Selected Titles in This Series

671 **Andrzej Rosłanowski and Saharon Shelah,** Norms on possibilities I: Forcing with trees and creatures, 1999

670 **Steve Jackson,** A computation of δ^1_5, 1999

669 **Seán Keel and James McKernan,** Rational curves on quasi-projective surfaces, 1999

668 **E. N. Dancer and P. Poláčik,** Realization of vector fields and dynamics of spatially homogeneous parabolic equations, 1999

667 **Ethan Akin,** Simplicial dynamical systems, 1999

666 **Mark Hovey and Neil P. Strickland,** Morava K-theories and localisation, 1999

665 **George Lawrence Ashline,** The defect relation of meromorphic maps on parabolic manifolds, 1999

664 **Xia Chen,** Limit theorems for functionals of ergodic Markov chains with general state space, 1999

663 **Ola Bratteli and Palle E. T. Jorgensen,** Iterated function systems and permutation representation of the Cuntz algebra, 1999

662 **B. H. Bowditch,** Treelike structures arising from continua and convergence groups, 1999

661 **J. P. C. Greenlees,** Rational S^1-equivariant stable homotopy theory, 1999

660 **Dale E. Alspach,** Tensor products and independent sums of $\mathcal{L}_p$-spaces, $1 < p < \infty$, 1999

659 **R. D. Nussbaum and S. M. Verduyn Lunel,** Generalizations of the Perron-Frobenius theorem for nonlinear maps, 1999

658 **Hasna Riahi,** Study of the critical points at infinity arising from the failure of the Palais-Smale condition for n-body type problems, 1999

657 **Richard F. Bass and Krzysztof Burdzy,** Cutting Brownian paths, 1999

656 **W. G. Bade, H. G. Dales, and Z. A. Lykova,** Algebraic and strong splittings of extensions of Banach algebras, 1999

655 **Yuval Z. Flicker,** Matching of orbital integrals on $GL(4)$ and $GSp(2)$, 1999

654 **Wancheng Sheng and Tong Zhang,** The Riemann problem for the transportation equations in gas dynamics, 1999

653 **L. C. Evans and W. Gangbo,** Differential equations methods for the Monge-Kantorovich mass transfer problem, 1999

652 **Arne Meurman and Mirko Primc,** Annihilating fields of standard modules of $\mathfrak{sl}(2,\mathbb{C})^\sim$ and combinatorial identities, 1999

651 **Lindsay N. Childs, Cornelius Greither, David J. Moss, Jim Sauerberg, and Karl Zimmermann,** Hopf algebras, polynomial formal groups, and Raynaud orders, 1998

650 **Ian M. Musson and Michel Van den Bergh,** Invariants under Tori of rings of differential operators and related topics, 1998

649 **Bernd Stellmacher and Franz Georg Timmesfeld,** Rank 3 amalgams, 1998

648 **Raúl E. Curto and Lawrence A. Fialkow,** Flat extensions of positive moment matrices: Recursively generated relations, 1998

647 **Wenxian Shen and Yingfei Yi,** Almost automorphic and almost periodic dynamics in skew-product semiflows, 1998

646 **Russell Johnson and Mahesh Nerurkar,** Controllability, stabilization, and the regulator problem for random differential systems, 1998

645 **Peter W. Bates, Kening Lu, and Chongchun Zeng,** Existence and persistence of invariant manifolds for semiflows in Banach space, 1998

644 **Michael David Weiner,** Bosonic construction of vertex operator para-algebras from symplectic affine Kac-Moody algebras, 1998

643 **Józef Dodziuk and Jay Jorgenson,** Spectral asymptotics on degenerating hyperbolic 3-manifolds, 1998

642 **Chu Wenchang,** Basic almost-poised hypergeometric series, 1998

(*Continued in the back of this publication*)

Norms on Possibilities I: Forcing with Trees and Creatures

MEMOIRS
of the
American Mathematical Society

Number 671

Norms on Possibilities I: Forcing with Trees and Creatures

Andrzej Rosłanowski
Saharon Shelah

September 1999 • Volume 141 • Number 671 (first of 4 numbers) • ISSN 0065-9266

American Mathematical Society
Providence, Rhode Island

1991 *Mathematics Subject Classification.*
Primary 03E35; Secondary 03E40, 03E05.

Library of Congress Cataloging-in-Publication Data

Rosłanowski, Andrzej, 1960–
Norms on possibilities I : forcing with trees and creatures / Andrzej Rosłanowski, Saharon Shelah.
p. cm. — (Memoirs of the American Mathematical Society, ISSN 0065-9266 ; no. 671)
"Volume 141, number 671 (first of 4 numbers)."
Includes bibliographical references.
ISBN 0-8218-1180-0
1. Set theory. 2. Forcing (Model theory) I. Shelah, Saharon. II. Title. III. Series.
QA3.A57 no. 671
[QA248]
510 s–dc21
[511.3′22] 99-27220
CIP

Memoirs of the American Mathematical Society

This journal is devoted entirely to research in pure and applied mathematics.

Subscription information. The 1999 subscription begins with volume 137 and consists of six mailings, each containing one or more numbers. Subscription prices for 1999 are $448 list, $358 institutional member. A late charge of 10% of the subscription price will be imposed on orders received from nonmembers after January 1 of the subscription year. Subscribers outside the United States and India must pay a postage surcharge of $30; subscribers in India must pay a postage surcharge of $43. Expedited delivery to destinations in North America $35; elsewhere $130. Each number may be ordered separately; *please specify number* when ordering an individual number. For prices and titles of recently released numbers, see the New Publications sections of the *Notices of the American Mathematical Society.*

Back number information. For back issues see the *AMS Catalog of Publications.*

Subscriptions and orders should be addressed to the American Mathematical Society, P. O. Box 5904, Boston, MA 02206-5904. *All orders must be accompanied by payment.* Other correspondence should be addressed to Box 6248, Providence, RI 02940-6248.

Memoirs of the American Mathematical Society is published bimonthly (each volume consisting usually of more than one number) by the American Mathematical Society at 201 Charles Street, Providence, RI 02904-2294. Periodicals postage paid at Providence, RI. Postmaster: Send address changes to Memoirs, American Mathematical Society, P. O. Box 6248, Providence, RI 02940-6248.

This publication is indexed in *Science Citation Index®*, *SciSearch®*, *Research Alert®*, *CompuMath Citation Index®*, *Current Contents®/Physical, Chemical & Earth Sciences.*
Printed in the United States of America.

∞ The paper used in this book is acid-free and falls within the guidelines established to ensure permanence and durability.
Visit the AMS home page at URL: `http://www.ams.org/`

10 9 8 7 6 5 4 3 2 1 04 03 02 01 00 99

Contents

Annotated Content

3.2 **Preserving non-meager sets** [We deal with preservation of being a non-meager set. We show that if a tree–creating pair (K, Σ) is T-omittory then the forcing notion $\mathbb{Q}_1^{\rm tree}(K, \Sigma)$ preserves non-meager sets. We formulate a weaker property (being of the **NMP**–type) which in the finitary case implies that forcing notions $\mathbb{Q}_1^{\rm tree}(K, \Sigma)$, $\mathbb{Q}_f^*(K, \Sigma)$ preserve non-meager sets. We get a similar conclusion for $\mathbb{Q}_{\rm w\infty}^*(K, \Sigma)$ when (K, Σ) is a finitary creating pair which captures singletons.]

3.3 **Preserving non-null sets** [We formulate a property of tree–creating pairs which implies that the forcing notion $\mathbb{Q}_e^{\rm tree}(K, \Sigma)$ preserves non-null sets.]

3.4 **(No) Sacks Property** [An easy condition ensuring "no Sacks property" for forcing notions of our type.]

3.5 **Examples** [We build a tree–creating pair (K, Σ) such that the forcing notion $\mathbb{Q}_1^{\rm tree}(K, \Sigma)$ is proper, ω^ω-bounding, preserves the outer measure, preserves non-meager sets but does not have the Sacks property.]

CHAPTER 4: OMITTORY WITH HALVING We explain how omittory creating pairs with the weak Halving Property produce almost ω^ω-bounding forcing notions.

4.1 **What omittory may easily do** [We show why natural examples of forcing notions $\mathbb{Q}_{\rm s\infty}^*(K, \Sigma)$ (for an omittory creating pair (K, Σ)) add a Cohen real and make ground model reals meager.]

4.2 **More operations on weak creatures** [Just what the title says: we present more ways to put weak creatures together.]

4.3 **Old reals are unbounded** [We say when a creating pair (K, Σ) is of the **AB**–type and we show that this property may be concluded from easier–to–check properties. We show that $\mathbb{Q}_{\rm s\infty}^*(K, \Sigma)$ is almost ω^ω-bounding if (K, Σ) is growing condensed and of the **AB**–type. For omittory creating pairs we do not have to assume "condensed" but then we require a stronger variant of the **AB**, **AB**$^+$.]

4.4 **Examples** [We generalize the forcing notions from [**Sh 207**], [**RoSh 501**] building examples for properties investigated before.]

CHAPTER 5: AROUND NOT ADDING COHEN REALS We try to ensure that the forcing notions built according to our schema do not add Cohen reals even if iterated. We generalize "(f, g)–bounding" and further we arrive to a more general iterable condition implying "no Cohen reals".

5.1 **(f, g)–bounding** [We present easy ways to make sure that our method results in (f, g)–bounding forcing notions.]

5.2 **$(\bar{t}, \bar{\mathcal{F}})$–bounding** [We introduce a natural (in our context) generalization of (f, g)–bounding property. For the sake of completeness we show that the new property is preserved in CS iterations.]

5.3 **Quasi–generic Γ and preserving them** [We formulate a reasonably weak but still iterable condition for not adding Cohen reals. We define $\bar{t}$–systems, we say when Γ is quasi-W-generic and when a forcing notion is Γ–genericity preserving. These notions will be crucial in the next section too.]

5.4 **Examples** [We construct a sequence $\langle W_{\ell,h}^k : h \in \mathcal{F}_\ell, k, \ell \rangle$ such that $W_{\ell,h}^k$ are $\bar{t}$–systems and various forcing notions (including the random algebra) are $(\Gamma, W_{\ell,h}^k)$–genericity preserving (for quasi-generic Γ). We build a forcing notion $\mathbb{Q}_{\rm w\infty}^*(K, \Sigma)$ which is proper, ω^ω-bounding, (f, g)-bounding, makes ground model reals null and we use the technology of "Γ–genericity" to conclude that its CS iterations with Miller's forcing, Laver's forcing and random algebra do not add Cohen reals.]

CHAPTER 6: PLAYING WITH ULTRAFILTERS Our aim here is to build a model in which there is a p-point generated by $\aleph_1$ elements which is not a q-point and $\mathfrak{m}_1 = \aleph_2$.

6.1 **Generating an ultrafilter** [We say when and how quasi-W-generic Γ determines an ultrafilter on ω.]

6.2 **Between Ramsey and p-points** [We define semi–Ramsey and almost–Ramsey ultrafilters and we have a short look at them.]

6.3 **Preserving ultrafilters** [We give conditions on a tree–creating pair (K, Σ) which imply that the forcing notion $\mathbb{Q}_1^{\rm tree}(K, \Sigma)$ preserves "$\mathcal{D}$ is an ultrafilter" for $\mathcal{D}$ which is Ramsey, almost Ramsey. We say when the filter generated by $\mathcal{D}$ in the extension is almost Ramsey.]

6.4 **Examples** [We construct $\bar{t}$–systems W_L^n such that if a quasi-W_L^N-generic Γ generates a semi–Ramsey ultrafilter then it generates an almost–Ramsey ultrafilter, and we build a suitable quasi-generic Γ. For a function $\psi \in \omega^\omega$ we give a tree–creating pair (K, Σ) such that the forcing notion $\mathbb{Q}_1^{\rm tree}(K, \Sigma)$ preserves "$\mathcal{D}$ is an almost–Ramsey ultrafilter" and it adds a function $\dot{W}$ (with $\dot{W}(m) \in [\psi(m)]^{m+1}$) such that for each partial function $h \in \prod\limits_{m \in {\rm dom}(h)} \psi(m)$ infinitely often $h(m) \in \dot{W}(m)$. Next we apply it to get an answer to Matet's problem.]

CHAPTER 7: FRIENDS AND RELATIVES OF PP We deal with Balcerzak–Plewik number and various properties resembling PP–property.

7.1 **Balcerzak–Plewik number** [We recall the definition of $\kappa_{\rm BP}$ and we show that it is bounded by the dominating number of the relation determined by the strong PP–property.]

7.2 **An iterable friend of strong PP–property** [We introduce a property slightly stronger than the strong PP–property but which can be easily handled in CS iterations. We show that this property is natural for forcings $\mathbb{Q}_0^{\rm tree}(K, \Sigma)$, $\mathbb{Q}_{\rm w\infty}^*(K, \Sigma)$ (in finitary cases).]

7.3 **Bounded relatives of PP** [We define various PP-like properties for localizing functions below a given function. We say how one gets them for our forcing notions and how we may handle them in iterations.]

7.4 **Weakly non-reducible p-filters in iterations** [We show that a property of filters, crucial for getting PP-like properties for our forcing notions, is easy to preserve in CS iterations.]

7.5 **Examples** [For a perfect set $P \subseteq 2^\omega$ we build a creating pair (K, Σ) such that the forcing notion $\mathbb{Q}_f^*(K, \Sigma)$ is proper, ω^ω-bounding and adds a perfect subset Q of P whith property that $(\forall K \in [\omega]^\omega)(Q{\restriction}K \neq 2^K)$. We use this forcing notion to get consistency of $\mathfrak{d} < \kappa_{\rm BP}$. We show how forcing notions from other parts of the paper may be used to distinguish PP-like properties (and the corresponding cardinal invariants). We build an example of a forcing notion which is ω^ω–bounding and preserves non-meager sets but which does not have the strong PP-property.]

Abstract

In this Memoir [1] we present a systematic study of the method of *norms on possibilities* of building forcing notions with keeping their properties under full control. This technique allows us to answer several open problems, but on our way to get the solutions we develop various ideas interesting *per se*. These include a new iterable condition for "not adding Cohen reals" (which has a flavour of preserving special properties of p-points), new intriguing properties of ultrafilters (weaker than being Ramsey but stronger than p–point) and some new applications of variants of the PP–property.

The first author thanks the Hebrew University of Jerusalem and the Lady Davis Foundation for the Golda Meir Postdoctoral Fellowship, and KBN (State Committee for Scientific Research, Poland) for partial support through grant 2P03A01109.

The research of the second author was partially supported by "Basic Research Foundation" founded by the Israel Academy of Sciences and Humanities. Publication 470.

1991 Mathematics Subject Classification: Primary 03E35; Secondary 03E40, 03E05

Key words and phrases: Forcing, proper forcing notions, norms on possibilities, cardinal characteristics of the continuum.

[1]Received by the editor October 4, 1996.

CHAPTER 0

Introduction

Set Theory[1] began with Georg Cantor's work when he was studying some special sets of reals in connection with the theory of trigonometric series. This study led Cantor to the following fundamental question: does there exist a bijection between the natural numbers and the set of real numbers? He answered this question negatively by showing that there is no such function. Cantor's work did not stop here and with his sharp intuition he discovered new concepts like the aleph's scale:

$$0, 1, \dots, \aleph_0, \aleph_1, \dots, \aleph_\omega, \aleph_{\omega+1}, \dots.$$

Thus Cantor's theorem says that $\aleph_0 < 2^{\aleph_0}$ and Cantor's question was: is $2^{\aleph_0}$ equal to $\aleph_1$?

A real advance on Cantor's question was given by Kurt Gödel when he showed that it is (relatively) consistent that $2^{\aleph_0} = \aleph_1$. In 1963 Paul Cohen showed that if the ZF–axioms for Set Theory are consistent then there is a model for Set Theory where the continuum is bigger than $\aleph_1$. Cohen's work is the end of classical set theory and is beginning of a new era.

When the cardinality of the continuum is $\aleph_1$ (i.e. CH holds) most of the combinatorial problems are solved. When the continuum is at least $\aleph_3$ then most of the known technology fails and we meet very strong limitations and barriers.

When the continuum is $\aleph_2$ there are many independence results; moreover there are reasonably well developed techniques for getting them by sewing the countable support iterations of proper forcing notions together with theorems on preservation of various properties.

The aim of this paper is to present some tools applicable in the last case. We present here a technique of constructing of (proper) forcing notions that was introduced by Shelah for solving problems related to cardinal invariants like the unbounded number or the splitting number as well as questions of existence of special kinds of P-points (see Blass Shelah [**BsSh 242**] and Shelah [**Sh 207**], [**Sh 326**]). That method was successfully applied in Fremlin Shelah [**FrSh 406**], Rosłanowski Shelah [**RoSh 501**], Ciesielski Shelah [**CiSh 653**] and other papers. The first attempt to present a systematic study of the technique was done in the late eighties when the second author started work on preparation of a new edition of [**Sh:b**]. For a long time the new book, [**Sh:f**], was supposed to contain 19 chapters. The last chapter, *Norms on possibilities*, contained a series of general definitions and statements of some basic results. However, there was no new application (or: a good question to solve) and the author of the book decided to put this chapter

[1] A large part of the beginning of this introduction is based on notes of Haim Judah. I really think that they fit to the present paper, though Saharon Shelah is not convinced — Andrzej Rosłanowski.

aside. Several years later, when the first author started his cooperation with Shelah some new applications of *Norms on possibilities* appeared. But the real shape was given to the work due to questions of Tomek Bartoszyński and Pierre Matet. The answers were very stimulating for the development of the general method.

This paper is meant as the first one in a series of works presenting applicability of the method of *norms on possibilities.* In [**RoSh 670**] we will present more applications of this technique – for example we develop the ideas of Ciesielski Shelah [**CiSh 653**] to build models without magic sets and their relatives. Though one can get an impression here that our method results in non-ccc forcing notions, we managed to generalize it slightly and get a tool for constructing ccc forcing notions. That was successfully applied in [**RoSh 628**] to answer a problem of Kunen by constructing a ccc Borel ideal on 2^ω which is translation–invariant index–invariant and is distinct from the null ideal, the meager ideal and their intersection. It should be pointed out here, that already in Judah Rosłanowski Shelah [**JRSh 373**] an example of a ccc forcing notion built with the use of *norms on possibilities* was given (the forcing notion there can be presented as some $\mathbb{Q}_1^{\rm tree}(K,\Sigma)$ in the terminology here). Investigations of ccc forcing notions constructed according to our schema are continued in [**RoSh 672**]. There are serious hopes that the technique presented here might be used to deal with problems of large continuum due to special products. This would continue Goldstern Shelah [**GoSh 448**]. Another direction is study of σ-ideals related to forcing notions built according to the schema.

Let us note that most of the forcing notions constructed here fall into the category of snep–forcing notions of [**Sh 630**]. Consequently, the general machinery of definable forcing notions is applicable here. We may use it to improve some of our results, and also to get more tools for handling iterations (see [**Sh 630**] and [**Sh 669**] for more details).

We want to emphasize that though the aim of the paper is strongly related to independence proofs it should have some value for those firmly committed to unembellished ZFC, too. This is nicely expressed by the following:

THESIS 0.0.1. *We cannot discover the (candidates for) Theorems of ZFC without having good forcing techniques to show they are hard nuts.*

0.1. The content of the paper

Most of the results of the paper originated in answering a particular question by constructing an example of a forcing notion. However, the general idea of the paper is to extract those properties of the example which are responsible for the fact that it works, with the hope that it may help in further applications of the method. That led us to separation of "the general theory" from its applications, and caused us to introduce a large number of definitions specifying various properties of weak creating pairs. Each chapter ends with a section presenting examples and applications of the tools developed in previous sections. Moreover, at the end of the paper we present the list of all definitions which appeared in it. This is not a real index, but should be helpful. The first chapter introduces basic definitions and the general scheme. In the next chapter we deal with the fundamental question of when our forcing notions are proper. The first two chapters are a basis for the rest of the paper. After reading them one can jump to any of the following chapters.

In the third chapter we show how we may control some basic properties of forcing notions built according to our scheme. The properties we deal with here are related to measure and category and they lead us to example 3.5.1 of a proper forcing notion which is ω^ω–bounding, preserves non-meager sets and the outer measure, but does not have the Sacks Property. This answers Bartoszyński's request [**Ba94**, Problem 5].

The fourth part continues [**RoSh 501**], dealing with localizations of subsets of ω. Though the example constructed here is a minor modification of the one built there, it is presented according to our general setting. We show explicitly how the weak Halving Property works in this type of examples.

A serious problem in getting models of ZFC with given properties of measure and category is that of not adding Cohen reals. What is disturbing here is that we do not have any good (meaning: sufficiently weak but iterable) conditions for this. In the fifth chapter we show how one can ensure that our general scheme results in forcing notions not adding Cohen reals. A new iterable condition for this appears here and quite general tools are developed (see 5.3.6, 5.4.2). Finally, in 5.4.3, 5.4.4, we fully answer another request of Bartoszyński formulated in [**Ba94**, Problem 4]. We build a proper ω^ω–bounding forcing notion which preserves non-meager sets, makes ground model reals null, is (f,g)–bounding and such that countable support iterations of this forcing with Laver forcing, Miller forcing and random real forcing do not add Cohen reals.

The next chapter leads to answering a question of Matet and Pawlikowski. In 6.4.6 we show that it is consistent that there exists a p-point generated by $\aleph_1$ elements which is not a q-point and that for every $\psi\in\omega^\omega$ and a family $\mathcal{F}$ of $\aleph_1$ partial infinite functions $h:\mathrm{dom}(h)\longrightarrow\omega$ such that $h(n)<\psi(n)$ for $n\in\mathrm{dom}(h)\subseteq\omega$ there is $W\in\prod\limits_{n\in\omega}[\psi(n)]^{n+1}$ with $(\forall h\in\mathcal{F})(\exists^\infty n\in\mathrm{dom}(h))(h(n)\in W(n))$. Several general results on preserving special properties of ultrafilters are presented on the way to this solution.

A starting point for chapter 7 was a problem of Balcerzak and Plewik. We show that the Balcerzak–Plewik number $\kappa_{\rm BP}$ (see 7.1.1) is bounded by a cardinal invariant related to the strong PP–property (in 7.1.3). Next we show the consistency of both "$\mathfrak{d}<\kappa_{\rm BP}$" (in 7.5.2) and "$\kappa_{\rm BP}<\mathfrak{c}$" (in 7.5.3). We treat our solution as a good opportunity to look at various properties of forcing notions related to the PP–property (and corresponding cardinal invariants).

0.2. Notation

Most of our notation is standard and compatible with that of classical textbooks on Set Theory (like Bartoszyński Judah [**BaJu95**] or Jech [**J**]). However in forcing we keep the convention that *a stronger condition is the larger one.*

NOTATION 0.2.1.
1. $\mathbb{R}^{\geq 0}$ stands for the set of non-negative reals. The integer part of a real $r\in\mathbb{R}^{\geq 0}$ is denoted by $\lfloor r\rfloor$.
2. For two sequences η,ν we write $\nu\vartriangleleft\eta$ whenever ν is a proper initial segment of η, and $\nu\trianglelefteq\eta$ when either $\nu\vartriangleleft\eta$ or $\nu=\eta$. The length of a sequence η is denoted by $\ell g(\eta)$.
3. A *tree* is a family of finite sequences closed under initial segments. (In 1.3.1 we will define more general objects.) For a tree T the family of all ω–branches through T is denoted by $[T]$. We may use the notation $\lim(T)$ for this object too (see 1.3.1, note that a tree is a quasi tree).

4. The quantifiers $(\forall^\infty n)$ and $(\exists^\infty n)$ are abbreviations for
$$(\exists m \in \omega)(\forall n > m) \quad \text{and} \quad (\forall m \in \omega)(\exists n > m),$$
respectively.
5. For a function $h : X \longrightarrow X$ and an integer k we define $h^{(k)}$ as the kth–iteration of h: $h^{(1)} = h$, $h^{(k+1)} = h \circ h^{(k)}$.
6. For a set X, $[X]^{\leq\omega}$, $[X]^{<\omega}$ and $\mathcal{P}(X)$ will stand for families of countable, finite and all, respectively, subsets of the set X. The family of k-element subsets of X will be denoted by $[X]^k$. The set of all finite sequences with values in X is called $X^{<\omega}$ (so domains of elements of $X^{<\omega}$ are integers). The collection of all *finite partial* functions from ω to X is $X^{\underline{\omega}}$.
7. The Cantor space 2^ω and the Baire space ω^ω are the spaces of all functions from ω to 2, ω, respectively, equipped with natural (Polish) topology.
8. For a forcing notion $\mathbb{P}$, $\Gamma_{\mathbb{P}}$ stands for the canonical $\mathbb{P}$–name for the generic filter in $\mathbb{P}$. With this one exception, all $\mathbb{P}$–names for objects in the extension via $\mathbb{P}$ will be denoted with a dot above (e.g. $\dot\tau$, $\dot X$).
9. $\mathfrak{c}$ stands for the cardinality of the continuum. The dominating number (the minimal size of a dominating family in ω^ω in the ordering of eventual dominance) is denoted by $\mathfrak{d}$ and the unbounded number (the minimal size of an unbounded family in that order) is called $\mathfrak{b}$. $\mathcal{M}$, $\mathcal{N}$ stand for the σ–ideals of meager and null sets on the real line, respectively.

General Definitions 0.2.2. 1. For an ideal $\mathcal{J}$ of subsets of a space X we define its cardinal characteristics (called *additivity*, *covering number*, *uniformity* and *cofinality*, respectively):
- $\mathbf{add}(\mathcal{J}) = \min\{|\mathcal{A}| : \mathcal{A} \subseteq \mathcal{J} \ \&\ \bigcup\mathcal{A} \notin \mathcal{J}\}$,
- $\mathbf{cov}(\mathcal{J}) = \min\{|\mathcal{A}| : \mathcal{A} \subseteq \mathcal{J} \ \&\ \bigcup\mathcal{A} = X\}$,
- $\mathbf{non}(\mathcal{J}) = \min\{|Y| : Y \subseteq X \ \&\ Y \notin \mathcal{J}\}$,
- $\mathbf{cof}(\mathcal{J}) = \min\{|\mathcal{A}| : \mathcal{A} \subseteq \mathcal{J} \ \&\ (\forall A \in \mathcal{J})(\exists B \in \mathcal{A})(A \subseteq B)\}$.

2. Assume that X, Y are Polish spaces and $R \subseteq X \times Y$ is a Borel relation. Suppose that $\mathbf{V} \subseteq \mathbf{V}'$ are models of ZFC and that all parameters we need are in $\mathbf{V}$. We say that the extension $(\mathbf{V}, \mathbf{V}')$ has *the R-localization property* if
$$(\forall x \in X \cap \mathbf{V}')(\exists y \in Y \cap \mathbf{V})((x, y) \in R).$$
If $x \in X \cap \mathbf{V}'$, $y \in Y \cap \mathbf{V}$ and $(x, y) \in R$ then we say that y *R-localizes* x.
We say that a forcing notion $\mathbb{P}$ has the R–localization property if every generic extension of $\mathbf{V}$ via $\mathbb{P}$ has this property.
3. For a relation $R \subseteq X \times Y$ we define two cardinal numbers (the unbounded and the dominating number for R):
$$\mathfrak{b}(R) = \min\{|B| : (\forall y \in Y)(\exists x \in B)((x, y) \notin R)\}$$
$$\mathfrak{d}(R) = \min\{|D| : (\forall x \in X)(\exists y \in D)((x, y) \in R)\}.$$

CHAPTER 1

Basic definitions

In this chapter we introduce our heroes: forcing notions built of weak creatures. The Prologue is intended to give the reader some intuitions needed to get through a long list of definitions. A general scheme is presented in the second part, where we define weak creatures, sub–composition operations, weak creating pairs (K, Σ) and corresponding forcing notions $\mathbb{Q}_{\mathcal{C}(\mathbf{nor})}(K, \Sigma)$. However, in practice (at least in this paper) we will be interested in two special cases of the scheme. The first main family of weak creating pairs (and related forcing notions) are creating pairs determined by composition operations on creatures. The second family consists of tree–creating pairs coming from tree compositions on tree–creatures. These two options are introduced in the following two parts of the chapter. It should be underlined here that the rest of the paper will deal with these two (essentially disjoint and parallel) cases of the general scheme. In the last section we give some justifications for our work in the next chapter, showing that without extra care our schema may result in forcing notions collapsing $\aleph_1$.

Note: Our terminology (weak creatures, creatures, tree–creatures etc) might be slightly confusing, but it was developed during a long period of time (see introduction) and large parts of it are established in literature already.

Basic Notation: In this paper $\mathbf{H}$ will stand for a function with domain ω such that $(\forall m \in \omega)(|\mathbf{H}(m)| \geq 2)$. We usually assume that $0 \in \mathbf{H}(m)$ (for all $m \in \omega$); if it is not the case then we fix an element of $\mathbf{H}(m)$ and we use it whenever appropriate notions refer to 0. Moreover we fix "a sufficiently large" uncountable regular cardinal χ and we assume that at least $\mathbf{H} \in \mathcal{H}(\chi)$ (the family of sets of cardinality hereditarily less than χ) or, what is more natural, even $\mathbf{H} \in \mathcal{H}(\aleph_1)$.

1.0. Prologue

If one looks at forcing notions appearing naturally in the Set Theory of Reals (i.e. the forcing notions adding a real with certain properties and preserving various properties of the ground model reals) then one realizes that they often have a common pattern. A condition in such a forcing notion determines an initial segment of the real we want to add and it puts some restrictions on possible further extensions of the initial segment. When we pass to a stronger condition we extend the determined part of the generic real and we put more restrictions on possible extensions. But we usually demand that the amount of freedom which is left by the restrictions still goes to infinity in a sense. The basic part of the definition of such a forcing notion is to describe the way a condition puts a restriction on possible initial segments of the generic real. Typically the restriction can be described locally by use of "atoms" or "black boxes", in our terminology called *weak creatures*. So a weak creature t has a domain (contained in finite sequences) and for each sequence w

from the domain it gives a family of extensions of w (this is described by a relation $\mathbf{val}[t]$: if $\langle u,v\rangle \in \mathbf{val}[t]$ then v is an allowed extension of u). Moreover, such a t has a norm $\mathbf{nor}[t]$ which measures the amount of freedom it leaves. Further, we are told what we are allowed to do with weak creatures: typically we may shrink them, glue together or just forget about them (i.e. *omit* them). The results of permitted operations on a family $\mathcal{S}$ of weak creatures are elements of $\Sigma(\mathcal{S})$ in our notation (where Σ is a *sub-composition operation* on the considered family K of weak creatures), see 1.1.4. Now a condition in our forcing notion can be viewed as $(w,\mathcal{S})$, where w is a finite sequence (the determined part of the generic real) and $\mathcal{S}$ is a countable family of weak creatures satisfying some demands on its structure and requirements on $\mathbf{nor}[t]$ for $t\in\mathcal{S}$. When we want to build a stronger condition then we take $t\in\mathcal{S}$ such that w is in the domain of t and we pick up one of the possible extensions of w allowed by t. We may repeat this procedure finitely many times and we get a sequence w^* extending w. Next we choose a family $\mathcal{S}^*$ of weak creatures such that each $s\in\mathcal{S}^*$ is obtained by permitted procedures from some $\mathcal{S}_s\subseteq\mathcal{S}$ (i.e. $s\in\Sigma(\mathcal{S}_s)$). The pair $(w^*,\mathcal{S}^*)$ is an extension of $(w,\mathcal{S})$ provided $\mathcal{S}^*$ satisfies the structure demands and norm requirements.

However, this general schema breaks to two cases, which, though very similar, are of different flavors. In the first case we demand that the family $\mathcal{S}$ in a condition $(w,\mathcal{S})$ has a linear structure. Then we usually represent the condition as $(w,t_0,t_1,t_2,\dots)$, where for some sequence $0\le m_0<m_1<m_2<\dots<\omega$

> w is a sequence of length m_0 and for each $i<\omega$:
>
> > t_i is a weak creature saying in which way sequences of length m_i may be extended to sequences of length m_{i+1}.

So it is natural in this context to consider only weak creatures t such that for some integers $m^t_{\rm dn}<m^t_{\rm up}$ (dn stands for "down"), the domain of t is contained in sequences of length $m^t_{\rm dn}$ and every extension of a sequence from the domain allowed by t is of length $m^t_{\rm up}$. In other words we require that if $\langle u,v\rangle\in\mathbf{val}[t]$ then $\ell g(u)=m^t_{\rm dn}$ and $\ell g(v)=m^t_{\rm up}$. In applications the domain of the relation $\mathbf{val}[t]$ consist of *all legal* sequences of length $m^t_{\rm dn}$. Let us describe a simple example of this kind. Consider the Silver forcing $\mathbb{Q}$ "below 2^n": a condition in $\mathbb{Q}$ is a function $p:\mathrm{dom}(p)\longrightarrow\omega$ such that

$$\mathrm{dom}(p)\subseteq\omega \quad \& \quad |\omega\setminus\mathrm{dom}(p)|=\omega \quad \& \quad (\forall n\in\mathrm{dom}(p))(p(n)<2^n).$$

Let us look at this forcing in a different way.

Let K consist of all triples $t=(\mathbf{nor}[t],\mathbf{val}[t],\mathbf{dis}[t])$ such that

- $\mathbf{dis}[t]=(m^t,k^t)$ where $m^t<\omega$ and $k^t\in\{*\}\cup 2^{m^t}$,
- $\mathbf{nor}[t]=m^t$ if $k^t=*$ and $\mathbf{nor}[t]=0$ otherwise,
- $\mathbf{val}[t]=\{\langle u,v\rangle\in\prod\limits_{i<m^t}2^i\times\prod\limits_{i\le m^t}2^i : u\vartriangleleft v \ \ \& \ \ (k^t\neq * \ \Rightarrow\ v(m^t)=k^t)\}$.

For $\mathcal{S}\subseteq K$ we let $\Sigma(\mathcal{S})=\emptyset$ if $|\mathcal{S}|\neq 1$ and

if $t\in K$, $k^t\neq *$ then $\Sigma(\{t\})=\{t\}$,

if $t\in K$, $k^t=*$ then $\Sigma(\{t\})$ consists of all $s\in K$ with $m^s=m^t$.

Now, a condition p in $\mathbb{Q}$ may be represented as a sequence $(w^p,t^p_0,t^p_1,\dots)$, where

(a) each t^p_i is from K,

(b) w^p is a finite sequence of length $m^{t^p_0}$ such that $w^p(n)<2^n$ for $n<m^{t^p_0}$,

(c) $m^{t^p_i}=m^{t^p_0}+i$,

(d) $\limsup_{i\to\omega} \mathbf{nor}[t^p_i] = \infty$.

The order $\leq$ of $\mathbb{Q}$ is defined by $(w^p, t^p_0, t^p_1, t^p_2, \dots) \leq (w^q, t^q_0, t^q_1, t^q_2, \dots)$ if and only if for some $N < \omega$:

$w^p \trianglelefteq w^q$, $\ell g(w^q) = \ell g(w^p) + N$,
$\langle w^q{\restriction}\ell g(w^p) + i, w^q{\restriction}(w^p) + i + 1\rangle \in \mathbf{val}[t^p_i]$ for each $i < N$, and
$t^q_j \in \Sigma(t^p_{N+j})$ for every $j < \omega$.

The pair (K, Σ) is an example of a *creating pair* and the forcing notion $\mathbb{Q}$ (represented as above) is $\mathbb{Q}^*_{\mathrm{w}\infty}(K, \Sigma)$ (see 1.2.2 and 1.2.4).

On the other pole of possible weak creatures we have those which provide possible extensions for only one sequence η (i.e. those t for which $|\mathrm{dom}(\mathbf{val}[t])| = 1$). Weak creatures of this type are called *tree–creatures* and they say to us simply:

I know what the restrictions on extensions of a single sequence η are, and I do not look at other sequences at all.

Tree–creatures are fundamental for building forcing notions in which conditions are trees of a special kind. In these forcing notions a condition $p = (w, \mathcal{S})$ is such that w is a root (stem) of a tree T^p and each $t \in \mathcal{S}$ is a part of the tree T^p; usually such a t describes how one passes from an element $\eta \in T^p$ to its extensions in T^p (not necessarily immediate successors). It is natural to put some requirements on sub–composition operations Σ when the weak creatures we consider are tree–creatures, and this leads to the definition of *tree composition* and *tree–creating pair*, see 1.3.3. Moreover, it turns out to be very practical to consider special demands on the norms $\mathbf{nor}[t]$ to take an advantage of the tree–form of a condition, see 1.3.5. (Note that in further definitions we do not require that T^p is a tree but we demand that it is a *quasi tree* only. This will simplify the notation a little bit.) Let us illustrate this by a suitable representation of the Laver forcing $\mathbb{L}$. Recall that a condition in $\mathbb{L}$ is a tree $T \subseteq \omega^{<\omega}$ such that if $\eta \in T$, $\mathrm{root}(T) \trianglelefteq \eta$ then $|\mathrm{succ}_T(\eta)| = \omega$.

Let K consist of all triples $t = (\mathbf{nor}[t], \mathbf{val}[t], \mathbf{dis}[t])$ such that

- $\mathbf{dis}[t] = (\eta^t, A^t)$, where $A^t \in [\omega]^\omega$ and $\eta^t \in \omega^{<\omega}$,
- $\mathbf{nor}[t] = \ell g(\eta^t)$,
- $\mathbf{val}[t] = \{\langle \eta^t, \nu\rangle : \eta^t \vartriangleleft \nu \ \&\ \ell g(\nu) = \ell g(\eta^t) + 1 \ \&\ \nu(\ell g(\eta^t)) \in A^t\}$.

For $t \in K$ we let $\Sigma(\{t\}) = \{s \in K : \eta^s = \eta^t \ \&\ A^s \subseteq A^t\}$ and for $\mathcal{S} \subseteq K$ with $|\mathcal{S}| \neq 1$ we declare $\Sigma(\mathcal{S}) = \emptyset$. Now, a condition p in $\mathbb{L}$ can be represented as $(\eta^p, \langle t^p_\nu : \nu \in T^p\rangle)$, where $T^p \subseteq \omega^{<\omega}$ is a tree such that $\mathrm{root}(T^p) = \eta^p$ and for each $\nu \in T^p$, $\mathrm{root}(T^p) \trianglelefteq \nu$ we have $\mathrm{succ}_{T^p}(\nu) = \{\rho : \langle \nu, \rho\rangle \in \mathbf{val}[t^p_\nu]\}$ (so $\mathcal{S}$ is $\{t^p_\nu : \nu \in T^p\}$ here). Moreover, we demand that for each infinite branch η through T^p the norms $\mathbf{nor}[t^p_{\eta\restriction i}]$ go to infinity. The order $\leq$ of $\mathbb{L}$ is given by $(\eta^p, \langle t^p_\nu : \nu \in T^p\rangle) \leq (\eta^q, \langle t^q_\nu : \nu \in T^q\rangle)$ if and only if $\eta^q \in T^p$ and

$$(\forall \nu \in T^q)(\nu \in T^p \ \&\ t^q_\nu \in \Sigma(t^p_\nu)).$$

The pair (K, Σ) defined above is a tree creating pair and the forcing notion $\mathbb{L}$ is $\mathbb{Q}^{\mathrm{tree}}_1(K, \Sigma)$.

1.1. Weak creatures and related forcing notions

DEFINITION 1.1.1. 1. A triple $t = (\mathbf{nor}, \mathbf{val}, \mathbf{dis})$ is *a weak creature for* $\mathbf{H}$ if:

(a) $\mathbf{nor} \in \mathbb{R}^{\geq 0}$,

(b) **val** is a non-empty subset of

$$\{\langle x,y\rangle \in \bigcup_{m_0<m_1<\omega} [\prod_{i<m_0} \mathbf{H}(i) \times \prod_{i<m_1} \mathbf{H}(i)] : x \lhd y\},$$

(c) **dis** $\in \mathcal{H}(\chi)$.

The family of all weak creatures for **H** is denoted by WCR[**H**].

2. In the above definition we write **nor** = **nor**$[t]$, **val** = **val**$[t]$ and **dis** = **dis**$[t]$. [**val** is for value, **nor** is for norm, **dis** is for distinguish.]

REMARK 1.1.2. The **dis**$[t]$ in a weak creature t plays the role of an additional parameter which allows as to have distinct creatures with the same values of **val** and **nor**. This may be sometimes important in defining sub-composition operations on K (see 1.1.4 below): we will be able to have distinct values of $\Sigma(t_0)$, $\Sigma(t_1)$ though **val**$[t_0]$ = **val**$[t_1]$ and **nor**$[t_0]$ = **nor**$[t_1]$. One may think that this additional parameter describes the way the weak creature t was constructed (while **val**$[t]$, **nor**$[t]$ give the final effect of the construction). We may sometimes "forget" to mention **dis**$[t]$ explicitly – in most of the results and applications **dis**$[t]$ might be arbitrary. In the examples we construct, if we do not mention **dis**$[t]$ we mean that either it is 0 or its form is clear.

DEFINITION 1.1.3. 1. If we omit **H** we mean for some **H** or the **H** is clear from the context, etc.

2. We say that **H** is *finitary* (or *of a countable character*, respectively) if $\mathbf{H}(n)$ is finite (countable, resp.) for each $n \in \omega$. We say that $K \subseteq$ WCR[**H**] is *finitary* if **H** is finitary and **val**$[t]$ is finite for each $t \in K$.

DEFINITION 1.1.4. Let $K \subseteq$ WCR[**H**].

1. A function $\Sigma : [K]^{\leq\omega} \longrightarrow \mathcal{P}(K)$ is a *sub-composition operation on* K if:
   (a) *(transitivity)* if $\mathcal{S} \in [K]^{\leq\omega}$ and for each $s \in \mathcal{S}$ we have $s \in \Sigma(\mathcal{S}_s)$ then $\Sigma(\mathcal{S}) \subseteq \Sigma(\bigcup_{s\in\mathcal{S}} \mathcal{S}_s)$,
   (b) $r \in \Sigma(r)$ for each $r \in K$ and $\Sigma(\emptyset) = \emptyset$.

   [Note that $\Sigma(\mathcal{S})$ may be empty for non-empty $\mathcal{S}$; in future defining Σ we will describe it only for the cases it provides a non-empty result, in all other cases we will assume that $\Sigma(\mathcal{S}) = \emptyset$.]
2. In the situation described above the pair (K, Σ) is called *a weak creating pair for* **H**.
3. Suppose that (K, Σ) is a weak creating pair, $t_0, t_1 \in K$. We say that t_0, t_1 are Σ–equivalent (and we write then $t_0 \sim_\Sigma t_1$) if **nor**$[t_0]$ = **nor**$[t_1]$, **val**$[t_0]$ = **val**$[t_1]$ and for each $\mathcal{S} \subseteq [K \setminus \{t_0, t_1\}]^{\leq\omega}$ we have $\Sigma(\mathcal{S} \cup \{t_0\}) = \Sigma(\mathcal{S} \cup \{t_1\})$.

REMARK 1.1.5. Note that the relation $\sim_\Sigma$ as defined in 1.1.4(3) does not have to be transitive in a general case (so, perhaps, we should not use the name Σ–*equivalent*). However, if (K, Σ) is either a creating pair (see 1.2.2) or a tree–creating pair (see 1.3.3) then $\sim_\Sigma$ is an equivalence relation. Then, if additionally $\Sigma(\mathcal{S})$ is non-empty for finite $\mathcal{S}$ only, the value of $\Sigma(\mathcal{S})$ depends on the $\sim_\Sigma$–equivalence classes of elements of $\mathcal{S}$ only. Therefore we will tend to think in these situations that we identify all $\sim_\Sigma$–equivalent elements of K (or just consider a selector $K^* \subseteq K$ of $K/{\sim_\Sigma}$). If $\Sigma(\mathcal{S})$ may be non-empty for an infinite $\mathcal{S} \subseteq K$ (which may happen for tree–creating pairs), then we have to be more careful before we consider this identification: we should check that the values of Σ depend on $\sim_\Sigma$–equivalence classes only.

DEFINITION 1.1.6. Let (K,Σ) be a weak creating pair for $\mathbf{H}$.

1. For a weak creature $t \in K$ we define its basis (with respect to (K,Σ)) as

$$\mathbf{basis}(t) = \{w \in \bigcup_{m<\omega} \prod_{i<m} \mathbf{H}(i) : (\exists s \in \Sigma(t))(\exists u)(\langle w,u\rangle \in \mathbf{val}[s])\}.$$

2. For $w \in \bigcup\limits_{m<\omega} \prod\limits_{i<m} \mathbf{H}(i)$ and $\mathcal{S} \in [K]^{\leq\omega}$ we define the set $\mathrm{pos}(w,\mathcal{S})$ of possible extensions of w from the point of view of $\mathcal{S}$ (with respect to (K,Σ)) as:

$$\mathrm{pos}^*(w,\mathcal{S}) = \{u : (\exists s \in \Sigma(\mathcal{S}))(\langle w,u\rangle \in \mathbf{val}[s])\},$$

$$\begin{aligned}\mathrm{pos}(w,\mathcal{S}) = \{u :\ & \text{there are disjoint sets } \mathcal{S}_i \text{ (for } i<m<\omega) \text{ with } \bigcup_{i<m} \mathcal{S}_i = \mathcal{S} \\ & \text{and a sequence } 0<\ell_0<\ldots<\ell_{m-1}<\ell g(u) \text{ such that} \\ & u{\restriction}\ell_0 \in \mathrm{pos}^*(w,\mathcal{S}_0) \text{ and} \\ & u{\restriction}\ell_1 \in \mathrm{pos}^*(u{\restriction}\ell_0,\mathcal{S}_1)\ \&\ \ldots\ \&\ u \in \mathrm{pos}^*(u{\restriction}\ell_{m-1},\mathcal{S}_{m-1})\}.\end{aligned}$$

3. Whenever we use **basis** or pos we assume that the weak creating pair (K,Σ) with respect to which these notions are defined is understood.

DEFINITION 1.1.7. Suppose (K,Σ) is a weak creating pair for $\mathbf{H}$ and $\mathcal{C}(\mathbf{nor})$ is a property of ω-sequences of weak creatures from K (i.e. $\mathcal{C}(\mathbf{nor})$ can be thought of as a subset of K^ω). We define the forcing notion $\mathbb{Q}_{\mathcal{C}(\mathbf{nor})}(K,\Sigma)$ by

conditions are pairs (w,T) such that for some $k_0<\omega$:

(a) $w \in \prod\limits_{i<k_0} \mathbf{H}(i)$

(b) $T = \langle t_i : i<\omega\rangle$ where:

 (i) $t_i \in K$ for each i,

 (ii) $w \in \mathbf{basis}(t_i)$ for some $i<\omega$ and for each $u \in \mathrm{pos}(w,\{t_i : i \in I_0\})$, $I_0 \subseteq \omega$ there is $i \in \omega \setminus I_0$ such that $u \in \mathbf{basis}(t_i)$,

(c) the sequence $\langle t_i : i<\omega\rangle$ satisfies the condition $\mathcal{C}(\mathbf{nor})$;

the order is given by: $(w_1,T^1) \leq (w_2,T^2)$ if and only if for some disjoint sets $\mathcal{S}_0,\mathcal{S}_1,\mathcal{S}_2,\ldots \subseteq \omega$ we have:

$$w_2 \in \mathrm{pos}(w_1,\{t^1_\ell : \ell \in \mathcal{S}_0\}) \quad\text{and}\quad t^2_i \in \Sigma(\{t^1_\ell : \ell \in \mathcal{S}_{i+1}\}) \text{ for each } i<\omega$$

(where $T^\ell = \langle t^\ell_i : i<\omega\rangle$).

If $p=(w,T)$ we let $w^p=w$, $T^p=T$ and if $T^p=\langle t_i : i<\omega\rangle$ then we let $t^p_i=t_i$. We may write $(w,t_0,t_1,\ldots)$ instead of (w,T) (when $T=\langle t_i : i<\omega\rangle$).

PROPOSITION 1.1.8. *If (K,Σ) is a weak creating pair and $\mathcal{C}(\mathbf{nor})$ is a property of sequences of elements of K then $\mathbb{Q}_{\mathcal{C}(\mathbf{nor})}(K,\Sigma)$ is a forcing notion.*

REMARK 1.1.9. The reason for our notation $\mathcal{C}(\mathbf{nor})$ for the property relevant for **(c)** of 1.1.7 is that in the applications this conditions will say that the norms $\mathbf{nor}[t_i]$ go to the infinity in some sense. Some of the possibilities here are listed in 1.1.10 below.

DEFINITION 1.1.10. For a weak creature t let us denote

$$m_{\mathrm{dn}}(t) = \min\{\ell g(u) : u \in \mathrm{dom}(\mathbf{val}[t])\}.$$

We introduce the following (basic) properties of sequences of weak creatures which may serve as $\mathcal{C}(\mathbf{nor})$ in 1.1.7:

(s∞) A sequence $\langle t_i : i < \omega \rangle$ satisfies $\mathcal{C}^{\mathrm{s}\infty}(\mathbf{nor})$ if and only if

$$(\forall i < \omega)(\mathbf{nor}[t_i] > \max\{i, m_{\mathrm{dn}}(t_i)\}).$$

(∞) A sequence $\langle t_i : i < \omega \rangle$ satisfies $\mathcal{C}^{\infty}(\mathbf{nor})$ if and only if

$$\lim_{i \to \omega} \mathbf{nor}[t_i] = \infty.$$

(w∞) A sequence $\langle t_i : i < \omega \rangle$ satisfies $\mathcal{C}^{\mathrm{w}\infty}(\mathbf{nor})$ if and only if

$$\limsup_{i \to \omega} \mathbf{nor}[t_i] = \infty.$$

Let $f : \omega \times \omega \longrightarrow \omega$. We define the property introduced by f by

(f) A sequence $\langle t_i : i < \omega \rangle$ satisfies $\mathcal{C}^{f}(\mathbf{nor})$ if and only if

$$(\forall k < \omega)(\forall^\infty i)(\mathbf{nor}[t_i] > f(k, m_{\mathrm{dn}}(t_i))).$$

For notational convenience we will sometimes use the empty norm condition:

($\emptyset$) Each sequence $\langle t_i : i < \omega \rangle$ satisfies $\mathcal{C}^{\emptyset}(\mathbf{nor})$.

The forcing notions corresponding to the above properties (for a weak creating pair (K, Σ)) will be denoted by $\mathbb{Q}_{\mathrm{s}\infty}(K, \Sigma)$, $\mathbb{Q}_{\infty}(K, \Sigma)$, $\mathbb{Q}_{\mathrm{w}\infty}(K, \Sigma)$, $\mathbb{Q}_{f}(K, \Sigma)$ and $\mathbb{Q}_{\emptyset}(K, \Sigma)$, respectively.

REMARK 1.1.11. 1) Note that the second component of a pair $(w, T) \in \mathbb{Q}_{\mathcal{C}(\mathbf{nor})}(K, \Sigma)$ is *a sequence* of weak creatures, and in the most general case the order of its members may be important. For example the property $\mathcal{C}^{\mathrm{s}\infty}(\mathbf{nor})$ introduced in 1.1.10 is not permutation invariant and some changes of the order in the sequence $\langle t_i : i < \omega \rangle$ may produce a pair (w, T') which is not a legal condition. This is not what we would like to have here, so in applications in which this kind of problems appears we will restrict ourselves to suborders $\mathbb{Q}^*_{\mathcal{C}(\mathbf{nor})}(K, \Sigma)$ of $\mathbb{Q}_{\mathcal{C}(\mathbf{nor})}(K, \Sigma)$ in which we put additional structure demands on the sequences $\langle t_i : i < \omega \rangle$ (see 1.2.6). Moreover, to get properness for forcing notions $\mathbb{Q}^*_{\mathrm{s}\infty}(K, \Sigma)$ we will have to put some demands on (K, Σ) (see 2.1.6). These demands will cause that various variants of the norm condition $\mathcal{C}^{\mathrm{s}\infty}(\mathbf{nor})$ result in equivalent forcing notions (see 2.1.3). So, from the point of view of applications, the main reason for introducing $\mathcal{C}^{\mathrm{s}\infty}(\mathbf{nor})$ is a notational convenience.
2) Note that

$$\mathbb{Q}_{\mathrm{s}\infty}(K, \Sigma) \subseteq \mathbb{Q}_{\infty}(K, \Sigma) \subseteq \mathbb{Q}_{\mathrm{w}\infty}(K, \Sigma) \subseteq \mathbb{Q}_{\emptyset}(K, \Sigma)$$

where the inclusions mean "suborder" (but often not "complete suborder"). If we put some conditions on f (e.g. f is fast, see 1.1.12) then we may easily have $\mathbb{Q}_f(K, \Sigma) \subseteq \mathbb{Q}_\infty(K, \Sigma)$.
3) In our applications we will consider the forcing notions $\mathbb{Q}_f(K, \Sigma)$ only for functions $f : \omega \times \omega \longrightarrow \omega$ which are growing fast enough (see 1.1.12 below).

DEFINITION 1.1.12. A function $f : \omega \times \omega \longrightarrow \omega$ is *fast* if

$$(\forall k \in \omega)\big(\forall \ell \in \omega)(f(k, \ell) \leq f(k, \ell+1) \;\;\&\;\; 2 \cdot f(k, \ell) < f(k+1, \ell)\big).$$

The function f is $\mathbf{H}$-*fast* if additionally ($\mathbf{H}$ is finitary and) for each $k, \ell \in \omega$:

$$2^{\varphi_{\mathbf{H}}(\ell)} \cdot (f(k, \ell) + \varphi_{\mathbf{H}}(\ell) + 2) < f(k+1, \ell),$$

where $\varphi_{\mathbf{H}}(\ell) = |\prod\limits_{i<\ell} \mathbf{H}(i)|$.

DEFINITION 1.1.13. Suppose that (K,Σ) is a weak creating pair and $\mathcal{C}(\mathbf{nor})$ is a property of sequences of elements of K. Let $\dot{W}$ be a $\mathbb{Q}_{\mathcal{C}(\mathbf{nor})}(K,\Sigma)$-name such that

$$\Vdash_{\mathbb{Q}_{\mathcal{C}(\mathbf{nor})}(K,\Sigma)} \dot{W} = \bigcup\{w^p : p \in \Gamma_{\mathbb{Q}_{\mathcal{C}(\mathbf{nor})}(K,\Sigma)}\}.$$

PROPOSITION 1.1.14. *Suppose that (K,Σ) is a weak creating pair and $\mathcal{C}(\mathbf{nor})$ is a property such that the forcing notion $\mathbb{Q}_{\mathcal{C}(\mathbf{nor})}(K,\Sigma)$ is non-empty. Then:*

1. $\Vdash_{\mathbb{Q}_{\mathcal{C}(\mathbf{nor})}(K,\Sigma)}$ *"$\dot{W}$ is a member of $\prod\limits_{i<\omega}\mathbf{H}(i)$".*
2. *If $(\forall i\in\omega)(\mathbf{H}(i)=2)$ then $\Vdash_{\mathbb{Q}_{\mathcal{C}(\mathbf{nor})}(K,\Sigma)}$ "$\dot{W}$ is a real".*
3. *If for every $t\in K$, $u\in\mathbf{basis}(t)$ the set $\mathrm{pos}(u,t)$ has at least two elements then $\Vdash_{\mathbb{Q}_{\mathcal{C}(\mathbf{nor})}(K,\Sigma)}$ "$\dot{W}\notin\mathbf{V}$".*

REMARK 1.1.15. 1) We will always assume that the considered weak creating pairs (K,Σ) (and norm conditions $\mathcal{C}(\mathbf{nor})$) are such that $\mathbb{Q}_{\mathcal{C}(\mathbf{nor})}(K,\Sigma)\neq\emptyset$. Usually, it will be enough that K contains enough creatures with large norms and in each particular example this requirement will be easy to verify.
2) In general, the $\dot{W}$ defined in 1.1.13 does not have to encode the generic filter. We may formulate a condition ensuring this. Let (K,Σ) be a weak creating pair and $\mathcal{C}(\mathbf{nor})$ be a norm condition such that $\mathbb{Q}_{\mathcal{C}(\mathbf{nor})}(K,\Sigma)$ is not empty. For $p\in\mathbb{Q}_{\mathcal{C}(\mathbf{nor})}(K,\Sigma)$ define

$$S(p)\stackrel{\rm def}{=}\{w\in\bigcup_{n\in\omega}\prod_{i<n}\mathbf{H}(i) : (\exists q\geq p)(w\trianglelefteq w^q)\}$$

Clearly $S(p)$ is a subtree of $\bigcup\limits_{n\in\omega}\prod\limits_{i<n}\mathbf{H}(i)$. Moreover, for each $w\in\bigcup\limits_{n\in\omega}\prod\limits_{i<n}\mathbf{H}(i)$ and $p,q\in\mathbb{Q}_{\mathcal{C}(\mathbf{nor})}(K,\Sigma)$:

$$\begin{array}{lll} p\nVdash_{\mathbb{Q}_{\mathcal{C}(\mathbf{nor})}(K,\Sigma)} \text{“}w\ntrianglelefteq\dot{W}\text{”} & \text{if and only if} & w\in S(p)\quad\text{and}\\ p\Vdash_{\mathbb{Q}_{\mathcal{C}(\mathbf{nor})}(K,\Sigma)} \text{“}\dot{W}\in[S(q)]\text{”} & \text{if and only if} & S(p)\subseteq S(q).\end{array}$$

Now we may define a $\mathbb{Q}_{\mathcal{C}(\mathbf{nor})}(K,\Sigma)$–name $\dot{H}$ by

$$\Vdash_{\mathbb{Q}_{\mathcal{C}(\mathbf{nor})(K,\Sigma)}} \dot{H}=\{p\in\mathbb{Q}_{\mathcal{C}(\mathbf{nor})}(K,\Sigma) : \dot{W}\in[S(p)]\}$$

and we may want to claim that $\Vdash \dot{H}=\Gamma_{\mathbb{Q}_{\mathcal{C}(\mathbf{nor})}(K,\Sigma)}$. But for this we need to know that any two conditions in $\dot{H}$ are compatible. A sufficient and necessary requirement for this is:

($\boxdot$) *if* $p,q\in\mathbb{Q}_{\mathcal{C}(\mathbf{nor})}(K,\Sigma)$ and $S(p)\subseteq S(q)$
then $p\Vdash q\in\Gamma_{\mathbb{Q}_{\mathcal{C}(\mathbf{nor})}(K,\Sigma)}$ (or in other words $p\geq q$ modulo the equivalence of conditions).

In most of our examples and applications, the condition ($\boxdot$) will be easy to check. We will not mention it in future as we will not use its consequences.
Note however that it is very easy to build examples of weak creating pairs (K,Σ) (even creating pairs or tree creating pairs) for which ($\boxdot$) fails. Some of these examples might appear naturally.

1.2. Creatures

Now we will deal with the first specific case of the general scheme: creating pairs and forcing notions $\mathbb{Q}^*_{\mathcal{C}(\mathbf{nor})}(K,\Sigma)$. Notation and definitions introduced here are applicable to this case only and should not be confused with that for tree–creating pairs.

DEFINITION 1.2.1. Let t be a weak creature for $\mathbf{H}$.

1. If there is $m<\omega$ such that $(\forall \langle u,v\rangle \in \mathbf{val}[t])(\ell g(u)=m)$ then this unique m is called $m^t_{\rm dn}$.
2. If there is $m<\omega$ such that $(\forall \langle u,v\rangle \in \mathbf{val}[t])(\ell g(v)=m)$ then this unique m is called $m^t_{\rm up}$.
3. If both $m^t_{\rm dn}$ and $m^t_{\rm up}$ are defined then t is called *an* $(m^t_{\rm dn},m^t_{\rm up})$*-creature* (or just *a creature*).
4. $\mathrm{CR}_{m_{\rm dn},m_{\rm up}}[\mathbf{H}]=\{t\in \mathrm{WCR}[\mathbf{H}]: m^t_{\rm dn}=m_{\rm dn} \text{ and } m^t_{\rm up}=m_{\rm up}\}$,
$$\mathrm{CR}[\mathbf{H}]=\bigcup_{m_{\rm dn}<m_{\rm up}<\omega}\mathrm{CR}_{m_{\rm dn},m_{\rm up}}[\mathbf{H}].$$

DEFINITION 1.2.2. Suppose that $K\subseteq \mathrm{CR}[\mathbf{H}]$ and Σ is a sub-composition operation on K. We say that Σ is *a composition* on K (and we say that (K,Σ) is *a creating pair for* $\mathbf{H}$) if:

1. if $\mathcal{S}\in [K]^{\leq\omega}$ and $\Sigma(\mathcal{S})\neq\emptyset$ then $\mathcal{S}$ is finite and for some enumeration $\mathcal{S}=\{t_0,\ldots,t_{m-1}\}$ we have $m^{t_i}_{\rm up}=m^{t_{i+1}}_{\rm dn}$ for all $i<m-1$, and
2. for each $s\in\Sigma(t_0,\ldots,t_{m-1})$ we have $m^s_{\rm dn}=m^{t_0}_{\rm dn}$ and $m^s_{\rm up}=m^{t_{m-1}}_{\rm up}$.

In this paper we will always assume that the creating pair under considerations is additionally *nice* and *smooth* (see 1.2.5 below) and we will not repeat this demand later.

REMARK 1.2.3. Sets of creatures with pairwise distinct $m^t_{\rm dn}$'s might be naturally ordered according to this value and therefore in similar situations we identify sets of creatures with the corresponding sequences of creatures.

DEFINITION 1.2.4. 1. For $K\subseteq \mathrm{CR}[\mathbf{H}]$ and a composition operation Σ on K we define *finite candidates* (FC) and *pure finite candidates* (PFC) with respect to (K,Σ):

$$\begin{aligned}\mathrm{FC}(K,\Sigma)=\ &\{(w,t_0,\ldots,t_n): w\in \mathbf{basis}(t_0) \text{ and for each } i\leq n\\ &\quad t_i\in K, m^{t_i}_{\rm up}=m^{t_{i+1}}_{\rm dn} \text{ and } \mathrm{pos}(w,t_0,\ldots,t_i)\subseteq \mathbf{basis}(t_{i+1})\},\end{aligned}$$

$$\mathrm{PFC}(K,\Sigma)=\{(t_0,\ldots,t_n):(\exists w\in\mathbf{basis}(t_0))((w,t_0,\ldots,t_n)\in \mathrm{FC}(K))\}.$$

2. We have a natural partial order $\leq$ on $\mathrm{FC}(K,\Sigma)$ (like in 1.1.7). The partial order $\leq$ on PFC is defined by
$(t_0,\ldots,t_{n-1})\leq(s_0,\ldots,s_{m-1})$ if and only if $m^{t_{n-1}}_{\rm up}=m^{s_{m-1}}_{\rm up}$, and
$$(\forall w\in\mathbf{basis}(t_0))\big(w\in\mathbf{basis}(s_0) \text{ and } (w,t_0,\ldots,t_{n-1})\leq(w,s_0,\ldots,s_{m-1})\big)$$
(so $(t_0)\leq(s_0)$ means that $s_0\in\Sigma(t_0)$ *and* $\mathbf{basis}(t_0)\subseteq\mathbf{basis}(s_0)$).
3. A sequence $\langle t_0,t_1,t_2,\ldots\rangle$ of creatures from K is *a pure candidate* with respect to a creating pair (K,Σ) if
$$(\forall i<\omega)(m^{t_i}_{\rm up}=m^{t_{i+1}}_{\rm dn})\quad\text{and}$$
$$(\exists w\in\mathbf{basis}(t_0))(\forall i<\omega)(\mathrm{pos}(w,t_0,\ldots,t_i)\subseteq\mathbf{basis}(t_{i+1})).$$

The set of pure candidates with respect to (K,Σ) is denoted by $\mathrm{PC}(K,\Sigma)$. The partial order $\leq$ on $\mathrm{PC}(K,\Sigma)$ is defined naturally.

4. For a norm condition $\mathcal{C}(\mathbf{nor})$ the family of *$\mathcal{C}(\mathbf{nor})$-normed pure candidates* is

$$\mathrm{PC}_{\mathcal{C}(\mathbf{nor})}(K,\Sigma) \stackrel{\mathrm{def}}{=} \{\langle t_0,t_1,\ldots\rangle \in \mathrm{PC}(K,\Sigma) : \langle t_0,t_1,\ldots,\rangle \text{ satisfies } \mathcal{C}(\mathbf{nor})\}.$$

DEFINITION 1.2.5. Let (K,Σ) be a creating pair for $\mathbf{H}$. We say that

1. (K,Σ) is *nice* if for all $t_0,\ldots,t_{n-1}\in K$ and $s\in\Sigma(t_0,\ldots,t_{n-1})$ we have $\mathbf{basis}(t_0)\subseteq\mathbf{basis}(s)$.
2. (K,Σ) is *smooth* provided that:
   if $(w,t_0,\ldots,t_{n-1})\in\mathrm{FC}(K,\Sigma)$, $m<n$ and $u\in\mathrm{pos}(w,t_0,\ldots,t_{n-1})$
   then $u\restriction m^{t_m}_{\mathrm{dn}}\in\mathrm{pos}(w,t_0,\ldots,t_{m-1})$ and $u\in\mathrm{pos}(u\restriction m^{t_m}_{\mathrm{dn}},t_m,\ldots,t_{n-1})$.
3. K is *forgetful* if for every creature $t\in K$ we have:

$$[\langle w,u\rangle\in\mathbf{val}[t]\ \&\ w'\in\prod_{n<m^t_{\mathrm{dn}}}\mathbf{H}(n)]\quad\Rightarrow\quad\langle w',w'{}^\frown u\restriction[m^t_{\mathrm{dn}},m^t_{\mathrm{up}})\rangle\in\mathbf{val}[t].$$

4. K is *full* if $\mathrm{dom}(\mathbf{val}[t])=\prod\limits_{n<m^t_{\mathrm{dn}}}\mathbf{H}(n)$ for every $t\in K$.

As we said in 1.2.2, we will always demand that a creating pair is nice and smooth (but these properties occur naturally in applications). The main reason for the first assumption is to have the effect presented in 1.2.8(2) below and the second demand is to get the conclusion of 1.2.10. Before we state these observations let us modify a little bit the forcing notions we are interested in.

DEFINITION 1.2.6. Let (K,Σ) be a creating pair and $\mathcal{C}(\mathbf{nor})$ be a property of ω-sequences of creatures. The forcing notion $\mathbb{Q}^*_{\mathcal{C}(\mathbf{nor})}(K,\Sigma)$ is a suborder of $\mathbb{Q}_{\mathcal{C}(\mathbf{nor})}(K,\Sigma)$ consisting of these conditions $(w,t_0,t_1,\ldots)$ for which additionally

$$(\boxtimes_{1.2.6})\qquad(\forall i<\omega)(m^{t_i}_{\mathrm{up}}=m^{t_{i+1}}_{\mathrm{dn}}).$$

REMARK 1.2.7. 1) The forcing notions introduced in 1.2.6 fit better to the idea of creatures and compositions on them. Moreover in most of the applications the forcing notions $\mathbb{Q}^*_{\mathcal{C}(\mathbf{nor})}(K,\Sigma)$ and $\mathbb{Q}_{\mathcal{C}(\mathbf{nor})}(K,\Sigma)$ will be equivalent. Even in the most general case they are not so far from each other; note that if $p\in\mathbb{Q}^*_{\mathcal{C}(\mathbf{nor})}(K,\Sigma)$ and $q\in\mathbb{Q}_{\mathcal{C}(\mathbf{nor})}(K,\Sigma)$, $p\leq q$ (in $\mathbb{Q}_{\mathcal{C}(\mathbf{nor})}(K,\Sigma)$) then $q\in\mathbb{Q}^*_{\mathcal{C}(\mathbf{nor})}(K,\Sigma)$. Of course it may happen that $\mathbb{Q}^*_{\mathcal{C}(\mathbf{nor})}(K,\Sigma)$ is trivial – this usually suggests that the tree–approach is more suitable (see 1.3.3).
2) Several notions simplify for the forcing notions $\mathbb{Q}^*_{\mathcal{C}(\mathbf{nor})}(K,\Sigma)$. For example if $t_0,\ldots,t_{n-1}\in K$ are such that $m=m^{t_0}_{\mathrm{dn}}$, $m^{t_i}_{\mathrm{up}}=m^{t_{i+1}}_{\mathrm{dn}}$ and $w\in\prod\limits_{i<m}\mathbf{H}(i)$ then

$$\begin{aligned}\mathrm{pos}(w,t_0,\ldots,t_{n-1})=\{u:\ &\text{for some } 0\leq k_1<\ldots<k_\ell<n-1 \text{ we have}\\ &u\restriction m^{t_{k_1}}_{\mathrm{up}}\in\mathrm{pos}^*(w,t_0,\ldots,t_{k_1})\ \&\\ &u\restriction m^{t_{k_2}}_{\mathrm{up}}\in\mathrm{pos}^*(u\restriction m^{t_{k_1}}_{\mathrm{up}},t_{k_1+1},\ldots,t_{k_2})\ \&\ \ldots\ \&\\ &u\in\mathrm{pos}^*(u\restriction m^{t_{k_\ell}}_{\mathrm{up}},t_{k_\ell+1},\ldots,t_{n-1})\}.\end{aligned}$$

3) The norm condition $(\mathrm{s}\infty)$ (see 1.1.10) can be presented slightly simpler for $\mathbb{Q}^*_{\mathrm{s}\infty}(K,\Sigma)$. For $(w,t_0,t_1,\ldots)\in\mathbb{Q}^*_{\mathrm{s}\infty}(K,\Sigma)$ it says just that

$$(\forall i<\omega)(\mathbf{nor}[t_i]>m^{t_i}_{\mathrm{dn}}).$$

PROPOSITION 1.2.8. *Suppose (K,Σ) is a creating pair for $\mathbf{H}$.*

1. *Assume that K is full. Then (K,Σ) is nice and if $\mathcal{C}(\mathbf{nor})$ is one of the conditions* (s∞), (∞), *or* (f) *(where f is any fast function), then the forcing notion $\mathbb{Q}^*_{\mathcal{C}(\mathbf{nor})}(K,\Sigma)$ is a dense subset of $\mathbb{Q}_{\mathcal{C}(\mathbf{nor})}(K,\Sigma)$.*
2. *If (K,Σ) is nice, $(w,t_0,t_1,t_2,\dots)$ is a condition in $\mathbb{Q}^*_\emptyset(K,\Sigma)$ and $\langle s_n : n\in\omega\rangle$ is such that for some $0=k_0<k_1<\dots<\omega$, $s_n\in\Sigma(t_i : k_n\le i<k_{n+1})$ (for all $n\in\omega$) then $(w,s_0,s_1,s_2,\dots)$ is a condition in $\mathbb{Q}^*_\emptyset(K,\Sigma)$ (stronger than $(w,t_0,t_1,t_2,\dots)$).*
3. *If (K,Σ) is forgetful then it is full.*

DEFINITION 1.2.9. Let (K,Σ) be a creating pair, $\mathcal{C}(\mathbf{nor})$ be a norm condition, $p\in\mathbb{Q}^*_{\mathcal{C}(\mathbf{nor})}(K,\Sigma)$ and $\dot\tau$ be a $\mathbb{Q}^*_{\mathcal{C}(\mathbf{nor})}(K,\Sigma)$-name for an ordinal. We say that

1. *p essentially decides the name $\dot\tau$* if

$$(\exists m\in\omega)(\forall u\in \mathrm{pos}(w^p,t^p_0,\dots,t^p_{m-1}))((u,t^p_m,t^p_{m+1},\dots) \text{ decides the value of } \dot\tau),$$

2. *p approximates $\dot\tau$ at n* (or at t^p_n) whenever:
for each $w_1\in\mathrm{pos}(w^p,t^p_0,\dots,t^p_{n-1})$, if there is a condition $r\in\mathbb{Q}^*_{\mathcal{C}(\mathbf{nor})}(K,\Sigma)$ stronger than p and such that $w^r=w_1$ and r decides the value of $\dot\tau$ then the condition $(w_1,t^p_n,t^p_{n+1},\dots)$ decides the value of $\dot\tau$.

LEMMA 1.2.10. *Suppose that (K,Σ) is a smooth creating pair, $\mathcal{C}(\mathbf{nor})$ is a norm condition and $\dot\tau$ is a $\mathbb{Q}^*_{\mathcal{C}(\mathbf{nor})}(K,\Sigma)$–name for an ordinal. Assume that a condition $p\in\mathbb{Q}^*_{\mathcal{C}(\mathbf{nor})}(K,\Sigma)$ essentially decides $\dot\tau$ (approximates $\dot\tau$ at each n, respectively). Then each $q\ge p$ essentially decides $\dot\tau$ (approximates $\dot\tau$ at each n, respectively).*

PROOF. Immediate by smoothness. □

DEFINITION 1.2.11. Let (K,Σ) be a creating pair for $\mathbf{H}$.

1. For a property $\mathcal{C}(\mathbf{nor})$ of ω-sequences of creatures from K and conditions $p,q\in\mathbb{Q}^*_{\mathcal{C}(\mathbf{nor})}(K,\Sigma)$ we define
   $p\le_{\mathrm{apr}} q$ (in $\mathbb{Q}^*_{\mathcal{C}(\mathbf{nor})}(K,\Sigma)$) if and only if
   $p\le q$ and for some k we have $(\forall i<\omega)(t^p_{i+k}=t^q_i)$
   (so then $w^q\in\mathrm{pos}(w^p,t^p_0,\dots,t^p_{k-1})$ too).
2. We define relations $\le^{\mathrm{s}\infty}_n$ (for $n<\omega$) on $\mathbb{Q}^*_{\mathrm{s}\infty}(K,\Sigma)$ by:
   (α) $p\le^{\mathrm{s}\infty}_0 q$ (in $\mathbb{Q}^*_{\mathrm{s}\infty}(K,\Sigma)$) if $p\le q$ and $w^p=w^q$,
   (β) $p\le^{\mathrm{s}\infty}_{n+1} q$ (in $\mathbb{Q}^*_{\mathrm{s}\infty}(K,\Sigma)$) if $p\le^{\mathrm{s}\infty}_0 q$ and $t^p_i=t^q_i$ for $i<n+1$.
3. Relations $\le^\infty_n$ on $\mathbb{Q}^*_\infty(K,\Sigma)$ (for $n<\omega$) are defined by:
   (α) $p\le^\infty_0 q$ (in $\mathbb{Q}^*_\infty(K,\Sigma)$) if $p\le q$ and $w^p=w^q$,
   (β^*) $p\le^\infty_{n+1} q$ (in $\mathbb{Q}^*_\infty(K,\Sigma)$) if $p\le^\infty_0 q$ and

$$t^p_j=t^q_j \text{ for all } j\le\min\{i<\omega:\mathbf{nor}[t^p_i]>n+1\} \text{ and}$$

$$\{t^q_i : i<\omega \ \&\ \mathbf{nor}[t^q_i]\le n+1\}\subseteq\{t^p_i : i<\omega\}.$$

4. Relations $\le^{\mathrm{w}\infty}_n$ on $\mathbb{Q}^*_{\mathrm{w}\infty}(K,\Sigma)$ are defined by
   (α) $p\le^{\mathrm{w}\infty}_0 q$ (in $\mathbb{Q}^*_{\mathrm{w}\infty}(K,\Sigma)$) if $p\le q$ and $w^p=w^q$,
   (β^+) $p\le^{\mathrm{w}\infty}_{n+1} q$ (in $\mathbb{Q}^*_{\mathrm{w}\infty}(K,\Sigma)$) if $p\le^{\mathrm{w}\infty}_0 q$ and

$$t^p_j=t^q_j \text{ for all } j\le\min\{i<\omega:\mathbf{nor}[t^p_i]>n+1\}.$$

5. Let $f:\omega\times\omega\longrightarrow\omega$ be a fast function. Relations $\le^f_n$ on $\mathbb{Q}^*_f(K,\Sigma)$ are defined by:
   (α) $p\le^f_0 q$ (in $\mathbb{Q}^*_\infty(K,\Sigma)$) if $p\le q$ and $w^p=w^q$,

(β^f) $p \leq^f_{n+1} q$ (in $\mathbb{Q}^*_f(K,\Sigma)$) if $p \leq^f_0 q$ and

$$t^p_j = t^q_j \text{ for all } j \leq \min\{i < \omega : \mathbf{nor}[t^p_i] > f(n+1, m^{t^p_i}_{\rm dn})\} \qquad \text{and}$$

$$\{t^q_i : i < \omega \ \&\ \mathbf{nor}[t^q_i] \leq f(n+1, m^{t^q_i}_{\rm dn})\} \subseteq \{t^p_i : i < \omega\}.$$

6. We may omit superscripts in $\leq^{s\infty}_n$, $\leq^{\infty}_n$, $\leq^{w\infty}_n$ and $\leq^f_n$ if it is clear from the context in which forcing notion we are working (i.e. what is the norm condition we deal with).

REMARK 1.2.12. The difference between e.g. (3) and (4) is in the last condition of (3), of course.

PROPOSITION 1.2.13. *Suppose (K,Σ) is a creating pair for* $\mathbf{H}$. *Let $\mathcal{C}(\mathbf{nor})$ be one of the following properties of ω-sequences: $\mathcal{C}^{s\infty}(\mathbf{nor})$, $\mathcal{C}^{\infty}(\mathbf{nor})$, $\mathcal{C}^{w\infty}(\mathbf{nor})$ or $\mathcal{C}^f(\mathbf{nor})$ for some fast function f (see 1.1.10) and let $\leq_n$ be the corresponding relations (defined in 1.2.11). Then*

1. *$\leq_{\rm apr}$ is a partial order (stronger than $\leq$) on $\mathbb{Q}^*_{\mathcal{C}(\mathbf{nor})}(K,\Sigma)$.*
2. *$\leq_n$ (with superscripts) are partial orders (stronger than $\leq$) on the respective $\mathbb{Q}^*_{\mathcal{C}(\mathbf{nor})}(K,\Sigma)$ and $p \leq_{n+1} q$ implies $p \leq_n q$.*
3. *Suppose that $p_n \in \mathbb{Q}^*_{\mathcal{C}(\mathbf{nor})}(K,\Sigma)$ (for $n \in \omega$) are such that*
$$(\forall n \in \omega)(p_n \leq_{n+1} p_{n+1}).$$
Then the naturally defined limit condition $p = \lim\limits_n p_n$ satisfies:
$$p \in \mathbb{Q}^*_{\mathcal{C}(\mathbf{nor})}(K,\Sigma) \quad \text{and} \quad (\forall n < \omega)(p_n \leq_{n+1} p).$$

REMARK 1.2.14. A natural property one could ask for in the context of creating pairs is some kind of monotonicity:

$$\mathbf{basis}(t) = \mathrm{dom}(\mathbf{val}[t]) \quad \text{and} \quad \mathrm{pos}(u,t) = \{v : \langle u, v\rangle \in \mathbf{val}[t]\},$$

for $t \in K$ and $u \in \mathbf{basis}(t)$. However, there is no real need for it, as all our demands and assumptions on creating pairs will refer to pos (and not **val**). But for tree–creating pairs we will postulate the respective demand, mainly to simplify notation (and have explicit tree–representations of conditions), see 1.3.3(3).

1.3. Tree creatures and tree–like forcing notions

Here we introduce the second option for our general scheme: forcing notions in which conditions are trees with norms. This case, though parallel to the one of creating pairs, is of different character and therefore we reformulate all general definitions for this particular context.

DEFINITION 1.3.1. 1. *A quasi tree* is a set T of finite sequences with the $\vartriangleleft$-smallest element denoted by $\mathrm{root}(T)$.

2. A quasi tree T is *a tree* if it is closed under initial segments. If T is a quasi tree then $\mathrm{dcl}(T)$ is the smallest tree containing T (the downward closure of T).
3. For a quasi tree T and $\eta \in T$ we define *the successors of η in T, the restriction of T to η, the splitting points of T* and *maximal points of T* by:
$$\mathrm{succ}_T(\eta) = \{\nu \in T : \eta \vartriangleleft \nu \ \&\ \neg(\exists \rho \in T)(\eta \vartriangleleft \rho \vartriangleleft \nu)\},$$
$$T^{[\eta]} = \{\nu \in T : \eta \trianglelefteq \nu\},$$

$$\mathrm{split}(T) = \{\eta \in T : |\mathrm{succ}_T(\eta)| \geq 2\},$$

$$\max(T) = \{\nu \in T : \text{ there is no } \rho \in T \text{ such that } \nu \vartriangleleft \rho\}.$$

We put $\hat{T} = T \setminus \max(T)$.

4. The set of all limit infinite branches through a quasi tree T is

$$\lim(T) \stackrel{\mathrm{def}}{=} \{\eta : \eta \text{ is an } \omega\text{–sequence} \quad \text{and } (\exists^\infty n)(\eta{\restriction}n \in T)\}.$$

The quasi tree T is *well founded* if $\lim(T) = \emptyset$.

5. A subset F of a quasi tree T is *a front* in T if no two distinct members of F are $\vartriangleleft$-comparable and

$$(\forall \eta \in \lim(T) \cup \max(T))(\exists n \in \omega)(\eta \restriction n \in F).$$

REMARK 1.3.2. Note the difference between $\lim(T)$ and $\lim(\mathrm{dcl}(T))$ for a quasi tree T. In particular, it is possible that a quasi tree T is well-founded but there is an infinite branch through $\mathrm{dcl}(T)$. Moreover, a front in T does not have to be a front in $\mathrm{dcl}(T)$.

DEFINITION 1.3.3. 1. A weak creature $t \in \mathrm{WCR}[\mathbf{H}]$ is a *tree–creature* if $\mathrm{dom}(\mathbf{val}[t])$ is a singleton $\{\eta\}$ and no two distinct elements of $\mathrm{rng}(\mathbf{val}[t])$ are $\vartriangleleft$–comparable;

$\mathrm{TCR}[\mathbf{H}]$ is the family of all tree–creatures for $\mathbf{H}$.

2. $\mathrm{TCR}_\eta[\mathbf{H}] = \{t \in \mathrm{TCR}[\mathbf{H}] : \mathrm{dom}(\mathbf{val}[t]) = \{\eta\}\}$.
3. A sub-composition operation Σ on $K \subseteq \mathrm{TCR}[\mathbf{H}]$ is *a tree composition* (and then (K, Σ) is called *a tree–creating pair* (for $\mathbf{H}$)) if:
   (a) if $\mathcal{S} \in [K]^{\leq\omega}$, $\Sigma(\mathcal{S}) \neq \emptyset$ then $\mathcal{S} = \{s_\nu : \nu \in \hat{T}\}$ for some well founded quasi tree $T \subseteq \bigcup\limits_{n<\omega} \prod\limits_{i<n} \mathbf{H}(i)$ and a system $\langle s_\nu : \nu \in \hat{T}\rangle \subseteq K$ such that for each finite sequence $\nu \in \hat{T}$

   $$s_\nu \in \mathrm{TCR}_\nu[\mathbf{H}] \quad \text{and} \quad \mathrm{rng}(\mathbf{val}[s_\nu]) = \mathrm{succ}_T(\nu),$$

   and
   (b) if $t \in \Sigma(s_\nu : \nu \in \hat{T})$ then $t \in \mathrm{TCR}_{\mathrm{root}(T)}[\mathbf{H}]$ and $\mathrm{rng}(\mathbf{val}[t]) \subseteq \max(T)$.

   If $\hat{T} = \{\mathrm{root}(T)\}$, $t = t_{\mathrm{root}(T)} \in \mathrm{TCR}_{\mathrm{root}(T)}[\mathbf{H}]$ and $\mathrm{rng}(\mathbf{val}[t]) = \max(T)$ then we will write $\Sigma(t)$ instead of $\Sigma(t_\nu : \nu \in \hat{T})$.
4. A tree-composition Σ on K is *bounded* if for each $t \in \Sigma(s_\nu : \nu \in \hat{T})$ we have

$$\mathbf{nor}[t] \leq \max\{\mathbf{nor}[s_\nu] : (\exists \eta \in \mathrm{rng}(\mathbf{val}[t]))(\nu \vartriangleleft \eta)\}.$$

REMARK 1.3.4. 1) Note that sets of tree creatures relevant for tree compositions have a natural structure: we identify here $\mathcal{S}$ with $\{s_{\nu(s)} : s \in \mathcal{S}\}$ where $\nu(s)$ is such that $s \in \mathrm{TCR}_{\nu(s)}$ and $s_{\nu(s)} = s$.
2) To check consistency of our notation for tree creatures with that of 1.1.7 note that in 1.3.3(3), if $s_\nu \in \Sigma(s_{\nu,\eta} : \eta \in \hat{T}_\nu)$ for each $\nu \in \hat{T}$, T is a well founded quasi tree as in (3)**(a)** of 1.3.3 then $T^* \stackrel{\mathrm{def}}{=} \bigcup\limits_{\nu \in \hat{T}} T_\nu$ is a well founded quasi tree, $\hat{T}^* = \bigcup\limits_{\nu \in \hat{T}} \hat{T}_\nu$ and $\langle s_{\nu,\eta} : \nu \in \hat{T}, \eta \in \hat{T}_\nu\rangle$ is a system for which Σ may be non-empty, i.e. it satisfies the requirements of 1.3.3(3)**(a)**.
3) Note that if (K, Σ) is a tree–creating pair for $\mathbf{H}$, $t \in \mathrm{TCR}_\eta[\mathbf{H}]$ then $\mathbf{basis}(t) = \{\eta\}$ and $\mathrm{pos}(\eta, t) = \mathrm{rng}(\mathbf{val}[t])$ (see 1.1.6). For this reason we will write $\mathrm{pos}(t)$ for $\mathrm{pos}(\eta, t)$ and $\mathrm{rng}(\mathbf{val}[t])$ in the context of tree–creating pairs.

4) Tree–creating pairs have the properties corresponding to the niceness and smoothness of creating pairs (see 1.2.5, compare with 1.3.9).

When dealing with tree–creating pairs it seems to be more natural to consider both very special norm conditions and some restrictions on conditions of the forcing notions we consider. The second is not very serious: the forcing notions $\mathbb{Q}^{\rm tree}_e(K,\Sigma)$ (for $e<5$) introduced in 1.3.5 below are dense subsets of the general forcing notions $\mathbb{Q}_{\mathcal{C}(\mathbf{nor})}(K,\Sigma)$ (for suitable conditions $\mathcal{C}(\mathbf{nor})$). We write the definition of $\mathbb{Q}^{\rm tree}_e(K,\Sigma)$ fully, not referring the reader to 1.1.7, to show explicitly the way tree creating pairs work.

Definition 1.3.5. Let (K,Σ) be a tree–creating pair for $\mathbf{H}$.

1. We define the forcing notion $\mathbb{Q}^{\rm tree}_1(K,\Sigma)$ by letting:
   conditions be sequences $p=\langle t_\eta:\eta\in T\rangle$ such that
   (a) $T\subseteq\bigcup\limits_{n\in\omega}\prod\limits_{i<n}\mathbf{H}(i)$ is a non-empty quasi tree with $\max(T)=\emptyset$,
   (b) $t_\eta\in{\rm TCR}_\eta[\mathbf{H}]\cap K$ and ${\rm pos}(t_\eta)={\rm succ}_T(\eta)$ (see 1.3.4(3)),
   $({\rm c})_1$ for every $\eta\in\lim(T)$ we have:
   $$\text{the sequence } \langle\mathbf{nor}[t_{\eta\restriction k}]:k<\omega,\eta\restriction k\in T\rangle \text{ diverges to infinity;}$$
   the order be given by:
   $\langle t^1_\eta:\eta\in T^1\rangle\leq\langle t^2_\eta:\eta\in T^2\rangle$ if and only if
   $T^2\subseteq T^1$ and for each $\eta\in T^2$ there is a well founded quasi tree $T_{0,\eta}\subseteq(T^1)^{[\eta]}$ such that $t^2_\eta\in\Sigma(t^1_\nu:\nu\in\hat{T}_{0,\eta})$.
   If $p=\langle t_\eta:\eta\in T\rangle$ then we write ${\rm root}(p)={\rm root}(T)$, $T^p=T$, $t^p_\eta=t_\eta$ etc.
2. Similarly we define forcing notions $\mathbb{Q}^{\rm tree}_e(K,\Sigma)$ for $e=0,2,3,4$ replacing the condition $(\mathbf{c})_1$ by $(\mathbf{c})_e$ respectively, where:
   $({\rm c})_0$ for every $\eta\in\lim(T)$:
   $$\limsup\langle\mathbf{nor}[t_{\eta\restriction k}]:k<\omega,\eta\restriction k\in T\rangle=\infty,$$
   $({\rm c})_2$ for every $\eta\in T$ and $n<\omega$ there is ν such that $\eta\vartriangleleft\nu\in T$ and $\mathbf{nor}[t_\nu]\geq n$,
   $({\rm c})_3$ for every $\eta\in T$ and $n<\omega$ there is ν such that $\eta\vartriangleleft\nu\in T$ and
   $$(\forall\rho\in T)(\nu\vartriangleleft\rho\quad\Rightarrow\quad\mathbf{nor}[t_\rho]\geq n),$$
   $({\rm c})_4$ for every $n<\omega$, the set
   $$\{\nu\in T:(\forall\rho\in T)(\nu\vartriangleleft\rho\quad\Rightarrow\quad\mathbf{nor}[t_\rho]\geq n)\}$$
   contains a front of the quasi tree T.
3. If $p\in\mathbb{Q}^{\rm tree}_e(K,\Sigma)$ then we let $p^{[\eta]}=\langle t^p_\nu:\nu\in(T^p)^{[\eta]}\rangle$ for $\eta\in T^p$.
4. For the sake of notational convenience we define partial order $\mathbb{Q}^{\rm tree}_\emptyset(K,\Sigma)$ in the same manner as $\mathbb{Q}^{\rm tree}_e(K,\Sigma)$ above but we omit the requirement $(\mathbf{c})_e$ (like in 1.1.10; so this is essentially $\mathbb{Q}_\emptyset(K,\Sigma)$).

Remark 1.3.6. 1) In the definition above we do not follow exactly the notation of 1.1.7: we omit the first part w^p of a condition p as it can be clearly read from the rest of the condition. Of course the missing item is ${\rm root}(p)$. In this new notation the name $\dot{W}$ of 1.1.13 may be defined by

$$\Vdash_{\mathbb{Q}^{\rm tree}_e(K,\Sigma)}\dot{W}=\bigcup\{{\rm root}(p):p\in\Gamma_{\mathbb{Q}^{\rm tree}_e(K,\Sigma)}\}.$$

2) Note that

$$\mathbb{Q}_4^{\rm tree}(K,\Sigma)\subseteq\mathbb{Q}_1^{\rm tree}(K,\Sigma)\subseteq\mathbb{Q}_0^{\rm tree}(K,\Sigma)\subseteq\mathbb{Q}_2^{\rm tree}(K,\Sigma)\qquad\text{and}$$
$$\mathbb{Q}_1^{\rm tree}(K,\Sigma)\subseteq\mathbb{Q}_3^{\rm tree}(K,\Sigma)\subseteq\mathbb{Q}_2^{\rm tree}(K,\Sigma)$$

but in general these inclusions do not mean "complete suborders". If the tree–creating pair is t-omittory (see 2.3.4) then $\mathbb{Q}_4^{\rm tree}(K,\Sigma)$ is dense in $\mathbb{Q}_2^{\rm tree}(K,\Sigma)$ and thus all these forcing notions are equivalent. If (K,Σ) is $\bar{2}$–big (see 2.3.2) then $\mathbb{Q}_4^{\rm tree}(K,\Sigma)$ is dense in $\mathbb{Q}_1^{\rm tree}(K,\Sigma)$ (see 2.3.12).

Let us give two simple examples of tree–creating pairs.

Let $\mathbf{H}(i)=\omega$ (for $i\in\omega$).

Let $K_0\subseteq\mathrm{TCR}[\mathbf{H}]$ consists of these tree–creatures s that if $s\in\mathrm{TCR}_\eta[\mathbf{H}]$ then $\mathrm{rng}(\mathbf{val}[s])\subseteq\{\eta^\frown\langle k\rangle : k\in\omega\}$ and

$$\mathbf{nor}[s]=\begin{cases}\ell g(\eta) & \text{if } \mathbf{val}[s] \text{ is infinite,}\\ 0 & \text{otherwise.}\end{cases}$$

The operation Σ_0 gives non-empty values for singletons only; for $s\in K_0$ we let $\Sigma_0(s)=\{t\in K_0:\mathbf{val}[t]\subseteq\mathbf{val}[s]\}$ (an operation Σ defined in this manner will be further called *trivial*). Clearly (K_0,Σ_0) is a tree–creating pair. Note that:

(a) the forcing notions $\mathbb{Q}_2^{\rm tree}(K_0,\Sigma_0)$ and $\mathbb{Q}_0^{\rm tree}(K_0,\Sigma_0)$ are equivalent to Miller's Rational Perfect Set Forcing;

(b) the forcing notions $\mathbb{Q}_1^{\rm tree}(K_0,\Sigma_0)$, $\mathbb{Q}_3^{\rm tree}(K_0,\Sigma_0)$, $\mathbb{Q}_4^{\rm tree}(K_0,\Sigma_0)$ are equivalent to the Laver forcing

(thus $\mathbb{Q}_1^{\rm tree}(K_0,\Sigma_0)$ is not a complete suborder of $\mathbb{Q}_0^{\rm tree}(K_0,\Sigma_0)$, and $\mathbb{Q}_3^{\rm tree}(K_0,\Sigma_0)$ is not a complete suborder of $\mathbb{Q}_2^{\rm tree}(K_0,\Sigma_0)$).

Let us modify the norms on the tree–creatures a little. For this we define a function $f:\omega^{<\omega}\longrightarrow\omega$ by

$$f(\langle\rangle)=0,\quad f(\eta^\frown\langle k\rangle)=\begin{cases}f(\eta)+1 & \text{if } k=0\\ f(\eta) & \text{otherwise.}\end{cases}$$

Now, let K_1 consist of tree creatures $s\in\mathrm{TCR}[\mathbf{H}]$ such that $\mathrm{rng}(\mathbf{val}[s])\subseteq\{\eta^\frown\langle k\rangle : k\in\omega\}$ (where $s\in\mathrm{TCR}_\eta[\mathbf{H}]$) and

$$\mathbf{nor}[s]=\begin{cases}f(\eta) & \text{if } \mathbf{val}[s] \text{ is infinite,}\\ 0 & \text{otherwise.}\end{cases}$$

Let Σ_1 be the trivial tree-composition on K_1, so it is nonempty for singletons only and then $\Sigma_1(s)=\{t\in K_1:\mathbf{val}[t]\subseteq\mathbf{val}[s]\}$ (clearly (K_1,Σ_1) is a tree creating pair). Then

(a) the forcing notion $\mathbb{Q}_0^{\rm tree}(K_1,\Sigma_1)$ is a dense suborder of $\mathbb{Q}_2^{\rm tree}(K_1,\Sigma_1)$,

(b) the partial orders $\mathbb{Q}_1^{\rm tree}(K_1,\Sigma_1)$ and $\mathbb{Q}_4^{\rm tree}(K_1,\Sigma_1)$ are empty, but

(c) $\mathbb{Q}_3^{\rm tree}(K_1,\Sigma_1)$ is not–trivial (it adds a new real) and it is not a complete suborder of $\mathbb{Q}_2^{\rm tree}(K_1,\Sigma_1)$ (e.g. incompatibility is not preserved) and it is disjoint from $\mathbb{Q}_0^{\rm tree}(K_1,\Sigma_1)$.

DEFINITION 1.3.7. Let (K,Σ) be a tree creating pair, $e<5$, $p\in\mathbb{Q}_e^{\rm tree}(K,\Sigma)$. A set $A\subseteq T^p$ is called *an e-thick antichain* (or just *a thick antichain*) if it is an antichain in $(T^p,\vartriangleleft)$ and for every condition $q\in\mathbb{Q}_e^{\rm tree}(K,\Sigma)$ stronger than p the intersection $A\cap\mathrm{dcl}(T^q)$ is non-empty.

PROPOSITION 1.3.8. *Suppose that (K,Σ) is a tree-creating pair for $\mathbf{H}$, $e<5$, $p\in\mathbb{Q}_e^{\rm tree}(K,\Sigma)$ and $\eta\in T^p$. Then:*

1. *$\mathbb{Q}^{\rm tree}_e(K,\Sigma)$ is a partial order.*
2. *Each e-thick antichain in T^p is a maximal antichain. Every front of T^p is an e-thick antichain in T^p.*
3. *If $e\in\{1,3,4\}$, $n<\omega$ then the set*

$$B_n(p)\stackrel{\rm def}{=}\{\eta\in T^p:\ \ (i)\ \ (\forall\nu\in T^p)(\eta\trianglelefteq\nu\ \ \Rightarrow\ \ \mathbf{nor}[t_\nu]>n)\ \ but\\ (ii)\ \ no\ \eta'\vartriangleleft\eta,\eta'\in T^p\ satisfies\ (i)\}$$

is a maximal $\vartriangleleft$-antichain in T^p. If $e=4$ then $B_n(p)$ is a front of T^p.

4. *For every $m,n<\omega$ the set*

$$F^m_n(p)\stackrel{\rm def}{=}\{\eta\in T^p:\ \ (i)\ \ \mathbf{nor}[t_\eta]>n\ and\\ (ii)\ |\{\eta'\in T^p:\eta'\vartriangleleft\eta\ \&\ \mathbf{nor}[t_{\eta'}]>n\}|=m\}$$

is a maximal $\vartriangleleft$-antichain of T^p. If $e\in\{0,1,4\}$ then $F^m_n(p)$ is a front of T^p.

5. *If K is finitary (so $|\mathbf{val}[t]|<\omega$ for $t\in K$, see 1.1.3) then every front of T^p is a front of $\mathrm{dcl}(T^p)$ and hence it is finite.*
6. *If Σ is bounded then each $F^m_n(p)$ is a thick antichain of T^p.*
7. *$p\le p^{[\eta]}\in\mathbb{Q}^{\rm tree}_e(K,\Sigma)$ and $\mathrm{root}(p^{[\eta]})=\eta$.*

REMARK 1.3.9. One of the useful properties of tree–creating pairs (K,Σ) and forcing notions $\mathbb{Q}^{\rm tree}_e(K,\Sigma)$ is the following:

$(*)_{1.3.9}$ *Suppose that $p,q\in\mathbb{Q}^{\rm tree}_e(K,\Sigma)$, $p\le q$ (so in particular $T^q\subseteq T^p$), $\eta\in T^q$ and $\nu\vartriangleleft\eta$, $\nu\in T^p$.*
Then $p^{[\nu]}\le q^{[\eta]}$.

DEFINITION 1.3.10. Let $p,q\in\mathbb{Q}^{\rm tree}_e(K,\Sigma)$, $e<3$ (and (K,Σ) a tree–creating pair). We define relations $\le^e_n$ for $n\in\omega$ by:

1. If $e\in\{0,2\}$ then:
   $p\le^e_0 q$ (in $\mathbb{Q}^{\rm tree}_e(K,\Sigma)$) if $p\le q$ and $\mathrm{root}(p)=\mathrm{root}(q)$,
   $p\le^e_{n+1} q$ (in $\mathbb{Q}^{\rm tree}_e(K,\Sigma)$) if $p\le^e_0 q$ and
   if $\eta\in F^0_n(p)$ (see 1.3.8(4)) and $\nu\in T^p$, $\nu\trianglelefteq\eta$ then $\nu\in T^q$ and $t^q_\nu=t^p_\nu$.
2. The relations $\le^1_n$ (on $\mathbb{Q}^{\rm tree}_1(K,\Sigma)$) are defined by:
   $p\le^1_0 q$ (in $\mathbb{Q}^{\rm tree}_1(K,\Sigma)$) if and only if $p\le q$ and $\mathrm{root}(p)=\mathrm{root}(q)$,
   $p\le^1_{n+1} q$ (in $\mathbb{Q}^{\rm tree}_1(K,\Sigma)$) if $p\le^1_0 q$ and
   if $\eta\in F^0_n(p)$ (see 1.3.8(4)) and $\nu\in T^p$, $\nu\trianglelefteq\eta$ then $\nu\in T^q$, $t^p_\nu=t^q_\nu$, and

$$\{t^q_\eta:\eta\in T^q\ \&\ \mathbf{nor}[t^q_\eta]\le n\}\subseteq\{t^p_\eta:\eta\in T^p\}.$$

3. We may omit the superscript e in $\le^e_n$ if it is clear in which of the forcing notions $\mathbb{Q}^{\rm tree}_e(K,\Sigma)$ we are working.

PROPOSITION 1.3.11. *Let (K,Σ) be a tree–creating pair for $\mathbf{H}$, $e<3$.*

1. *The relations $\le^e_n$ are partial orders on $\mathbb{Q}^{\rm tree}_e(K,\Sigma)$ stronger than $\le$. The partial order $\le^e_{n+1}$ is stronger than $\le^e_n$.*
2. *Suppose that conditions $p_n\in\mathbb{Q}^{\rm tree}_e(K,\Sigma)$ are such that $p_n\le^e_{n+1}p_{n+1}$.*
   Then $\lim\limits_{n\in\omega}p_n\in\mathbb{Q}^{\rm tree}_e(K,\Sigma)$ and $(\forall n\in\omega)(p_n\le^e_{n+1}\lim\limits_{n\in\omega}p_n)$ (where the limit condition $p=\lim\limits_{n\in\omega}p_n$ is defined naturally; $T^p=\bigcap\limits_{n\in\omega}T^{p_n}$).

1.4. Non proper examples

In the next chapter we will see that if one combines a norm condition with suitable properties of a weak creating pair then the resulting forcing notion is proper. In particular we will see that (with the norm conditions defined in 1.3.5) getting properness in the case of tree–creating pairs is relatively easy. Here, however, we show that one cannot expect a general theorem like "$\mathbb{Q}^{\rm tree}_{C(\mathbf{nor})}(K,\Sigma)$ is always proper" and that we should be always careful a little bit. The forcing notions resulting from our general schema may collapse $\aleph_1$! For example, looking at the norm conditions introduced in 1.3.5(2) one could try to consider the following condition

$(\mathbf{c})_5$ $(\forall k\in\omega)(\forall^\infty n)(\forall\eta\in T^p)(\ell g(\eta)\geq n \ \Rightarrow\ \mathbf{nor}[t_\nu]\geq k)$.

If a creating pair (K,Σ) is finitary then, clearly, the forcing notions $\mathbb{Q}^{\rm tree}_5(K,\Sigma)$ and $\mathbb{Q}^{\rm tree}_4(K,\Sigma)$ are the same.

The forcing notion $\mathbb{Q}^{\rm tree}_5(K,\Sigma)$ might be even not proper. The following example shows this bad phenomenon which may be made quite general.

EXAMPLE 1.4.1. Let $\mathbf{H}(i)=\omega$ for $i\in\omega$.
There is a tree creating pair $(K_{1.4.1},\Sigma_{1.4.1})$ for $\mathbf{H}$ which is simple (see 2.1.7) and the forcing notion $\mathbb{Q}^{\rm tree}_5(K_{1.4.1},\Sigma_{1.4.1})$ is not proper (and collapses $\mathfrak{c}$ onto ω).

CONSTRUCTION. To define the family $K_{1.4.1}$ of tree creatures for $\mathbf{H}$ choose families $\emptyset\neq S^\ell_\eta\subseteq R_\eta\subseteq[\omega]^\omega$ and functions $h_\eta: R_\eta\longrightarrow\omega$ (for $\eta\in\omega^{<\omega}$, $0<\ell\leq \ell g(\eta)$) such that for every η,ℓ we have:

(α) $\omega\in R_\eta$, $h_\eta(\omega)=\ell g(\eta)+1$,

(β) if $F\in S^\ell_\eta$ then $h_\eta(F)=\ell$, each S^ℓ_η is infinite,

(γ) if $F_0,F_1\in S^\ell_\eta$, $F_0\neq F_1$ then $F_0\cap F_1=\emptyset$,

(δ) if $A\in R_\eta$, $h_\eta(A)\geq\ell+1$ then for each $F\in S^\ell_\eta$

$$A\cap F\in R_\eta \qquad\text{and}\qquad h_\eta(A\cap F)=\ell,$$

(ε) if $A_0,A_1\in R_\eta$, $A_0\subseteq A_1$ then $h_\eta(A_0)\leq h_\eta(A_1)$.

There are several possibilities to construct $S^\ell_\eta, R_\eta, h_\eta$ as above. One can do it for example in the following way. Fix $\eta\in\omega^n$. Take a system $\{K_\sigma:\sigma\in\omega^{\leq n}\}$ of infinite subsets of ω such that

a) $K_{\langle\rangle}=\omega$,

b) $\sigma_0\vartriangleleft\sigma_1\in\omega^{\leq n}\ \Rightarrow\ K_{\sigma_1}\subseteq K_{\sigma_0}$,

c) $\sigma_0,\sigma_1\in\omega^\ell\ \&\ \ell\leq n\ \&\ \sigma_0\neq\sigma_1\ \Rightarrow\ K_{\sigma_0}\cap K_{\sigma_1}=\emptyset$,

d) $\sigma\in\omega^{<n}\ \Rightarrow\ K_\sigma=\bigcup\limits_{m\in\omega}K_{\sigma^\frown\langle m\rangle}$.

Now put $R_\eta=\{\bigcup\limits_{\sigma\in I}K_\sigma:\emptyset\neq I\subseteq\omega^n\}$. For $A\in R_\eta$ we declare that

e) $h_\eta(A)\geq 1$, $h_\eta(\omega)=h_\eta(K_{\langle\rangle})=n+1$,

f) if for some $\sigma\in\omega^{n-\ell}$, $\ell\leq n$ the set A contains the set K_σ then $h_\eta(A)\geq \ell+1$,

g) if $A_0\subseteq A_1$, $A_0,A_1\in R_\eta$ then $h_\eta(A_0)\leq h_\eta(A_1)$.

Next we put

$$F^{\ell,m}_\eta=\bigcup\{K_{\sigma^\frown\langle m\rangle}:\sigma\in\omega^{n-\ell}\},\qquad\text{for } 0<\ell\leq n,\ m\in\omega\quad\text{and}$$
$$S^\ell_\eta=\{F^{\ell,m}_\eta:m\in\omega\}.$$

It should be easy to check that S^ℓ_η, R_η, h_η (for $\eta \in \omega^{<\omega}$, $0<\ell\leq \ell g(\eta)$) defined in this way satisfy the requirements (α)–(ε).

A tree creature $t \in \mathrm{TCR}_\eta[\mathbf{H}]$ is in $K_{1.4.1}$ if $\eta \in \omega^{<\omega}$, $\ell g(\eta) > 0$ and

$$\mathbf{dis}[t] \in R_\eta, \quad \mathbf{val}[t] = \{\langle \eta, \eta^\frown\langle m\rangle\rangle : m \in \mathbf{dis}[t]\} \quad \text{and} \quad \mathbf{nor}[t] = h_\eta(\mathbf{dis}[t]).$$

If $A \in R_\eta$ then the unique tree–creature $t \in \mathrm{TCR}_\eta[\mathbf{H}] \cap K_{1.4.1}$ such that $\mathbf{dis}[t] = A$ will be denoted by $t^{\eta,A}$.
The operation $\Sigma_{1.4.1}$ is trivial: it gives a non-empty result for singletons only and then $\Sigma_{1.4.1}(t) = \{s \in K_{1.4.1} : \mathbf{val}[s] \subseteq \mathbf{val}[t]\}$.

CLAIM 1.4.1.1. *The forcing notion $\mathbb{Q}^{\rm tree}_5(K_{1.4.1}, \Sigma_{1.4.1})$ collapses $\mathfrak{c}$ onto ω.*

Proof of the claim. Fix $\eta \in \omega^{<\omega} \setminus \{\langle\rangle\}$ for a moment.
Elements of S^ℓ_η are pairwise disjoint so we may naturally order them according to the smallest element. Say $S^\ell_\eta = \{F^{\ell,m}_\eta : m < \omega\}$. Let $f : [\ell g(\eta), \omega) \longrightarrow \omega$. We define a condition $p^{f,\eta} \in \mathbb{Q}^{\rm tree}_5(K_{1.4.1}, \Sigma_{1.4.1})$ putting (we keep the notation as for the forcing notions $\mathbb{Q}^{\rm tree}_e(K,\Sigma)$):

$\mathrm{root}(p^{f,\eta}) = \eta$;
let $k_0 = \ell g(\eta)$, $k_{\ell+1} = f(k_\ell) + k_\ell + 1$ (for $\ell < \omega$);
if $\nu \in T^{p^{f,\eta}}$, $k_{\ell-1} \leq \ell g(\nu) < k_\ell$ then $t^{p^{f,\eta}}_\nu = t^{\nu, F^{\ell, f(\ell g(\nu))}_\nu}$ and
$\mathrm{succ}_{T^{p^{f,\eta}}}(\nu) = \{\nu^\frown\langle m\rangle : m \in F^{\ell, f(\ell g(\nu))}_\nu\}$.

Clearly this defines $p^{f,\eta} \in \mathbb{Q}^{\rm tree}_5(K_{1.4.1}, \Sigma_{1.4.1})$. Note that
if $f, g : [\ell g(\eta), \omega) \longrightarrow \omega$ are distinct
then the conditions $p^{f,\eta}, p^{g,\eta}$ are incompatible in $\mathbb{Q}^{\rm tree}_5(K_{1.4.1}, \Sigma_{1.4.1})$
(by the requirement (γ)). Let $\dot{\tau}$ be a $\mathbb{Q}^{\rm tree}_5(K_{1.4.1}, \Sigma_{1.4.1})$–name for a function defined on $\omega^{<\omega}$ such that $\Vdash$"$\dot{\tau}(\eta) \in \mathbf{V}$ & $\dot{\tau}(\eta) : [\ell g(\eta), \omega) \longrightarrow \omega$" for $\eta \in \omega^{<\omega}$ and for $f : [\ell g(\eta), \omega) \longrightarrow \omega$ we have

$$p^{f,\eta} \Vdash_{\mathbb{Q}^{\rm tree}_5(K_{1.4.1}, \Sigma_{1.4.1})} \dot{\tau}(\eta) = f.$$

This definition is correct as $\{p^{f,\eta} : f : [\ell g(\eta), \omega) \to \omega\}$ is an antichain (of course it is not necessarily maximal in $\mathbb{Q}^{\rm tree}_5(K_{1.4.1}, \Sigma_{1.4.1})$). The claim will be shown if we prove that

$$\Vdash_{\mathbb{Q}^{\rm tree}_5(K_{1.4.1}, \Sigma_{1.4.1})} (\forall g \in \omega^\omega \cap \mathbf{V})(\exists \eta \in \omega^{<\omega})(\forall n \geq \ell g(\eta))(\dot{\tau}(\eta)(2n) = g(n)).$$

For this suppose that $g \in \omega^\omega$, $p \in \mathbb{Q}^{\rm tree}_5(K_{1.4.1}, \Sigma_{1.4.1})$. Choose an increasing sequence $\ell g(\mathrm{root}(p)) < k_0 < k_1 < \ldots$ of odd integers such that for each $\ell < \omega$

$$(\forall \nu \in T^p)(k_\ell \leq \ell g(\nu) \quad \Rightarrow \quad \mathbf{nor}[t^p_\nu] \geq \ell + 2).$$

Let $f : [k_0, \omega) \longrightarrow \omega$ be a function such that:

1. $f(k_\ell) = k_{\ell+1} - k_\ell - 1$
2. if $n \geq k_0$ then $f(2n) = g(n)$.

(Note that these clauses are compatible as the k_ℓ's are odd. Of course there is still much freedom left in defining f.)

Choose $\eta \in T^p \cap \omega^{k_0}$ and look at the condition $p^{f,\eta}$. Due to the requirement (δ) this condition is compatible with p: define $r \in \mathbb{Q}^{\rm tree}_5(K_{1.4.1}, \Sigma_{1.4.1})$ by

$\mathrm{root}(r) = \eta$, $T^r \subseteq T^p$ and
if $\nu \in T^r$, $k_{\ell-1} \leq \ell g(\nu) < k_\ell$ then

$$t^r_\nu = t^{\nu, A_\nu} \quad \text{and} \quad \mathrm{succ}_{T^r}(\nu) = \{\nu^\frown\langle m\rangle : m \in A_\nu\}$$

where $A_\nu = \mathbf{dis}[t^p_\nu] \cap F^{\ell, f(\ell g(\nu))}_\nu$.

Note that by the choice of k_ℓ and the requirement (δ) we have

$$t^{\nu, A_\nu} \in \Sigma_{1.4.1}(t^p_\nu) \cap \Sigma_{1.4.1}(t^{p^{f,\eta}}_\nu) \quad \text{and} \quad \mathbf{nor}[t^{\nu, A_\nu}] = \ell.$$

Consequently the definition of r is correct. Clearly r is stronger than both p and $p^{f,\eta}$. Thus

$$r \Vdash_{\mathbb{Q}^{\rm tree}_5(K_{1.4.1}, \Sigma_{1.4.1})} (\forall n \geq k_0)(\dot{\tau}(\eta)(n) = f(n))$$

which together with

$$(\forall n \geq k_0)(f(2n) = g(n)) \quad \text{and} \quad k_0 = \ell g(\eta)$$

finishes the proof of the claim. □

□

Our next example shows that the assumption that (K, Σ) is finitary in 2.1.6 is crucial.

EXAMPLE 1.4.2. Let $\mathbf{H}(i) = \omega$ for $i \in \omega$.
There is a creating pair $(K_{1.4.2}, \Sigma_{1.4.2})$ for $\mathbf{H}$ which is forgetful and growing (see 2.1.1(3)) but the forcing notion $\mathbb{Q}^*_\infty(K_{1.4.2}, \Sigma_{1.4.2})$ is not proper (and collapses $\mathfrak{c}$ onto ω).
(By 2.1.3 we may replace $\mathbb{Q}^*_\infty$ by either $\mathbb{Q}^*_{\rm w\infty}$ or $\mathbb{Q}^*_{\rm s\infty}$ or $\mathbb{Q}^*_f$ (for a fast function f).)

CONSTRUCTION. This is similar to 1.4.1: for $0 < \ell \leq i < \omega$ choose $\emptyset \neq S^\ell_i \subseteq R_i \subseteq [\omega]^\omega$ and functions $h_i : R_i \longrightarrow \omega$ satisfying the requirements (α)–(ε) of the construction of 1.4.1 (with i instead of η and $\ell g(\eta)$) and

(ζ) $\bigcup S^\ell_i = \omega$ for each $0 < \ell \leq i < \omega$

(this additional condition is satisfied by the example constructed there). Fix an enumeration $S^\ell_i = \{F^{\ell,m}_i : m \in \omega\}$.

A creature $t \in K_{1.4.2}$ may be described in the following way. For each $i \in [m^t_{\rm dn}, m^t_{\rm up})$ we have a set $A_i \in R_i$. Now:

$$\mathbf{dis}[t] = \langle A_i : m^t_{\rm dn} \leq i < m^t_{\rm up} \rangle$$
$$\mathbf{val}[t] = \{\langle u, v\rangle \in \prod_{i < m^t_{\rm dn}} \mathbf{H}(i) \times \prod_{i < m^t_{\rm up}} \mathbf{H}(i) : u \vartriangleleft v \ \& \ (\forall i \in [m^t_{\rm dn}, m^t_{\rm up}))(v(i) \in A_i)\},$$
$$\mathbf{nor}[t] = \max\{h_i(A_i) : i \in [m^t_{\rm dn}, m^t_{\rm up})\}.$$

If creatures $t_0, \ldots, t_{n-1} \in K_{1.4.2}$ are determined by sets $A^j_i \in R_i$ (for $j < n$, $i \in [m^{t_j}_{\rm dn}, m^{t_j}_{\rm up})$) in the way described above and $m^{t_{j+1}}_{\rm dn} = m^{t_j}_{\rm up}$ (for $j < n-1$) then $\Sigma_{1.4.2}(t_0, \ldots, t_{n-1})$ consists of all creatures $t \in K_{1.4.2}$ which are determined (in the way described above) by some sets $A_i \in R_i$ (for $i \in [m^{t_0}_{\rm dn}, m^{t_{n-1}}_{\rm up})$) such that $A_i \subseteq A^j_i$ whenever $m^{t_j}_{\rm dn} \leq i < m^{t_j}_{\rm up}$, $j < n$.

It is easy to check that $\Sigma_{1.4.2}$ is a composition operation on $K_{1.4.2}$. The creating pair $(K_{1.4.2}, \Sigma_{1.4.2})$ is forgetful and growing.

CLAIM 1.4.2.1. *The forcing notion $\mathbb{Q}^*_\infty(K_{1.4.2}, \Sigma_{1.4.2})$ collapses $\mathfrak{c}$ onto ω.*

Proof of the claim. We proceed like in 1.4.1.1 (with small modifications however). Let $\pi : \omega \longrightarrow \omega \times \omega : n \mapsto (\pi_0(n), \pi_1(n))$ be a bijection. Let $\pi^*_n : \omega^{[n,\omega)} \longrightarrow \omega^\omega$ (for $n \in \omega$) be mappings defined in the following manner. Let $f : [n, \omega) \longrightarrow \omega$;

inductively define $n_0 = n$, $n_{\ell+1} = n_\ell + \pi_0(f(n_\ell)) + \pi_1(f(n_\ell)) + 2$ (for $\ell < \omega$) and then $m_\ell = n_\ell + \pi_0(f(n_\ell)) + 1$ (so $n_\ell < m_\ell < n_{\ell+1}$); now put $\pi^*_n(f)(\ell) = f(m_\ell)$ (for $\ell \in \omega$).

For $0 < n < \omega$, $u \in \prod\limits_{i<n} \mathbf{H}(i)$ and a function $f : [n, \omega) \longrightarrow \omega$ define a condition $p^{u,f} \in \mathbb{Q}^*_\infty(K_{1.4.2}, \Sigma_{1.4.2})$:

$k_0 = n$, $k_{2\ell+1} = k_{2\ell} + \pi_0(f(k_{2\ell})) + 1$, $k_{2\ell+2} = k_{2\ell+1} + \pi_1(f(k_{2\ell})) + 1$,

$w^{p^{u,f}} = u$,

if $k_{2\ell} \le i+n < k_{2\ell+2}$ then $t^{p^{u,f}}_i \in K_{1.4.2}$ is such that $m^{t^{p^{u,f}}_i}_{\rm dn} = n+i$, $m^{t^{p^{u,f}}_i}_{\rm up} = n+i+1$, $\mathbf{dis}[t^{p^{u,f}}_i] = \langle F^{\ell+1, f(i+n)}_{i+n}\rangle$.

As in 1.4.1.1, if $f, g : [n, \omega) \longrightarrow \omega$ are distinct, $u \in \prod\limits_{i<n} \mathbf{H}(i)$ then the conditions $p^{u,f}, p^{u,g}$ are incompatible. Consequently we may choose a $\mathbb{Q}^*_\infty(K_{1.4.2}, \Sigma_{1.4.2})$–name $\dot\tau$ for a function on ω such that $\Vdash (\forall n \in \omega)(\dot\tau(n) : [n, \omega) \to \omega)$ and $p^{u,f} \Vdash \dot\tau(n) = f$. To finish it is enough to show that

$$\Vdash_{\mathbb{Q}^*_\infty(K_{1.4.2}, \Sigma_{1.4.2})} (\forall g \in \omega^\omega \cap \mathbf{V})(\exists n \in \omega)(\pi^*_n(\dot\tau(n)) = g).$$

Suppose that $p \in \mathbb{Q}^*_\infty(K_{1.4.2}, \Sigma_{1.4.2})$, $g \in \omega^\omega$. Choose $2 < i_0 < i_1 < \ldots < \omega$ such that $\mathbf{nor}[t^p_{i_\ell}] \ge \ell + 2$ and next choose $k_0 < k_1 < k_2 < \ldots < \omega$ such that for each $\ell \in \omega$:

$m^{t^p_{i_\ell}}_{\rm dn} \le k_\ell < m^{t^p_{i_\ell}}_{\rm up}$ and for some set $A_\ell \in R_{k_\ell}$ we have

$$h_{k_\ell}(A_\ell) \ge \ell + 2 \quad \text{and} \quad (\forall n \in A_\ell)(\exists \langle u, v\rangle \in \mathbf{val}[t^p_{i_\ell}])(v(k_\ell) = n)$$

(possible by the way we defined $(K_{1.4.2}, \Sigma_{1.4.2})$). Choose any $v \in \mathrm{pos}(w^p, t^p_0, \ldots, t^p_{i_0})$ and let $u = v{\restriction}k_0$. Next choose $f : [k_0, \omega) \longrightarrow \omega$ such that for each $\ell \in \omega$:

$$\pi_0(f(k_{2\ell})) = k_{2\ell+1} - k_{2\ell} - 1, \quad \pi_1(f(k_{2\ell})) = k_{2\ell+2} - k_{2\ell+1} - 1, \quad f(k_{2\ell+1}) = g(\ell),$$

and if $k \in (k_{2\ell}, k_{2\ell+2}) \setminus \{k_{2\ell+1}\}$, $m^{t^p_i}_{\rm dn} \le k < m^{t^p_i}_{\rm up}$ ($\ell < \omega$, $i < \omega$) then

$$(\exists \langle u, v\rangle \in \mathbf{val}[t^p_i])(v(k) \in F^{\ell+1, f(k)}_k).$$

One easily checks that the choice of f is possible (remember the additional requirement (ζ)) and that the conditions $p^{u,f}$ and p are compatible in $\mathbb{Q}^*_\infty(K_{1.4.2}, \Sigma_{1.4.2})$. As

$$p^{u,f} \Vdash_{\mathbb{Q}^*_\infty(K_{1.4.2}, \Sigma_{1.4.2})} \pi^*_n(\dot\tau(n)) = \pi^*_n(f) = g,$$

we finish the proof of the claim. □

□

One could expect that the main reason for collapsing $\mathfrak{c}$ in the two examples constructed above is that the (K, Σ)'s there are not finitary. But this is not the case. Using similar ideas we may build a finitary creating pair (K, Σ) for which the forcing notion $\mathbb{Q}^*_\infty(K, \Sigma)$ collapses $\mathfrak{c}$ onto ω as well. This is the reason why we have to use forcing notions $\mathbb{Q}^*_f(K, \Sigma)$ with (K, Σ) satisfying extra demands (including Halving and bigness, see 2.2.12) and why $\mathbb{Q}^*_\infty(K, \Sigma)$ is used only for growing (K, Σ) (so then $\mathbb{Q}^*_{\rm s\infty}(K, \Sigma)$ is dense in $\mathbb{Q}^*_\infty(K, \Sigma)$, see 2.1). This bad effect can be made quite general and we will present it in this way, trying to show the heart of the matter. One could try to cover the previous examples by our "negative theory" too, but this would involve much more complications.

DEFINITION 1.4.3. We say that a weak creating pair (K,Σ) is *local* if for every $t\in K$, $w\in \mathbf{basis}(t)$ and $u\in \mathrm{pos}(w,t)$ we have $\ell g(u)=\ell g(w)+1$.

DEFINITION 1.4.4. Let (K,Σ) be a (nice and smooth) creating pair for $\mathbf{H}$ which is local (so $t\in K\ \Rightarrow\ m^t_{\rm up}=m^t_{\rm dn}+1$) and simple (which means that $\Sigma(\mathcal{S})\neq\emptyset\ \Rightarrow\ |\mathcal{S}|=1$; see 2.1.7(1)).
We say that (K,Σ) is *definitely bad* if there are a perfect tree $T\subseteq\omega^{<\omega}$ and mappings F_0,F_1 such that

1. $T\cap\omega^m$ is finite for each $m\in\omega$, $\quad \mathrm{dom}(F_0)=\mathrm{dom}(F_1)=\bigcup\limits_{m<\omega}\prod\limits_{i<m}\mathbf{H}(i)$,
2. if $v\in\prod\limits_{i\leq m}\mathbf{H}(i)$, $m<\omega$ then $F_1(v):T\cap\omega^m\longrightarrow 2$ and $F_0(v):T\cap\omega^m\longrightarrow T$ is such that $(\forall\eta\in T\cap\omega^m)(F_0(v)(\eta)\in\mathrm{succ}_T(\eta))$,
3. *if* $t\in K$, $\mathbf{nor}[t]\geq 2$, $m=m^t_{\rm dn}>0$, $i<2$ *and* $F^*:T\cap\omega^m\longrightarrow T$ *is such that* $F^*(\nu)\in\mathrm{succ}_T(\nu)$ *for* $\nu\in T\cap\omega^m$ *then there is* $s\in\Sigma(t)$ *such that* $\mathbf{nor}[s]\geq\mathbf{nor}[t]-1$ *and for each* $\eta\in T\cap\omega^{m-1}$ *there is* $\nu\in\mathrm{succ}_T(\eta)$ *with*

$$(\forall u\in\mathbf{basis}(s))(\forall v\in\mathrm{pos}(u,s))(F_0(v)(\nu)=F^*(\nu)\ \ \&\ \ F_1(v)(\nu)=i).$$

PROPOSITION 1.4.5. *Suppose that* (K,Σ) *is a local, simple and definitely bad creating pair for* $\mathbf{H}$ *such that* $\Sigma(t)$ *is finite for each* $t\in K$. *Then*

(a) *the forcing notion* $\mathbb{Q}^*_\infty(K,\Sigma)$ *collapses* $\mathfrak{c}$ *onto* ω,
(b) *if* $f:\omega\times\omega\longrightarrow\omega$ *is a fast function then the forcing notion* $\mathbb{Q}^*_f(K,\Sigma)$ *collapses* $\mathfrak{c}$ *onto* ω.

PROOF. In both cases the proof is exactly the same, so let us deal with **(a)** only. So suppose that a finitely branching perfect tree $T\subseteq\omega^{<\omega}$ and functions F_0,F_1 witness that (K,Σ) is definitely bad. Let $G^0:T\times\prod\limits_{i<\omega}\mathbf{H}(i)\longrightarrow[T]$ and $G^1:[T]\times\prod\limits_{i<\omega}\mathbf{H}(i)\longrightarrow 2^\omega$ be defined by

$$\begin{array}{ll} G^0(\eta,W)\restriction\ell g(\eta)=\eta, & \text{and}\\ G^0(\eta,W)\restriction(m+1)=F_0(W\restriction(m+1))(G^0(\eta,W)\restriction m) & \text{for } m\geq\ell g(\eta),\\ G^1(\rho,W)(n)=F_1(W\restriction(n+1))(\rho\restriction n) & \text{for } n\in\omega.\end{array}$$

We are going to show that

$$\Vdash_{\mathbb{Q}^*_\infty(K,\Sigma)}(\forall r\in 2^\omega\cap\mathbf{V})(\exists\eta\in T)(\forall^\infty n\in\omega)(r(n)=G^1(G^0(\eta,\dot{W}),\dot{W})(n)),$$

where $\dot{W}$ is the $\mathbb{Q}^*_\infty(K,\Sigma)$–name defined in 1.1.13. To this end suppose that $r\in 2^\omega$ and $p\in\mathbb{Q}^*_\infty(K,\Sigma)$. We may assume that $\ell g(w^p)>0$ and $(\forall i\in\omega)(\mathbf{nor}[t^p_i]\geq 3)$.

Fix $i_0\in\omega$ for a moment. By downward induction on $i\leq i_0$ we choose $s^{i_0}_i\in\Sigma(t^p_i)$ and $F^*_{i_0,i}:T\cap\omega^{\ell g(w^p)+i-1}\longrightarrow T$ such that

(α) $\mathbf{nor}[s_{i_0,i}]\geq\mathbf{nor}[t^p_i]-1$,
(β) $F^*_{i_0,i}(\nu)\in\mathrm{succ}_T(\nu)$ for $\nu\in T$, $\ell g(\nu)=\ell g(w^p)+i-1$,
(γ) for all sequences $u\in\mathbf{basis}(s_{i_0,i})$ and $v\in\mathrm{pos}(u,s_{i_0,i})$ and every $\nu\in T$ of length $\ell g(\nu)=\ell g(w^p)+i-1$ we have

$$F_0(v)(F^*_{i_0,i}(\nu))=F^*_{i_0,i+1}(\nu)\quad\text{ and }\quad F_1(v)(F^*_{i_0,i}(\nu))=r(\ell g(w^p)+i)$$

[for $i=i_0$ we omit the first part of the above demand].

It should be clear that the choice of the $s_{i_0,i}$'s and $F^*_{i_0,i}$'s as above is possible by 1.4.4(3). All levels of the tree T are finite, so for each $i\in\omega$ there are finitely many mappings $F^*: T \cap \omega^{\ell g(w^p)+i-1} \longrightarrow T \cap \omega^{\ell g(w^p)+i}$. Moreover, each $\Sigma(t^p_i)$ is finite (for $i \in \omega$). Hence, by König's lemma, we find a sequence $\langle t^q_i : i \in \omega \rangle$ and a mapping $\mathbf{F}^*: T \longrightarrow T$ such that

(i) $(\forall \nu \in T)(\mathbf{F}^*(\nu) \in \mathrm{succ}_T(\nu))$ and
(ii) for each $i \in \omega$ there is $j(i) > i$ such that for every $j \leq i$

$$s^{j(i)}_j = t^q_j \quad \text{and} \quad (\forall \nu \in T \cap \omega^{\ell g(w^p)+j-1})(\mathbf{F}^*(\nu) = F^*_{j(i),j}(\nu)).$$

By (α) we have that $q = (w^p, t^q_0, t^q_1, \dots) \in \mathbb{Q}^*_\infty(K,\Sigma)$ and it is stronger than p. Take any $\eta \in T$ with $\ell g(\eta) = \ell g(w^p) - 1$ and let $\eta_0 = \mathbf{F}^*(\eta)$, $\eta_{i+1} = \mathbf{F}^*(\eta_i)$, and $\eta^+ = \lim(\eta_i) \in [T]$. It follows from (γ) and (ii) (e.g. inductively using smoothness) that for each $n \in \omega$ and $v \in \mathrm{pos}(w^p, t^q_0, \dots, t^q_{n-1})$ we have

$$(v, t^q_n, t^q_{n+1}, \dots) \Vdash \text{“} G^0(\eta_0, \dot{W})\restriction \ell g(v) = \eta^+ \restriction \ell g(v) \text{ and } G^1(G^0(\eta_0, \dot{W}), \dot{W})\restriction[\ell g(\eta_0), \ell g(v)) = r\restriction[\ell g(\eta_0), \ell g(v))\text{”}.$$

Hence we conclude

$$q \Vdash_{\mathbb{Q}^*_\infty(K,\Sigma)} (\forall n \geq \ell g(\eta_0))(G^1(G^0(\eta_0, \dot{W}), \dot{W})(n) = r(n)),$$

finishing the proof. □

EXAMPLE 1.4.6. Let $f: \omega \times \omega \longrightarrow \omega$ be a fast function (for example $f(k,\ell) = 2^{2k}(\ell+1)$). There are a finitary function $\mathbf{H}$ and a creating pair $(K_{1.4.6}, \Sigma_{1.4.6})$ for $\mathbf{H}$ such that

(a) $(K_{1.4.6}, \Sigma_{1.4.6})$ is local, simple, forgetful and definitely bad (and smooth),
(b) $\Sigma_{1.4.6}(t)$ is finite for each $t \in K_{1.4.6}$,
(c) the forcing notions $\mathbb{Q}^*_\infty(K_{1.4.6}, \Sigma_{1.4.6})$ and $\mathbb{Q}^*_f(K_{1.4.6}, \Sigma_{1.4.6})$ are not trivial and thus collapse $\mathfrak{c}$ onto ω.

CONSTRUCTION. Let $f: \omega \times \omega \longrightarrow \omega$ be fast. For $n \in \omega$ let $k_n = 2^{f(n+1,n+1)}$. Next, for $n \in \omega$, let $\mathbf{H}(n)$ consist of all pairs $\langle z_0, z_1 \rangle$ such that

$$z_0 : \prod_{i<n} k_i \longrightarrow k_n \quad \text{and} \quad z_1 : \prod_{i<n} k_i \longrightarrow 2.$$

Immediately by the definition, one sees that $\mathbf{H}$ is finitary. Now we define the creating pair $(K_{1.4.6}, \Sigma_{1.4.6})$ for $\mathbf{H}$. A creature $t \in \mathrm{CR}[\mathbf{H}]$ with $m^t_{\mathrm{dn}} > 0$ is in $K_{1.4.6}$ if:

- $\mathbf{dis}[t] = \langle m^t_{\mathrm{dn}}, \langle A^t_\nu : \nu \in \prod\limits_{i<m^t_{\mathrm{dn}}-1} k_i \rangle, F^t_0, F^t_1 \rangle$, where
  $A^t_\nu \subseteq k_{m^t_{\mathrm{dn}}-1}$ for $\nu \in \prod\limits_{i<m^t_{\mathrm{dn}}-1} k_i$,
  $F^t_0 : \{\nu \in \prod\limits_{i<m^t_{\mathrm{dn}}} k_i : \nu(m^t_{\mathrm{dn}}-1) \in A^t_{\nu\restriction(m^t_{\mathrm{dn}}-1)}\} \longrightarrow k_{m^t_{\mathrm{dn}}}$,
  $F^t_1 : \{\nu \in \prod\limits_{i<m^t_{\mathrm{dn}}} k_i : \nu(m^t_{\mathrm{dn}}-1) \in A^t_{\nu\restriction(m^t_{\mathrm{dn}}-1)}\} \longrightarrow 2$,
- $\mathbf{val}[t]$ consists of all pairs $\langle w, u \rangle \in \prod\limits_{i<m^t_{\mathrm{dn}}} \mathbf{H}(i) \times \prod\limits_{i\leq m^t_{\mathrm{dn}}} \mathbf{H}(i)$ such that $w \vartriangleleft u$ and if $u(m^t_{\mathrm{dn}}) = \langle z_0, z_1 \rangle$ then $z_0 \supseteq F^t_0$ and $z_1 \supseteq F^t_1$.
- $\mathbf{nor}[t] = k_{m^t_{\mathrm{dn}}} - \max\{|A^t_\nu| : \nu \in \prod\limits_{i<m^t_{\mathrm{dn}}-1} k_i\}$.

If $t \in \mathrm{CR}[H]$, $m^t_{\mathrm{dn}} = 0$ then we take t to $K_{1.4.6}$ if:
$\mathbf{nor}[t] = 0$, $\mathbf{dis}[t] = \langle 0, x^t, i^t \rangle$, where $x^t \in k_0$, $i^t < 2$, $\mathbf{val}[t] = \{\langle\, \langle\rangle, u \,\rangle\}$, where

$u \in \prod\limits_{i\le 0} \mathbf{H}(i)$, $u(0) = \langle x^t, i^t\rangle$ (both x^t and i^t are treated here as functions from $\{\langle\rangle\}$).

The composition operation $\Sigma_{1.4.6}$ is the trivial one (so $\Sigma_{1.4.6}(\mathcal{S})$ is non-empty for singletons only and $\Sigma_{1.4.6}(t) = \{s \in K_{1.4.6} : \mathbf{val}[s] \subseteq \mathbf{val}[t]\}$). Easily $(K_{1.4.6}, \Sigma_{1.4.6})$ is a local, simple and forgetful creating pair. Note that if $n > 0$ and $t \in K_{1.4.6}$ is such that $m^t_{\rm dn} = n$ and $A^t_\nu = \emptyset$ for each $\nu \in \prod\limits_{i<n-1} k_i$ then $\mathbf{nor}[t] = k_{n-1} > f(n,n)$, so the forcing notions $\mathbb{Q}^*_\infty(K_{1.4.6}, \Sigma_{1.4.6})$, $\mathbb{Q}^*_f(K_{1.4.6}, \Sigma_{1.4.6})$ are non-trivial. Finally let $T = \bigcup\limits_{n\in\omega} \prod\limits_{i<n} k_i$ and for $v \in \prod\limits_{i\le m} \mathbf{H}(i)$ let $F_0(v)$, $F_1(v)$ be such that $v(m) = \langle F_0(v), F_1(v)\rangle$. It should be clear that T, F_0, F_1 witness that $(K_{1.4.6}, \Sigma_{1.4.6})$ is definitely bad. □

CHAPTER 2

Properness and the reading of names

This chapter is devoted to getting the basic property: properness. The first two sections deal with forcing notions determined by creating pairs. We define properties of creating pairs implying that appropriate forcing notions are proper. Some of these properties may look artificial, but in applications they appear naturally. The third part deals with forcing notions $\mathbb{Q}^{\rm tree}_e(K,\Sigma)$ (determined by a tree creating pair). Here, properness is an almost immediate consequence of our choice of norm conditions. In most cases, proving properness of a forcing notion we get much stronger property: continuous reading of names for ordinals. This property will be intensively used in the rest of the paper. Finally, in the last part of the chapter we give several examples for properties introduced and studied before.

2.1. Forcing notions $\mathbf{Q}^*_{\rm s\infty}(K,\Sigma)$, $\mathbf{Q}^*_{\rm w\infty}(K,\Sigma)$

DEFINITION 2.1.1. Let (K,Σ) be a creating pair for $\mathbf{H}$.

1. For $t\in K$, $m_0\leq m^t_{\rm dn}$, $m^t_{\rm up}\leq m_1$ we define the creature $s\stackrel{\rm def}{=} t\restriction^+ [m_0,m_1)$ by:

$$\begin{array}{rcl}\mathbf{nor}[s] & = & \mathbf{nor}[t],\\ \mathbf{dis}[s] & = & \langle 4,m_0,m_1\rangle^\frown\langle\mathbf{dis}[t]\rangle,\\ \mathbf{val}[s] & = & \{\langle w,u\rangle\in\prod\limits_{i<m_0}\mathbf{H}(i)\times\prod\limits_{i<m_1}\mathbf{H}(i): \langle u\restriction m^t_{\rm dn}, u\restriction m^t_{\rm up}\rangle\in\mathbf{val}[t]\ \&\\ & & \quad w\vartriangleleft u\ \&\ (\forall i\in[m_0,m^t_{\rm dn})\cup[m^t_{\rm up},m_1))(u(i)=0)\}.\end{array}$$

   [Note that $t\restriction^+ [m_0,m_1)$ is well defined only if $\mathbf{val}[s]\neq\emptyset$ above and then $m^s_{\rm dn}=m_0$, $m^s_{\rm up}=m_1$.]
2. The creating pair (K,Σ) is *omittory* if:
   ($\boxtimes_0$) if $t\in K$ and $u\in\mathbf{basis}(t)$ then $u^\frown\mathbf{0}_{[m^t_{\rm dn},m^t_{\rm up})}\in\mathrm{pos}(u,t)$ but there is $v\in\mathrm{pos}(u,t)$ such that $v\restriction[m^t_{\rm dn},m^t_{\rm up})\neq\mathbf{0}_{[m^t_{\rm dn},m^t_{\rm up})}$,
   ($\boxtimes_1$) for every $(t_0,\ldots,t_{n-1})\in\mathrm{PFC}(K,\Sigma)$ and $i<n$:
   $$t_i\restriction^+ [m^{t_0}_{\rm dn},m^{t_{n-1}}_{\rm up})\in\Sigma(t_0,\ldots,t_{n-1}),$$
   ($\boxtimes_2$) if $t,t\restriction^+ [m_0,m_1)\in K$ then for every $u\in\mathbf{basis}(t\restriction^+ [m_0,m_1))$ and $v\in\mathrm{pos}(u,t\restriction^+ [m_0,m_1))$ we have
   $$v(n)\neq 0\ \&\ n\in[\ell g(u),\ell g(v))\quad\Rightarrow\quad n\in[m^t_{\rm dn},m^t_{\rm up}).$$
   [Note that ($\boxtimes_0$) implies that in the cases relevant for ($\boxtimes_1$), $t_i\restriction^+ [m^{t_0}_{\rm dn},m^{t_{n-1}}_{\rm up})$ is well defined.]
3. (K,Σ) is *growing* if for any $(t_0,\ldots,t_{n-1})\in\mathrm{PFC}(K,\Sigma)$ there is a creature $t\in\Sigma(t_0,\ldots,t_{n-1})$ such that $\mathbf{nor}[t]\geq\max\limits_{i<n}\mathbf{nor}[t_i]$.

PROPOSITION 2.1.2. *If (K,Σ) is omittory then it is growing.*

PROPOSITION 2.1.3. *Suppose that a creating pair (K,Σ) is growing.*

1. *Then $\mathbb{Q}^*_{\mathrm{s}\infty}(K,\Sigma)$ is a dense subset of both $\mathbb{Q}^*_{\infty}(K,\Sigma)$ and $\mathbb{Q}^*_{\mathrm{w}\infty}(K,\Sigma)$ and $\mathbb{Q}^*_f(K,\Sigma)$ (for every fast (see 1.1.12) function $f:\omega\times\omega\longrightarrow\omega$). Consequently whenever we work with growing creating pairs we may interchange the respective forcing notions as they are equivalent.*
2. *Moreover, if $g:\omega\times\omega\longrightarrow\omega$, $p\in\mathbb{Q}^*_{\mathrm{w}\infty}(K,\Sigma)$ then there is $q\in\mathbb{Q}^*_{\mathrm{s}\infty}(K,\Sigma)$ such that $p\leq^{\mathrm{w}\infty}_0 q$ and*

$$(\forall n\in\omega)(\mathbf{nor}[t^q_n]>g(n,m^{t^q_n}_{\mathrm{dn}})).$$

PROOF. Suppose that $g:\omega\times\omega\longrightarrow\omega$ and $(w,t_0,t_1,\dots)\in\mathbb{Q}^*_{\mathrm{w}\infty}(K,\Sigma)$. Choose an increasing sequence $k_0<k_1<k_2<\dots$ such that

$$\mathbf{nor}[t_{k_0}]>g(0,m^{t_0}_{\mathrm{dn}})\qquad\text{and}\qquad\mathbf{nor}[t_{k_{n+1}}]>g(n+1,m^{t_{k_n}}_{\mathrm{up}})$$

(exists by 1.1.10(w∞)) and choose $s_0\in\Sigma(t_0,\dots,t_{k_0})$, $s_{n+1}\in\Sigma(t_{k_n+1},\dots,t_{k_{n+1}})$ such that $\mathbf{nor}[s_n]\geq\mathbf{nor}[t_{k_n}]$ (exist by 2.1.1(3)). Hence (by 1.2.8(2); remember that we assume (K,Σ) is nice)

$$q\stackrel{\mathrm{def}}{=}(w,s_0,s_1,s_2,\dots)\in\mathbb{Q}^*_{\mathrm{s}\infty}(K,\Sigma)$$

and clearly $(w,t_0,t_1,t_2,\dots)\leq_0 q$. □

THEOREM 2.1.4. *Assume (K,Σ) is a finitary creating pair. Further assume that $p\in\mathbb{Q}^*_{\mathrm{s}\infty}(K,\Sigma)$ and for $n<\omega$ we have a $\mathbb{Q}^*_{\mathrm{s}\infty}(K,\Sigma)$-name $\dot\tau_n$ such that $\Vdash_{\mathbb{Q}^*_{\mathrm{s}\infty}(K,\Sigma)}$ "$\dot\tau_n$ is an ordinal" and $\ell<\omega$.* Then *there is $q=(w^p,s_0,s_1,s_2,\dots)$ such that:*

(a) *$p\leq^{\mathrm{s}\infty}_\ell q\in\mathbb{Q}^*_{\mathrm{s}\infty}(K,\Sigma)$ and*
(b) if *$\ell\leq n<\omega$, $m\leq m^{s_{n-1}}_{\mathrm{up}}$ then the condition q approximates $\dot\tau_m$ at s_n (see 1.2.9(2)).*

PROOF. Let $p=\langle w^p,t^p_0,t^p_1,t^p_2,\dots\rangle$. Let $s_i=t^p_i$ for $i<\ell$. Now, by induction on $n\geq\ell$ we define $q_n,s_n,t^n_{n+1},t^n_{n+2},\dots$ such that:

(i) $q_\ell=p$,
(ii) $q_{n+1}=(w^p,s_0,\dots,s_n,t^n_{n+1},t^n_{n+2},\dots)\in\mathbb{Q}^*_{\mathrm{s}\infty}(K,\Sigma)$
(iii) $q_n\leq^{\mathrm{s}\infty}_n q_{n+1}$
(iv) if $w_1\in\mathrm{pos}(w^p,s_0,\dots,s_{n-1})$, $m\leq m^{s_{n-1}}_{\mathrm{up}}$ and there is a condition $r\in\mathbb{Q}^*_{\mathrm{s}\infty}(K,\Sigma)$, $\leq^{\mathrm{s}\infty}_0$–stronger than $(w_1,s_n,t^n_{n+1},t^n_{n+2},\dots)$ which decides the value of $\dot\tau_m$
 then the condition $(w_1,s_n,t^n_{n+1},t^n_{n+2},\dots)$ does it.

Arriving at the stage $n+1>\ell$ we have defined

$$q_n=(w^p,s_0,\dots,s_{n-1},t^{n-1}_n,t^{n-1}_{n+1},\dots).$$

Fix an enumeration $\langle(w^n_i,m^n_i):i<k_n\rangle$ of

$$\mathrm{pos}(w^p,s_0,\dots,s_{n-1})\times(m^{s_{n-1}}_{\mathrm{up}}+1)$$

(since each $\mathbf{H}(m)$ is finite, k_n is finite). Next choose by induction on $k\leq k_n$ conditions $q_{n,k}\in\mathbb{Q}^*_{\mathrm{s}\infty}(K,\Sigma)$ such that:

(α) $q_{n,0}=q_n$
(β) $q_{n,k}$ is of the form $(w^p,s_0,\dots,s_{n-1},t^{n,k}_n,t^{n,k}_{n+1},t^{n,k}_{n+1},\dots)$
(γ) $q_{n,k}\leq_n q_{n,k+1}$

(δ) if, in $\mathbb{Q}^*_{s\infty}(K,\Sigma)$, there is a condition $r \geq_0 (w^n_k, t^{n,k}_n, t^{n,k}_{n+1}, t^{n,k}_{n+2}, \dots)$ which decides (in $\mathbb{Q}^*_{s\infty}(K,\Sigma)$) the value of $\dot{\tau}_{m^n_k}$, then

$$(w^n_k, t^{n,k+1}_n, t^{n,k+1}_{n+1}, t^{n,k+1}_{n+2}, \dots) \in \mathbb{Q}^*_{s\infty}(K,\Sigma)$$

is a condition which forces a value to $\dot{\tau}_{m^n_k}$.
(Note: $(w^n_k, t^{n,k}_n, t^{n,k}_{n+1}, t^{n,k}_{n+2}, \dots) \in \mathbb{Q}^*_{s\infty}(K,\Sigma)$.)

For this part of the construction we need our standard assumption that (K,Σ) is nice. Note that choosing $(w^n_k, t^{n,k+1}_n, t^{n,k+1}_{n+1}, t^{n,k+1}_{n+2}, \dots)$ we want to be sure that

$$(w^p, s_0, \dots, s_{n-1}, t^{n,k+1}_n, t^{n,k+1}_{n+1}, t^{n,k+1}_{n+2}, \dots) \in \mathbb{Q}^*_{s\infty}(K,\Sigma)$$

(remember that 1.1.7(b)(ii) might fail). But by 1.2.8(2) it is not a problem. Next, the condition $q_{n+1} \stackrel{\rm def}{=} q_{n,k_n} \in \mathbb{Q}^*_{s\infty}(K,\Sigma)$ satisfies (iv): the keys are the clause (δ) and the fact that

$$(w^n_k, t^{n,k+1}_n, t^{n,k+1}_{n+1}, t^{n,k+1}_{n+2}, \dots) \leq^{s\infty}_0 (w^n_k, t^{n,k_n}_n, t^{n,k_n}_{n+1}, t^{n,k_n}_{n+2}, \dots) \in \mathbb{Q}^*_{s\infty}(K,\Sigma).$$

Thus $s_n \stackrel{\rm def}{=} t^{n,k_n}_n$, q_{n+1} and $t^n_{n+k} \stackrel{\rm def}{=} t^{n,k_n}_{n+k}$ are as required.

Now, by 1.2.13:

$$q \stackrel{\rm def}{=} (w^p, s_0, s_1, \dots, s_l, s_{l+1}, \dots) = \lim_n q_n \in \mathbb{Q}^*_{s\infty}(K,\Sigma).$$

Easily it satisfies the assertions of the theorem. □

A small modification of the proof of 2.1.4 shows the corresponding result for the forcing notion $\mathbb{Q}^*_{w\infty}(K,\Sigma)$:

THEOREM 2.1.5. *Assume (K,Σ) is a finitary creating pair and $p \in \mathbb{Q}^*_{w\infty}(K,\Sigma)$. Let $\dot{\tau}_n$ be $\mathbb{Q}^*_{w\infty}(K,\Sigma)$-names for ordinals (for $n < \omega$), $\ell < \omega$. Then there is a condition $q = (w^p, s_0, s_1, \dots) \in \mathbb{Q}^*_{w\infty}(K,\Sigma)$ such that*

(a) *$p \leq_\ell q$ and*
(b) *there is an increasing sequence $\ell = k_0 < k_1 < k_2 < \dots < \omega$ such that if $n < \omega$ and $m \leq m^{s_{k_n-1}}_{\rm up}$ then the condition q approximates $\dot{\tau}_m$ at s_{k_n}.*

As an immediate corollary to theorem 2.1.4 we get the following.

COROLLARY 2.1.6. *Assume that (K,Σ) is a finitary creating pair.*

(a) *Suppose that $\dot{\tau}_n$ are $\mathbb{Q}^*_{s\infty}(K,\Sigma)$-names for ordinals and $q \in \mathbb{Q}^*_{s\infty}(K,\Sigma)$ is a condition satisfying* **(b)** *of 2.1.4 (for $\ell = 0$). Further assume that $q \leq r \in \mathbb{Q}^*_{s\infty}(K,\Sigma)$, $n < \ell g(w^r)$ and $r \Vdash$ "$\dot{\tau}_n = \alpha$" (for some ordinal α).*
Then *for some $q' \in \mathbb{Q}^*_{s\infty}(K,\Sigma)$, $q \leq_{\rm apr} q' \leq_0 r$, we have $q' \Vdash$ "$\dot{\tau}_n = \alpha$".*

(Note: $\{q' \in \mathbb{Q}^*_{s\infty}(K,\Sigma) \colon q \leq_{\rm apr} q'\}$ is countable provided $\bigcup\limits_{i<\omega} \mathbf{H}(i)$ is countable.)

(b) *The forcing notion $\mathbb{Q}^*_{s\infty}(K,\Sigma)$ is proper (and α-proper for $\alpha < \omega_1$).*

It should be underlined here that 2.1.6 applies to forcing notions $\mathbb{Q}^*_{\infty}(K,\Sigma)$ for finitary growing creating pairs (remember 2.1.3). To get the respective conclusion for $\mathbb{Q}^*_{w\infty}(K,\Sigma)$ we need to assume more.

DEFINITION 2.1.7. We say that:

1. A weak creating pair (K,Σ) is *simple* if $\Sigma(\mathcal{S})$ is non-empty for singletons only.

2. A creating pair (K,Σ) is *gluing* if it is full and for every $k<\omega$ there is $n_0<\omega$ such that for every $n_0\le n<\omega$, $(t_0,\ldots,t_n)\in \mathrm{PFC}(K,\Sigma)$, for some $s\in\Sigma(t_0,\ldots,t_n)$ we have
$$\mathbf{nor}[s]\ge\min\{k,\mathbf{nor}[t_0],\ldots,\mathbf{nor}[t_n]\}.$$
In this situation the integer n_0 is called *the gluing witness for k.*

The two properties defined above are, in a sense, two extremal situations under which we may say something on (K,Σ). The demand "either simple or gluing" (like in 2.2.11) should not be surprising if one realizes that then we know what may happen when Σ is applied, at least in terms of m^t_{dn}, m^t_{up}.

COROLLARY 2.1.8. *Assume that (K,Σ) is a finitary and simple creating pair.*

(a) *Suppose that $\dot\tau_n$ are $\mathbb{Q}^*_{\mathrm{w}\infty}(K,\Sigma)$–names for ordinals, $k_0<k_1<\ldots<\omega$ and $q\in\mathbb{Q}^*_{\mathrm{w}\infty}(K,\Sigma)$ are as in 2.1.5(b). Suppose that $q\le r\in\mathbb{Q}^*_{\mathrm{w}\infty}(K,\Sigma)$, and r decides the value of one of the names $\dot\tau_n$, say $r\Vdash$ "$\dot\tau_n=\alpha$".*
   Then *for some $q'\in\mathbb{Q}^*_{\mathrm{w}\infty}(K,\Sigma)$ we have*
$$q\le_{\mathrm{apr}} q',\quad q'\Vdash \text{“}\dot\tau_n=\alpha\text{”}\quad \textit{and } q',r \textit{ are compatible.}$$
(b) *The forcing notion $\mathbb{Q}^*_{\mathrm{w}\infty}(K,\Sigma)$ is proper (and even more).*

REMARK 2.1.9. Note the presence of "simple" in the assumptions of 2.1.8. In practical applications of forcing notions of the type $\mathbb{Q}^*_{\mathrm{w}\infty}$ we can get more, see 2.1.12 below.

DEFINITION 2.1.10. Let (K,Σ) be a creating pair for $\mathbf{H}$. We say that (K,Σ) *captures singletons* if (K,Σ) is forgetful and for every $(t_0,\ldots,t_n)\in\mathrm{PFC}(K,\Sigma)$ and for each $u\in\mathbf{basis}(t_0)(=\prod\limits_{m<m^{t_0}_{\mathrm{dn}}}\mathbf{H}(m))$ and $v\in\mathrm{pos}(u,t_0,\ldots,t_n)$ *there is* $(s_0,\ldots,s_k)\in\mathrm{PFC}(K,\Sigma)$ such that $(t_0,\ldots,t_n)\le(s_0,\ldots,s_k)$ (see 1.2.4) and
$$\mathrm{pos}(u,s_0,\ldots,s_k)=\{v\},\quad m^{t_0}_{\mathrm{dn}}=m^{s_0}_{\mathrm{dn}},\quad m^{s_k}_{\mathrm{up}}=m^{t_n}_{\mathrm{up}}.$$
[Note that we put no demands on the norms of the s_i's.]

PROPOSITION 2.1.11. *Suppose that (K,Σ) is a creating pair which captures singletons (so in particular it is forgetful), $p\in\mathbb{Q}^*_{\mathrm{w}\infty}(K,\Sigma)$ and $\dot\tau$ is a $\mathbb{Q}^*_{\mathrm{w}\infty}(K,\Sigma)$–name for an ordinal. Then there is $q\in\mathbb{Q}^*_{\mathrm{w}\infty}(K,\Sigma)$ such that*
$$p\le^{\mathrm{w}\infty}_0 q\qquad \textit{and}\qquad q \textit{ decides } \dot\tau.$$

PROOF. Take $r\in\mathbb{Q}^*_{\mathrm{w}\infty}(K,\Sigma)$ such that $p\le r$ and $r\Vdash\dot\tau=\alpha$ (for some α). Look at w^r: for some $n\in\omega$ we have $w^r\in\mathrm{pos}(w^p,t^p_0,\ldots,t^p_{n-1})$. By 2.1.10 we find $s_0,\ldots,s_k$ such that $\mathrm{pos}(w^p,s_0,\ldots,s_k)=\{w^r\}$ and
$$(w^p,t^p_0,t^p_1,\ldots)\le(w^p,s_0,\ldots,s_k,t^r_0,t^r_1,\ldots)\stackrel{\mathrm{def}}{=}q\in\mathbb{Q}^*_{\mathrm{w}\infty}(K,\Sigma).$$
Clearly $p\le^{\mathrm{w}\infty}_0 q$. To show that $q\Vdash\dot\tau=\alpha$ we use our standard assumption that (K,Σ) is smooth. Suppose that $q'\ge q$ is such that $q'\Vdash\dot\tau\ne\alpha$. We may assume that $\ell g(w^{q'})>m^{s_k}_{\mathrm{up}}$. By the smoothness we have $w^{q'}\restriction m^{s_k}_{\mathrm{up}}\in\mathrm{pos}(w^p,s_0,\ldots,s_k)$, and so $w^{q'}\restriction m^{s_k}_{\mathrm{up}}=w^r$, and $w^{q'}\in\mathrm{pos}(w^r,t^r_0,\ldots,t^r_\ell)$ (for a suitable $\ell<\omega$). Consequently $q'\ge r$ and this is a contradiction. □

COROLLARY 2.1.12. *Assume (K,Σ) is a finitary creating pair which captures singletons.*

1. *Let $p \in \mathbb{Q}^*_{\mathrm{w}\infty}(K,\Sigma)$, $\dot{\tau}_n$ be $\mathbb{Q}^*_{\mathrm{w}\infty}(K,\Sigma)$-names for ordinals (for $n<\omega$) and $\ell<\omega$. Then there is a condition $q \in \mathbb{Q}^*_{\mathrm{w}\infty}(K,\Sigma)$ such that*
   (a) *$p \le_\ell q$ and*
   (b) *the condition q essentially decides (see 1.2.9(1)) each name $\dot{\tau}_n$.*
2. *The forcing notion $\mathbb{Q}^*_{\mathrm{w}\infty}(K,\Sigma)$ is proper.*

PROOF. 1) Follows from 2.1.5 and 2.1.11.
2) Follows from 1). □

2.2. Forcing notion $\mathbf{Q}^*_f(K,\Sigma)$: bigness and halving

DEFINITION 2.2.1. Let $\bar{r} = \langle r_m : m < \omega\rangle$ be a non-decreasing sequence of integers ≥ 2. For a creating pair (K,Σ) for $\mathbf{H}$ we say that:

1. (K,Σ) is *big* if for every $k<\omega$ there is $m<\omega$ such that:
   if $t \in K$, $\mathbf{nor}[t] \geq m$, $u \in \mathbf{basis}(t)$ and $c : \mathrm{pos}(u,t) \longrightarrow \{0,1\}$
   then there is $s \in \Sigma(t)$ such that $\mathbf{nor}[s] \geq k$, and $c{\restriction}\mathrm{pos}(u,s)$ is constant.
   In this situation we call m *a bigness witness for k*.
2. (K,Σ) is *$\bar{r}$–big* if for each $t \in K$ such that $\mathbf{nor}[t] > 1$ and $u \in \mathbf{basis}(t)$ and $c : \mathrm{pos}(u,t) \longrightarrow r_{m^t_{\mathrm{dn}}}$ there is $s \in \Sigma(t)$ such that $\mathbf{nor}[s] \geq \mathbf{nor}[t] - 1$ and $c \restriction \mathrm{pos}(u,s)$ is constant.

REMARK 2.2.2. Clearly, for a creating pair (K,Σ), $\bar{r}$–big implies implies big.

To show how the notions introduced in definition 2.2.1 work we start with proving an application of 2.1.4 to the case when the creating pair is additionally big and growing.

PROPOSITION 2.2.3. *Assume (K,Σ) is a finitary, growing and big creating pair. If $p \in \mathbb{Q}^*_{\mathrm{s}\infty}(K,\Sigma)$, $p \Vdash$ "$\dot{\tau} < m$", $m<\omega$ then there is q, $p \le_0 q \in \mathbb{Q}^*_{\mathrm{s}\infty}(K,\Sigma)$ such that $q \Vdash$ "$\dot{\tau} = m_0$" for some m_0.*

PROOF. Let $h \in \omega^\omega$ be such that $h(k)$ is a bigness witness for k (remember that (K,Σ) is big, see 2.2.1(1)). Note that 2.1.4 + 2.1.3(2) give us a condition $p' = (w^p, s_0, s_1, s_2, \dots) \in \mathbb{Q}^*_{\mathrm{s}\infty}(K,\Sigma)$ such that $p \le^{\mathrm{s}\infty}_0 p'$ and

(α) $\mathbf{nor}[s_\ell] \geq h^{(m\cdot|\mathrm{pos}(w^p,s_0,\dots,s_{\ell-1})|)}(m^{s_\ell}_{\mathrm{dn}} + 1)$ for all $\ell<\omega$ and
(β) p' approximates the name $\dot{\tau}$ at each $n<\omega$.

Using iteratively the choice of $h(k)$ we will have then

(γ) for every $\ell<\omega$ and each function

$$d : \{\langle u,v\rangle : u \in \mathrm{pos}(w^p, s_0, \dots, s_{\ell-1}) \ \&\ v \in \mathrm{pos}(u,s_\ell)\} \longrightarrow m+1$$

there is a creature $s \in \Sigma(s_\ell)$ such that $\mathbf{nor}[s] > m^{s_\ell}_{\mathrm{dn}}$ and

$$d \restriction \{\langle u,v\rangle \in \mathrm{dom}(d) : v \in \mathrm{pos}(u,s)\}$$

depends on the first coordinate only.

(Since, as usual, we assume that (K,Σ) is nice, we have in (γ) above that $\mathbf{basis}(s) \supseteq \mathrm{pos}(w^p, s_0, \dots, s_{\ell-1})$.)

Now apply (γ) to find $s'_\ell \in \Sigma(s_\ell)$ (for $\ell<\omega$) such that $\mathbf{nor}[s'_\ell] > m^{s_\ell}_{\mathrm{dn}} = m^{s'_\ell}_{\mathrm{dn}}$ and for every $u \in \mathrm{pos}(w^p, s'_0, \dots, s'_{\ell-1})$ we have

(δ) for each $v_0, v_1 \in \mathrm{pos}(u, s'_\ell)$, $i<m$

$$(v_0, s'_{\ell+1}, s'_{\ell+2}, \dots) \Vdash \dot{\tau} = i \quad \text{iff} \quad (v_1, s'_{\ell+1}, s'_{\ell+2}, \dots) \Vdash \dot{\tau} = i.$$

Look at $q \stackrel{\rm def}{=} (w^p, s'_0, s'_1, s'_2, \dots)$. First it is a condition in $\mathbb{Q}^*_{s\infty}(K,\Sigma)$ as (K,Σ) is nice and $s'_\ell \in \Sigma(s_\ell)$, $\mathbf{nor}[s'_\ell] > m^{s'_\ell}_{\rm dn}$. To show that

$$(\exists m_0 < m)(q \Vdash \dot{\tau} = m_0)$$

take a condition $r = (w_1, t_0, t_1, \dots) \geq q$ such that r decides the value of $\dot{\tau}$, $w_1 \in \mathrm{pos}(w^p, s'_0, \dots, s'_{\ell-1})$ and ℓ is the smallest possible. By (β) we know that the condition $(w_1, s'_\ell, s'_{\ell+1}, \dots)$ forces a value to $\dot{\tau}$, say m_0. We claim that $\ell = 0$, i.e. $w_1 = w^p$ (which is enough as then q decides the value of $\dot{\tau}$). Why? Suppose that $\ell > 0$ and look at the requirement (δ) for $\ell - 1$, $u = w_1 \restriction m^{s'_{\ell-1}}_{\rm dn}$. By the smoothness $u \in \mathrm{pos}(w^p, s'_0, \dots, s'_{\ell-2})$ and consequently, by (δ), for each $v \in \mathrm{pos}(u, s'_{\ell-1})$ we have

$$(v, s'_\ell, s'_{\ell+1}, \dots) \Vdash \dot{\tau} = m_0.$$

Applying smoothness once again we note that for each $w \in \mathrm{pos}(u, s'_{\ell-1}, s'_\ell, \dots, s'_k)$

$$w \restriction m^{s'_\ell}_{\rm dn} \in \mathrm{pos}(u, s'_{\ell-1}) \quad \text{and} \quad w \in \mathrm{pos}(w \restriction m^{s'_\ell}_{\rm dn}, s'_\ell, \dots, s'_k).$$

Hence for each such w we have

$$(w \restriction m^{s'_\ell}_{\rm dn}, s'_\ell, s'_{\ell+1}, \dots,) \Vdash \dot{\tau} = m_0 \quad \text{and}$$

$$(w, s'_{k+1}, s'_{k+2}, \dots) \geq (w \restriction m^{s'_\ell}_{\rm dn}, s'_\ell, s'_{\ell+1}, \dots)$$

and so

$$(w, s'_{k+1}, s'_{k+2}, \dots,) \Vdash \dot{\tau} = m_0.$$

Hence we may conclude that $(u, s'_{\ell-1}, s'_\ell, \dots) \Vdash \dot{\tau} = m_0$ which contradicts the choice of ℓ. □

REMARK 2.2.4. One may notice that the assumptions of 2.2.3 are difficult to satisfy in most natural cases. First examples of growing creating pairs one has in mind are omittory creating pairs. However, if we demand that an omittory creating pair (K,Σ) is smooth then we get to

$$t \in K \ \ \& \ \ u \in \mathbf{basis}(t) \quad \Rightarrow \quad u^\frown \mathbf{0}_{[m^t_{\rm dn}, m^t_{\rm up})} \in \mathrm{pos}(u,t).$$

This excludes bigness as defined in 2.2.1. Thus it is desirable to consider in this case a weaker condition, which more fits to specific properties of omittory creating pairs.

DEFINITION 2.2.5. An omittory creating pair (K,Σ) is *omittory–big* if for every $k < \omega$ there is $m < \omega$ such that:
if $t \in K$, $\mathbf{nor}[t] \geq m$, $u \in \mathbf{basis}(t)$ and $c : \mathrm{pos}(u,t) \longrightarrow \{0,1\}$ *then* there is $s \in \Sigma(t)$ such that $\mathbf{nor}[s] \geq k$ and $c{\restriction}(\mathrm{pos}(u,s) \setminus \{\mathbf{0}_{[m^t_{\rm dn}, m^t_{\rm up})}\})$ is constant.
We may call m *an omittory-bigness witness for* k.

PROPOSITION 2.2.6. *Assume (K,Σ) is a finitary, omittory and omittory–big creating pair. Suppose that $p \in \mathbb{Q}^*_{s\infty}(K,\Sigma)$, $p \Vdash$ "$\dot{\tau} < m$", $m < \omega$. Then there is a condition $q \in \mathbb{Q}^*_{s\infty}(K,\Sigma)$ such that $p \leq^{s\infty}_0 q$ and q decides the value of $\dot{\tau}$.*

PROOF. We start as in the proof of 2.2.3, but in (γ) there we say that

$$d \restriction \{\langle u, v\rangle \in \mathrm{dom}(d) : v \in \mathrm{pos}(u,s) \ \& \ (\exists k \in [\ell g(u), \ell g(v)))(v(h) \neq 0)\}$$

depends on the first coordinate only, and therefore we get $s'_\ell \in \Sigma(s_\ell)$ as there but with (δ) replaced by

$(\delta)^-$ for each $v_0, v_1 \in \mathrm{pos}(u, s'_\ell) \setminus \{u^\frown \mathbf{0}_{[m^{s'_\ell}_{\mathrm{dn}}, m^{s'_\ell}_{\mathrm{up}})}\}$, $i<m$

$$(v_0, s'_{\ell+1}, s'_{\ell+2}, \dots) \Vdash \dot\tau = i \quad \text{iff} \quad (v_1, s'_{\ell+1}, s'_{\ell+2}, \dots) \Vdash \dot\tau = i.$$

Now comes the main modification of the proof of 2.2.3. We choose an infinite set $I = \{i_0, i_1, i_2, \dots\} \subseteq \omega$ such that for every $i<m$ we have

if: $k < \ell \leq \ell' < \omega$, $w \in \mathrm{pos}(w^p, s'_0, \dots, s'_{i_k})$, $v_0 \in \mathrm{pos}(w, s'_{i_\ell} \restriction [m^{s'_{i_k}}_{\mathrm{up}}, m^{s'_{i_\ell}}_{\mathrm{up}}))$, $v_1 \in \mathrm{pos}(w, s'_{i_{\ell'}} \restriction [m^{s'_{i_k}}_{\mathrm{up}}, m^{s'_{i_{\ell'}}}_{\mathrm{up}}))$, and

$$(\exists m \in [m^{s'_{i_\ell}}_{\mathrm{dn}}, m^{s'_{i_\ell}}_{\mathrm{up}}))(v_0(m) \neq 0) \quad \text{and} \quad (\exists m \in [m^{s'_{i_{\ell'}}}_{\mathrm{dn}}, m^{s'_{i_{\ell'}}}_{\mathrm{up}}))(v_1(m) \neq 0)$$

then: $(v_0, s'_{i_\ell+1}, s'_{i_\ell+2}, \dots) \Vdash \dot\tau = i$ iff $(v_1, s'_{i_{\ell'}+1}, s'_{i_{\ell'}+2}, \dots) \Vdash \dot\tau = i$

and a similar condition for the case of $w = w^p$, $0 \leq \ell \leq \ell' < \omega$. The construction of the set I is rather standard (by induction) and it goes like in the proof of the suitable property for the Mathias forcing (see e.g. [**BaJu95**, 7.4.6]). Next we look at

$$q \stackrel{\mathrm{def}}{=} (w^p, s'_{i_0} \restriction [m^{s'_0}_{\mathrm{dn}}, m^{s'_{i_0}}_{\mathrm{up}}), s'_{i_1} \restriction [m^{s'_{i_0}}_{\mathrm{up}}, m^{s'_{i_1}}_{\mathrm{up}}), s'_{i_2} \restriction [m^{s'_{i_1}}_{\mathrm{up}}, m^{s'_{i_2}}_{\mathrm{up}}), \dots).$$

It should be clear that it is a condition in $\mathbb{Q}^*_{\mathrm{s}\infty}(K,\Sigma)$ which is $\leq^{\mathrm{s}\infty}_0$–stronger than p. Note that, as (K,Σ) is omittory (remember the demand $(\boxtimes_0)$ of 2.1.1(2)), by the choice of the set I we have

if $w_1 \in \mathrm{pos}(w^q, t^q_0, \dots, t^q_k)$, $k<\omega$, $w_1 \restriction [m^{t^q_k}_{\mathrm{dn}}, m^{t^q_k}_{\mathrm{up}}) = \mathbf{0}_{[m^{t^q_k}_{\mathrm{dn}}, m^{t^q_k}_{\mathrm{up}})}$ and the condition $(w_1, t^q_{k+1}, t^q_{k+2}, \dots)$ decides the value of $\dot\tau$ then $(w_1 \restriction m^{t^q_k}_{\mathrm{dn}}, t^q_k, t^q_{k+1}, t^q_{k+2}, \dots)$ does so.

Now, like in 2.2.3 we show that the condition q decides the value of $\dot\tau$, using the remark above and the choice of the set I. □

DEFINITION 2.2.7. Let (K,Σ) be a creating pair.

1. We say that (K,Σ) has the *Halving Property* if there is a mapping

$$\mathrm{half} : K \longrightarrow K$$

such that
   (a) for each $t \in K$, $\mathrm{half}(t) \in \Sigma(t)$ and $\mathbf{nor}[\mathrm{half}(t)] \geq \frac{1}{2}\mathbf{nor}[t]$,
   (b) if $t_0, \dots, t_n \in K$, $\min\{\mathbf{nor}[t_i] : i \leq n\} \geq 2$ and a creature $t \in \Sigma(\mathrm{half}(t_0), \dots, \mathrm{half}(t_n))$ is such that $\mathbf{nor}[t] > 0$ then there is $s \in \Sigma(t_0, \dots, t_n)$ such that

$$\mathbf{nor}[s] \geq \min\{\frac{1}{2}\mathbf{nor}[t_i] : i \leq n\} \quad \text{and} \quad (\forall u \in \mathbf{basis}(t_0))(\mathrm{pos}(u,s) \subseteq \mathrm{pos}(u,t)).$$

2. We say that (K,Σ) has the *weak Halving Property* if there is a mapping $\mathrm{half} : K \longrightarrow K$ which satisfies **(a)** above and
   $(\mathrm{b})^-$ if $t_0 \in K$, $\mathbf{nor}[t_0] \geq 2$ and $t \in \Sigma(\mathrm{half}(t_0))$ is such that $\mathbf{nor}[t] > 0$ then there is a creature $s \in \Sigma(t_0)$ such that

$$\mathbf{nor}[s] \geq \frac{1}{2}\mathbf{nor}[t_0] \quad \text{and} \quad (\forall u \in \mathbf{basis}(t_0))(\mathrm{pos}(u,s) \subseteq \mathrm{pos}(u,t)).$$

3. Whenever we say that (K,Σ) has the (weak) Halving Property we assume that the function $\mathrm{half} : K \longrightarrow K$ witnessing this is fixed.

REMARK 2.2.8. Remember that we standardly assume that creating pairs are nice, so in 2.2.7(1b), 2.2.7(2b$^-$) we have $\mathbf{basis}(t_0) \subseteq \mathbf{basis}(t)$ and $\mathbf{basis}(t_0) \subseteq \mathbf{basis}(s)$. Of course, the Halving Property implies the weak Halving Property. Moreover, the two notions agree for simple creating pairs.

Our next lemma shows how we are going apply the Halving Property.

LEMMA 2.2.9. *Assume that (K,Σ) is a creating pair with the Halving Property (witnessed by a mapping* half $: K \longrightarrow K$*). Suppose that $f: \omega\times\omega \longrightarrow \omega$ is a fast function, $\dot\tau$ is a $\mathbb{Q}^*_f(K,\Sigma)$-name for an ordinal, $n<\omega$, $0<\varepsilon\leq 1$ and $p \in \mathbb{Q}^*_f(K,\Sigma)$ is a condition such that*

$$(\forall i\in\omega)(\mathbf{nor}[t^p_i] \geq \varepsilon\cdot f(n, m^{t^p_i}_{\rm dn})).$$

Further assume that there is a condition $r\in\mathbb{Q}^*_f(K,\Sigma)$ such that

$$(\forall i\in\omega)(\mathbf{nor}[t^r_i]>0) \quad \textit{and} \quad (w^p, \mathrm{half}(t^p_0), \mathrm{half}(t^p_1), \dots) \leq^f_0 r \quad \textit{and}$$

r essentially decides $\dot\tau$ (see 1.2.9). Then *there is a condition $q\in\mathbb{Q}^*_f(K,\Sigma)$ such that*

$$(\forall i\in\omega)(\mathbf{nor}[t^q_i] \geq \frac{\varepsilon}{2}\cdot f(n, m^{t^q_i}_{\rm dn})), \quad p\leq^f_0 q \quad \textit{and } q \textit{ essentially decides } \dot\tau.$$

PROOF. First note that the niceness implies that $(w^p, \mathrm{half}(t^p_0), \mathrm{half}(t^p_1), \dots)$ is a condition in $\mathbb{Q}^*_f(K,\Sigma)$ (by 2.2.7(1)(a) and 1.2.8(2); remember that f is fast). Now suppose that $r\in\mathbb{Q}^*_f(K,\Sigma)$ is as in the assumptions of the lemma. Take $m<\omega$ so large that

$$(\forall u\in \mathrm{pos}(w^p, t^r_0, \dots, t^r_{m-1}))((u, t^r_m, t^r_{m+1}, \dots) \text{ decides the value of } \dot\tau) \quad \text{and}$$

$$(\forall i\geq m)(\mathbf{nor}[t^r_i] \geq \frac{\varepsilon}{2} f(n, m^{t^r_i}_{\rm dn}))$$

(for the first requirement remember that (K,Σ) is smooth; the second is possible since $\varepsilon\leq 1$). Next choose integers $0=i_0<i_1<\ldots<i_{m-1}<i_m$ such that

$$(\forall \ell<m)(t^r_\ell \in \Sigma(\mathrm{half}(t^p_{i_\ell}), \dots, \mathrm{half}(t^p_{i_{\ell+1}-1}))).$$

Applying the Halving Property (see 2.2.7(1b); remember that we have assumed $\mathbf{nor}[t^r_\ell]>0$ for each $\ell\in\omega$) we find $s_\ell \in \Sigma(t^p_{i_\ell}, \dots, t^p_{i_{\ell+1}-1})$ (for $\ell<m$) such that

$$\mathbf{nor}[s_\ell] \geq \min\{\tfrac{1}{2}\mathbf{nor}[t^p_i] : i_\ell\leq i<i_{\ell+1}\} \quad \text{and}$$
$$(\forall u \in \mathbf{basis}(t^p_{i_\ell}))(\mathrm{pos}(u, s_\ell) \subseteq \mathrm{pos}(u, t^r_\ell)).$$

Then easily

$$q \stackrel{\rm def}{=} (w^p, s_0, \dots, s_{m-1}, t^r_m, t^r_{m+1}, \dots) \in \mathbb{Q}^*_f(K,\Sigma), \quad p\leq^f_0 q \quad \text{and}$$

$$(\forall i\in\omega)(\mathbf{nor}[t^q_i] \geq \frac{\varepsilon}{2} f(n, m^{t^q_i}_{\rm dn}))$$

(for the last statement remember that f is fast so $f(n,\cdot)$ is non-decreasing). Moreover q essentially decides the value of $\dot\tau$ as

$$\mathrm{pos}(w^p, s_0, \dots, s_{m-1}) \subseteq \mathrm{pos}(w^p, t^r_0, \dots, t^r_{m-1}).$$

□

REMARK 2.2.10. One could ask why we cannot in the conclusion of 2.2.9 require that simply $\mathbf{nor}[t^q_i] \geq \frac{1}{2}\mathbf{nor}[t^p_j]$ (for j chosen somehow suitably, e.g. such that $m^{t^q_i}_{\rm dn} = m^{t^p_j}_{\rm dn}$). The reason is that "the upgrading procedure" given by 2.2.7(1b) takes care of possibilities only: no other relation between s,t there is required. In particular we do not know if $\Sigma(t)\cap\Sigma(s)\neq\emptyset$. Consequently, if we apply this procedure to *all* m (replacing each t^r_m by suitable s_m) then we may get a condition incompatible with r.

THEOREM 2.2.11. *Assume that a creating pair (K,Σ) for $\mathbf{H}$ is finitary, $\bar{2}$-big and has the Halving Property. Further suppose that a function $f:\omega\times\omega\longrightarrow\omega$ is $\mathbf{H}$-fast and that (K,Σ) is either simple or gluing (see 2.1.7). Let $\dot{\tau}_m$ be $\mathbb{Q}^*_f(K,\Sigma)$-names for ordinals (for $m\in\omega$), $p\in\mathbb{Q}^*_f(K,\Sigma)$ and $n<\omega$.*
Then there is a condition $q\in\mathbb{Q}^*_f(K,\Sigma)$ such that $p\leq^f_n q$ and q essentially decides (see 1.2.9) all the names $\dot{\tau}_m$ (for $m\in\omega$).

PROOF. Let (K,Σ) and f be as in the assumptions of the theorem.

CLAIM 2.2.11.1. *Suppose that $\dot{\tau}$ is a $\mathbb{Q}^*_f(K,\Sigma)$-name for an ordinal, $n<\omega$, a condition $p\in\mathbb{Q}^*_f(K,\Sigma)$ and a real ε are such that*

$$2^{|\mathrm{pos}(w^p,t^p_0)|}\cdot 2^{-\varphi_{\mathbf{H}}(m^{t^p_1}_{\rm dn})}\leq\varepsilon\leq 1$$

(where $\varphi_{\mathbf{H}}(\ell)=|\prod\limits_{i<\ell}\mathbf{H}(i)|$, see 1.1.12), and

$$\mathbf{nor}[t^p_0]>1 \quad \textit{and} \quad (\forall i>0)(\mathbf{nor}[t^p_i]\geq\varepsilon\cdot f(n+1,m^{t^p_i}_{\rm dn})).$$

Then there is a condition $q\in\mathbb{Q}^*_f(K,\Sigma)$ such that $p\leq^f_0 q$, $t^q_0\in\Sigma(t^p_0)$, $\mathbf{nor}[t^q_0]\geq\mathbf{nor}[t^p_0]-1$, q essentially decides the value of $\dot{\tau}$ and

$$(\forall i>0)(\mathbf{nor}[t^q_i]\geq\frac{\varepsilon}{2^{|\mathrm{pos}(w^p,t^p_0)|}}\cdot f(n+1,m^{t^q_i}_{\rm dn})).$$

Proof of the claim: First note that our assumptions on ε (and the fact that f is $\mathbf{H}$-fast) imply that if $t\in K$ is such that $m^t_{\rm dn}\geq m^{t^p_1}_{\rm dn}$ and

$$\mathbf{nor}[t]\geq\frac{\varepsilon}{2^{|\mathrm{pos}(w^p,t^p_0)|}}\cdot f(n+1,m^t_{\rm dn})$$

then

$$\mathbf{nor}[t]>f(n,m^t_{\rm dn})+\varphi_{\mathbf{H}}(m^t_{\rm dn})+2.$$

Let $\{w^0_m: m<m_0\}$ enumerate $\mathrm{pos}(w^p,t^p_0)$. We inductively choose conditions p^0_m for $m\leq m_0$ such that

(α) $p^0_0=(w^0_0,t^p_1,t^p_2,\dots)$,

(β) if there is a condition $r\in\mathbb{Q}^*_f(K,\Sigma)$ such that $p^0_m\leq_0 r$ and

$$(\forall i\in\omega)(\mathbf{nor}[t^r_i]\geq\frac{\varepsilon}{2^{m+1}}\cdot f(n+1,m^{t^r_i}_{\rm dn}))$$

and r essentially decides $\dot{\tau}$ then we choose such an r and we put

$$p^0_{m+1}=(w^0_{m+1},t^r_0,t^r_1,\dots),$$

(γ) if we cannot apply the clause (β) (i.e. there is no r as above) then

$$p^0_{m+1}=(w^0_{m+1},\mathrm{half}(t^{p^0_m}_0),\mathrm{half}(t^{p^0_m}_1),\dots).$$

[For the sake of the uniformity of the inductive definition we let $w^0_{m_0}$ to be e.g. w^0_0.] Note that by the niceness there are no problems in the above construction (i.e. $p^0_m \in \mathbb{Q}^*_f(K,\Sigma)$). Let $c : \mathrm{pos}(w^p, t^p_0) \longrightarrow 2$ be such that

$$c(w^0_m) = \begin{cases} 0 & \text{if the clause } (\beta) \text{ was applied to choose } p^0_{m+1}, \\ 1 & \text{otherwise.} \end{cases}$$

Due to the bigness of (K,Σ) we find a creature $s_0 \in \Sigma(t^p_0)$ such that c is constant on $\mathrm{pos}(w^p, s_0)$ and $\mathbf{nor}[s_0] \geq \mathbf{nor}[t^p_0] - 1$. Note that

$$q \stackrel{\mathrm{def}}{=} (w^p, s_0, t_0^{p^0_{m_0}}, t_1^{p^0_{m_0}}, t_2^{p^0_{m_0}}, \ldots) \in \mathbb{Q}^*_f(K,\Sigma)$$

(the niceness applies here once again, see 1.2.8; the norm condition should be obvious as $p^0_{m_0} \in \mathbb{Q}^*_f(K,\Sigma)$). Moreover

$$\mathbf{nor}[t_i^{p^0_{m_0}}] \geq \frac{\varepsilon}{2^{|\mathrm{pos}(w^p, t^p_0)|}} \cdot f(n+1, m_{\mathrm{dn}}^{t_i^{p^0_{m_0}}})$$

and hence, in particular,

$$\mathbf{nor}[t_i^{p^0_{m_0}}] > f(n, m_{\mathrm{dn}}^{t_i^{p^0_{m_0}}}) + \varphi_{\mathbf{H}}(m_{\mathrm{dn}}^{t_i^{p^0_{m_0}}}) + 2.$$

If the constant value of $c \restriction \mathrm{pos}(w^p, s_0)$ is 0 then easily the condition q satisfies the requirements of the claim (use 1.2.10).

So we want to exclude the possibility that the constant value is 1. For this assume that it is the case. First note that then, due to the way we constructed $p^0_{m_0}$, we may apply lemma 2.2.9 and conclude that there are no $u \in \mathrm{pos}(w^p, s_0)$ and $r \in \mathbb{Q}^*_f(K,\Sigma)$ such that $\mathbf{nor}[t^r_i] > 0$ for all $i \in \omega$ and

$$(u, t_0^{p^0_{m_0}}, t_1^{p^0_{m_0}}, \ldots) \leq_0 r \quad \text{and} \quad r \text{ essentially decides the value of } \dot{\tau}.$$

Now we inductively choose an increasing sequence $\ell_0 < \ell_1 < \ldots$ of integers, creatures $s_1, s_2, \ldots \in K$ and conditions $p_0, p_1, \ldots \in \mathbb{Q}^*_f(K,\Sigma)$ such that

1. $p_0 = q$ (defined above), $\ell_0 = 0$,
2. ℓ_{k+1} is such that $\ell_{k+1} > \ell_k$ and
$$(\forall i \geq \ell_{k+1})(\mathbf{nor}[t^{p_k}_i] \geq f(n+k+2, m_{\mathrm{dn}}^{t^{p_k}_i})),$$
3. $s_i = t^{p_k}_i$ for $i \leq \ell_k$ and $\mathbf{nor}[t^{p_k}_i] > f(n+k, m_{\mathrm{dn}}^{t^{p_k}_i}) + \varphi_{\mathbf{H}}(m_{\mathrm{dn}}^{t^{p_k}_i}) + 2$ for $i > \ell_k$,
4. there are no $u \in \mathrm{pos}(w^p, s_0, \ldots, s_{\ell_k})$ and $r \in \mathbb{Q}^*_f(K,\Sigma)$ such that
$$(\forall i \in \omega)(\mathbf{nor}[t^r_i] > 0)$$
and $(u, t^{p_k}_{\ell_k+1}, t^{p_k}_{\ell_k+2}, \ldots) \leq_0 r$ and r essentially decides $\dot{\tau}$,
5. for $i \in (\ell_k, \ell_{k+1}]$, $s_i \in \Sigma(t^{p_k}_i)$ and $\mathbf{nor}[s_i] > f(n+k, m^{s_i}_{\mathrm{dn}})$,
6. $\mathbf{nor}[s_{\ell_{k+1}}] > f(n+k+1, m_{\mathrm{dn}}^{s_{\ell_{k+1}}})$ and $p_k \leq^f_k p_{k+1}$.

Suppose we have defined ℓ_k, p_k and s_i for $i \leq \ell_k$. Choose ℓ_{k+1} according to the requirement (2) above. Fix an enumeration

$$\{w^k_m : m < m_k\} = \mathrm{pos}(w^p, s_0, \ldots, s_{\ell_k}, t^{p_k}_{\ell_k+1}, \ldots, t^{p_k}_{\ell_{k+1}}).$$

We inductively choose conditions $p^k_m \in \mathbb{Q}^*_f(K,\Sigma)$ (in the way analogous to the construction of p^0_m's):

$(\alpha)_k$ $p^k_0 = (w^k_0, t^{p_k}_{\ell_{k+1}+1}, t^{p_k}_{\ell_{k+1}+2}, \ldots)$,

$(\beta)_k$ if there is a condition $r\in\mathbb{Q}^*_f(K,\Sigma)$ such that $p^k_m\leq_0 r$ and

$$(\forall i\in\omega)(\mathbf{nor}[t^r_i]\geq\frac{1}{2^{m+1}}\cdot f(n+k+2,m^{t^r_i}_{\mathrm{dn}}))$$

and r essentially decides $\dot\tau$ then we choose such an r and we put

$$p^k_{m+1}=(w^k_{m+1},t^r_0,t^r_1,\dots),$$

$(\gamma)_k$ if we cannot apply the clause $(\beta)_k$ then

$$p^k_{m+1}=(w^k_{m+1},\mathrm{half}(t^{p^k_m}_0),\mathrm{half}(t^{p^k_m}_1),\dots).$$

As previously, due to the assumptions about (K,Σ), we can carry out the construction. Let $c_k:\mathrm{pos}(w^p,s_0,\dots,s_{\ell_k},t^{p_k}_{\ell_k+1},\dots,t^{p_k}_{\ell_{k+1}})\longrightarrow 2$ be a function such that

$$c_k(w^k_m)=\begin{cases}0 & \text{if the clause } (\beta)_k \text{ was applied to choose } p^k_{m+1}\\ 1 & \text{otherwise.}\end{cases}$$

Applying successively $\bar{2}$-bigness (to each $t^{p_k}_i$ for $i=\ell_{k+1},\ell_{k+1}-1,\dots,\ell_k+1$) we find creatures $s_{\ell_k+1},\dots,s_{\ell_{k+1}}$ such that for each $i\in[\ell_k+1,\ell_{k+1}]$

$$s_i\in\Sigma(t^{p_k}_i)\quad\text{ and }\quad\mathbf{nor}[s_i]\geq\mathbf{nor}[t^{p_k}_i]-|\mathrm{pos}(w^p,s_0,\dots,s_{\ell_k},t^{p_k}_{\ell_k+1},\dots,t^{p_k}_{i-1})|$$

and for each $u\in\mathrm{pos}(w^p,s_0,\dots,s_{\ell_k},s_{\ell_k+1},\dots,s_{\ell_{k+1}})$ the value of $c_k(u)$ depends on $u\restriction m^{s_{\ell_k}}_{\mathrm{up}}$ only. If the constant value of $c_k\restriction\mathrm{pos}(v,s_{\ell_k+1},\dots,s_{\ell_{k+1}})$ (for some sequence $v\in\mathrm{pos}(w^p,s_0,\dots,s_{\ell_k})$) is 0 then easily

$$(v,s_{\ell_k+1},\dots,s_{\ell_{k+1}},t^{p^k_{m_k}}_0,t^{p^k_{m_k}}_1,\dots)\in\mathbb{Q}^*_f(K,\Sigma)$$

(remember that (K,Σ) is nice), it is $\leq_0$-stronger than $(v,t^{p^k}_{\ell_k+1},t^{p^k}_{\ell_k+2},\dots)$ and it essentially decides $\dot\tau$ (by 1.2.10). This contradicts the inductive assumption (4). So the constant value is always 1 and consequently $c_k\restriction\mathrm{pos}(w^p,s_0,\dots,s_{\ell_k},\dots,s_{\ell_{k+1}})$ is constantly 1. Put

$$p_{k+1}\stackrel{\mathrm{def}}{=}(w^p,s_0,\dots,s_{\ell_{k+1}},t^{p^k_{m_k}}_0,t^{p^k_{m_k}}_1,\dots)\in\mathbb{Q}^*_f(K,\Sigma).$$

Note that for all $i\in\omega$

$$\mathbf{nor}[t^{p^k_{m_k}}_i]>f(n+k+1,m^{t^{p^k_{m_k}}_i}_{\mathrm{dn}})+\varphi_{\mathbf{H}}(m^{t^{p^k_{m_k}}_i}_{\mathrm{dn}})+2$$

and thus p_{k+1} satisfies (3). By the construction (and 2.2.9) the condition p_{k+1} satisfies the inductive requirement (4) (for $k+1$). Since f is $\mathbf{H}$-fast we have that for $i\in[\ell_k+1,\ell_{k+1}]$ (by the inductive assumption (3))

$$\mathbf{nor}[t^{p_{k+1}}_i]\geq\mathbf{nor}[t^{p_k}_i]-|\mathrm{pos}(w^p,s_0,\dots,s_{\ell_k},t^{p_k}_{\ell_k+1},\dots t^{p_k}_{i-1})|>f(n+k,m^{t^{p_{k+1}}_i}_{\mathrm{dn}}),$$

and for $i>\ell_{k+1}$

$$\mathbf{nor}[t^{p_{k+1}}_i]>f(n+k+1,m^{t^{p_{k+1}}_i}_{\mathrm{dn}})$$

and (if $k>0$)

$$\mathbf{nor}[t^{p_k}_{\ell_k}]=\mathbf{nor}[s_{\ell_k}]>f(n+k,m^{t^{p_k}_{\ell_k}}_{\mathrm{dn}}).$$

Hence $p_k\leq^f_k p_{k+1}$. Moreover $\mathbf{nor}[s_{\ell_{k+1}}]>f(n+k+1,m^{s_{\ell_{k+1}}}_{\mathrm{dn}})$. Thus the requirements (5) and (6) are satisfied.

Finally look at the limit condition

$$p^* = (w^p, s_0, s_1, \dots) = \lim_{k\in\omega} p_k \in \mathbb{Q}^*_f(K,\Sigma).$$

Now we have to distinguish the two cases: (K,Σ) is simple and (K,Σ) is gluing.

If our creating pair is simple then we take a condition $r \geq p^*$ which decides the value of $\dot\tau$ and such that $\mathbf{nor}[t^r_i] > 0$ for all $i \in \omega$. We may assume that for some $k \in \omega$ we have $w^r \in \mathrm{pos}(w^p, s_0, \dots, s_{\ell_k})$. Then $(w^r, t^{p_k}_{\ell_k+1}, t^{p_k}_{\ell_k+2}, \dots) \leq_0 r$ and this contradicts the assumption (4) about p_k.

Now suppose that (K,Σ) is gluing. Note that choosing ℓ_{k+1} we may take it arbitrarily large. So we may assume that (additionally) $\ell_{k+1} - \ell_k$ is larger then the gluing witness for $f(k, m^{s_{\ell_k}}_{\mathrm{up}})$. Then we find $s^*_{k+1} \in \Sigma(s_{\ell_k+1}, \dots, s_{\ell_{k+1}})$ such that

$$\mathbf{nor}[s^*_{k+1}] \geq \min\{f(k, m^{s_{\ell_k}}_{\mathrm{up}}), \mathbf{nor}[s_{\ell_k+1}], \dots, \mathbf{nor}[s_{\ell_{k+1}}]\} = f(k, m^{s_{\ell_k}}_{\mathrm{up}}).$$

Put $s^*_0 = s_0$ and consider the condition

$$p^{**} = (w^p, s^*_0, s^*_1, s^*_2, \dots) \geq p^*.$$

Now we finish choosing the r as earlier above p^{**}. The claim is proved.

CLAIM 2.2.11.2. *Suppose that $\dot\tau$ is a $\mathbb{Q}^*_f(K,\Sigma)$-name for an ordinal, $n < \omega$ and $p \in \mathbb{Q}^*_f(K,\Sigma)$. Then there is $q \in \mathbb{Q}^*_f(K,\Sigma)$ such that $p \leq^f_n q$ and q essentially decides $\dot\tau$.*

Proof of the claim: Take $i_0 < \omega$ so large that

$$(\forall i \geq i_0)(\mathbf{nor}[t^p_i] > f(n+1, m^{t^p_i}_{\mathrm{dn}})).$$

Let $\{w_m : m < m^*\}$ enumerate $\mathrm{pos}(w^p, t^p_0, \dots, t^p_{i_0})$. Choose inductively q_m, ε_m (for $m < m^*$) such that

q_0 is given by 2.2.11.1 for $(w_0, t^p_{i_0+1}, t^p_{i_0+2}, \dots)$, $\varepsilon_0 = 1$

$\varepsilon_1 = \frac{1}{2^{|\mathrm{pos}(w_0, t^p_{i_0+1})|}}$, $\varepsilon_{m+1} = \frac{\varepsilon_m}{2^{|\mathrm{pos}(w_m, t^{q_{m-1}}_0)|}}$ (for $m > 0$),

q_{m+1} is given by claim 2.2.11.1 for $(w_{m+1}, t^{q_m}_0, t^{q_m}_1, \dots)$, ε_{m+1}.

Note that arriving at the stage $m+1 < m^*$ of this construction we have

$$|\mathrm{pos}(w_0, t^p_{i_0+1})| + \sum_{\ell \leq m} |\mathrm{pos}(w_{\ell+1}, t^{q_\ell}_0)| \leq \varphi_{\mathbf{H}}(m^{t^{q_m}_1}_{\mathrm{dn}}),$$

so ε_{m+1} satisfies the respective demand. Moreover,

$$\mathbf{nor}[t^{q_m}_i] \geq \varepsilon_{m+1} \cdot f(n+1, m^{t^{q_m}_i}_{\mathrm{dn}}) > f(n, m^{t^{q_m}_i}_{\mathrm{dn}}) \qquad \text{for } i > 0,\ m < m^*,$$

and for each $m < m^*$

$$\mathbf{nor}[t^{q_m}_0] \geq \mathbf{nor}[t^p_{i_0+1}] - m > f(n, m^{t^{q_m}_0}_{\mathrm{dn}}) + 2.$$

Consequently we may carry out the construction and finally letting

$$q \stackrel{\mathrm{def}}{=} (w^p, t^p_0, \dots, t^p_{i_0}, t^{q_{m^*-1}}_0, t^{q_{m^*-1}}_1, \dots)$$

we will clearly have a condition as required in the claim.

Applying inductively claim 2.2.11.2 to $\dot\tau_m$ and $n+m$ we finish the theorem (using 1.2.13 and 1.2.10). □

COROLLARY 2.2.12. *Suppose that a creating pair (K,Σ) is finitary, $\bar{2}$-big and has the Halving Property. Further assume that it is either simple or gluing. Let $f : \omega \times \omega \longrightarrow \omega$ be an $\mathbf{H}$-fast function. Then the forcing notion $\mathbb{Q}^*_f(K,\Sigma)$ is proper.*

2.3. Tree–creating (K,Σ)

PROPOSITION 2.3.1. *Let $e<3$, $n<\omega$, $p\in\mathbb{Q}^{\rm tree}_e(K,\Sigma)$ and let $A\subseteq T^p$ be an antichain in T^p such that $(\forall\eta\in A)(\exists\nu\in F^0_n(p))(\nu\vartriangleleft\eta)$. Assume that for each $\eta\in A$ we have a condition $q_\eta\in\mathbb{Q}^{\rm tree}_e(K,\Sigma)$ such that $p^{[\eta]}\leq_0 q_\eta$ and*

if $e=1$ then $\{t^{q_\eta}_\nu:\nu\in T^{q_\eta}\ \&\ \mathbf{nor}[t^{q_\eta}_\nu]\leq n\}\subseteq\{t^p_\nu:\nu\in T^p\}$.

Then *there exists $q\in\mathbb{Q}^{\rm tree}_e(K,\Sigma)$ such that $p\leq^e_{n+1} q$, $A\subseteq T^q$, $q^{[\eta]}=q_\eta$ for $\eta\in A$ and if $\nu\in T^p$ is such that there is no $\eta\in A$ with $\eta\trianglelefteq\nu$ then $\nu\in T^q$ and $t^p_\nu=t^q_\nu$.*

DEFINITION 2.3.2. Let (K,Σ) be a tree-creating pair for $\mathbf{H}$, $k<\omega$.

1. A tree creature $t\in K$ is called *k-big* if $\mathbf{nor}[t]>1$ and for every function $h:\mathrm{pos}(t)\longrightarrow k$ there is $s\in\Sigma(t)$ such that $h{\restriction}\mathrm{pos}(s)$ is constant and $\mathbf{nor}[s]\geq\mathbf{nor}[t]-1$.
2. We say that (K,Σ) is *k-big* if every $t\in K$ with $\mathbf{nor}[t]>1$ is k-big.

REMARK 2.3.3. The difference with 2.2.1 is not serious - we could have interfering.

DEFINITION 2.3.4. A tree creating pair (K,Σ) is t-omittory if for each system $\langle s_\nu:\nu\in\hat{T}\rangle\subseteq K$ such that $\hat{T}$ is a well founded quasi tree, $\mathrm{root}(s_\nu)=\nu$, $\mathrm{pos}(s_\nu)=\mathrm{succ}_T(\nu)$ (for $\nu\in\hat{T}$) and for every $\nu_0\in\hat{T}$ such that $\mathrm{pos}(s_{\nu_0})\subseteq\max(T)$ there is $s\in\Sigma(s_\nu:\nu\in\hat{T})$ such that

$$\mathbf{nor}[s]\geq\mathbf{nor}[s_{\nu_0}]-1\quad\text{ and }\quad\mathrm{pos}(s)\subseteq\mathrm{pos}(s_{\nu_0}).$$

REMARK 2.3.5. The name "t-omittory" comes from "tree–omittory": it is a natural notion corresponding to omittory creating pairs for the case of tree–creating pairs. The main point of being t-omittory is that if $p,q\in\mathbb{Q}^{\rm tree}_e(K,\Sigma)$, $p\leq q$ then we have a condition $r\in\mathbb{Q}^{\rm tree}_e(K,\Sigma)$ such that $p\leq^e_0 r$ and $\mathrm{dcl}(T^r)\subseteq\mathrm{dcl}(T^q)$ and $t^r_\nu=t^q_\nu$ for each $\nu\in T^r$, $\mathrm{root}(r)\vartriangleleft\nu$. [Why? Let $\eta=\mathrm{root}(q)$ and let $T^*\subseteq T^p$ be a well founded quasi tree such that

$$(\forall\nu\in\hat{T}^*)(\mathrm{succ}_{T^*}(\nu)=\mathrm{pos}(t^p_\nu)),\quad\text{ and }\quad\mathrm{root}(T^*)=\eta,\quad\text{ and }\quad t^q_\eta\in\Sigma(t^p_\nu:\nu\in\hat{T}^*).$$

Let $T^-=\{\mathrm{root}(p)\}\cup\bigcup\{\mathrm{pos}(t^p_\nu):\nu\vartriangleleft\eta\ \&\ \nu\in T^p\}\cup\mathrm{pos}(t^q_\eta)$. Clearly T^- is a well founded quasi tree and we may apply 2.3.4 to $\langle t^q_\eta,t^p_\nu:\nu\vartriangleleft\eta\ \&\ \nu\in T^p\rangle$ and η. Thus we get $t^r_{\mathrm{root}(p)}\in\Sigma(t^q_\eta,t^p_\nu:\nu\vartriangleleft\eta\ \&\ \nu\in T^p)$ such that $\mathrm{pos}(t^r_{\mathrm{root}(p)})\subseteq\mathrm{pos}(t^q_\eta)$. Note that, by transitivity of Σ, $t^r_{\mathrm{root}(p)}\in\Sigma(t^p_\nu:\nu\in\hat{T}^-\cup\hat{T}^*)$. For $\nu\in\mathrm{pos}(t^r_{\mathrm{root}(p)})$ let $t^r_\nu=t^q_\nu$ and so on. Easily, this defines a condition r as required.]

Moreover this property implies that $\mathbb{Q}^{\rm tree}_4(K,\Sigma)$ is dense in $\mathbb{Q}^{\rm tree}_2(K,\Sigma)$.

LEMMA 2.3.6. *Suppose (K,Σ) is a tree-creating pair, $e<3$, $p\in\mathbb{Q}^{\rm tree}_e(K,\Sigma)$, $n<\omega$ and $\dot\tau$ is a $\mathbb{Q}^{\rm tree}_e(K,\Sigma)$-name for an ordinal. Further assume that if $e=2$ then (K,Σ) is bounded (see 1.3.3(4)). Then:*

1. *There exist a condition $q\in\mathbb{Q}^{\rm tree}_e(K,\Sigma)$ and a maximal antichain $A\subseteq T^q$ of T^q such that:*
   - (α) *$p\leq^e_n q$,*
   - (β) *for every $\eta\in A$ the condition $q^{[\eta]}$ decides the value of $\dot\tau$,*
   - (γ) *A is an e-thick antichain of T^p (see definition 1.3.7).*
2. *Assume additionally that either $e=0$ and (K,Σ) is t-omittory or $e=1$ and (K,Σ) is 2-big. Then there are $q\in\mathbb{Q}^{\rm tree}_e(K,\Sigma)$ and a front F of T^q satisfying clauses (α) and (β) of (1) above.*

PROOF. 1) Put

$$A_0 = \{\nu \in T^p \colon \ (\exists \eta \in F^0_n(p))(\eta \lhd \nu) \text{ and there is } q \in \mathbb{Q}^{\rm tree}_e(K,\Sigma) \text{ such that } p^{[\nu]} \leq^e_0 q,\ q \text{ decides the value of } \dot\tau \text{ and if } e=1 \text{ then } (\forall \eta \in T^q)(\mathbf{nor}[t^q_\eta] > n)\}.$$

CLAIM 2.3.6.1. $(\forall r \in \mathbb{Q}^{\rm tree}_e(K,\Sigma))(p \leq r \ \Rightarrow \ A_0 \cap T^r \neq \emptyset)$.

Proof of the claim: Suppose $r \in \mathbb{Q}^{\rm tree}_e(K,\Sigma)$ is such that $p \leq r$. We may assume that r decides the value of $\dot\tau$ and if $e=1$ then additionally we have $(\forall \eta \in T^r)(\mathbf{nor}[t^r_\eta] > n)$ (by 1.3.8(3)). Now, as $F^0_n(p)$ is an e-thick antichain of T^p (see 1.3.8(4), (6); remember that if $e=2$ then we assume that (K,Σ) is bounded), we find $\nu \in T^r$ such that $(\exists \eta \in F^0_n(p))(\eta \lhd \nu)$. Look at the condition $r^{[\nu]}$: it witnesses that $\nu \in A_0$.

Thus

$$A \stackrel{\rm def}{=} \{\nu \in A_0 : \text{ there is no } \nu' \lhd \nu \text{ which is in } A_0\}$$

is an e-thick antichain of T^p and for each $\eta \in A$ we may take a witness q_η for $\eta \in A_0$. Now apply 2.3.1 to find $q \in \mathbb{Q}^{\rm tree}_e(K,\Sigma)$ such that $p \leq^e_n q$ and $q^{[\eta]} = q_\eta$ for $\eta \in A$.

2) If $e=0$ and (K,Σ) is t-omittory then for each $\nu \in F^1_n(p)$ we may choose a condition $q_\nu \geq_0 p^{[\nu]}$ deciding the value of $\dot\tau$ (see remark 2.3.5). Now apply 2.3.1.

Assume now that (K,Σ) is 2-big, $e=1$. Let

$$A_1 \stackrel{\rm def}{=} \{\nu \in T^p : \ (\exists \eta \in F^1_n(p))(\eta \unlhd \nu) \text{ and there is } q \geq_0 p^{[\nu]} \text{ such that } \{t^q_\rho : \rho \in T^q \ \&\ \mathbf{nor}[t^q_\rho] \leq n\} \subseteq \{t^p_\rho : \rho \in T^p\} \text{ and there is a front } F \text{ of } T^q \text{ with } (\forall \rho \in F)(q^{[\rho]} \text{ decides } \dot\tau)\}.$$

Our aim is to show that $F^1_n(p) \subseteq A_1$ which will finish the proof (applying 2.3.1 remember that $F^1_n(p)$ is a front of T^p "above" $F^0_n(p)$).

Fix $\eta_0 \in F^1_n(p)$. For each $\eta_0 \unlhd \eta \in T^p$ such that $\mathbf{nor}[t^p_\eta] > 1$ the creature t^p_η is 2-big so there is $s_\eta \in \Sigma(t^p_\eta)$ such that $\mathbf{nor}[s_\eta] \geq \mathbf{nor}[t^p_\eta] - 1$ and

$$\text{either} \quad \mathrm{pos}(s_\eta) \cap A_1 = \emptyset \quad \text{or} \quad \mathrm{pos}(s_\eta) \subseteq A_1.$$

CLAIM 2.3.6.2. *If $\eta_0 \unlhd \eta \in T^p$, $\mathbf{nor}[t^p_\eta] \geq n+2$ and $\mathrm{pos}(s_\eta) \cap A_1 \neq \emptyset$ then $\eta \in A_1$.*

Proof of the claim: By the the choice of s_η we know that then $\mathrm{pos}(s_\eta) \subseteq A_1$ so for $\rho \in \mathrm{pos}(s_\eta)$ we may choose a condition q_ρ and a front $F^\rho \subseteq T^{q_\rho}$ witnessing $\rho \in A_1$. Look at the quasi tree

$$T^r \stackrel{\rm def}{=} \{\eta\} \cup \mathrm{pos}(s_\eta) \cup \bigcup_{\rho \in \mathrm{pos}(s_\eta)} T^{q_\rho}.$$

It determines a condition $r \in \mathbb{Q}^{\rm tree}_1(K,\Sigma)$. It follows from the assumption $\mathbf{nor}[t^p_\eta] \geq n+2$ that $\mathbf{nor}[s_\eta] > n$ and therefore

$$\{t^r_\nu : \nu \in T^r \ \&\ \mathbf{nor}[t^r_\nu] \leq n\} \subseteq \{t^p_\nu : \nu \in T^{p^{[\eta]}}\}.$$

Hence the condition r together with $F^\eta = \bigcup\limits_{\rho \in \mathrm{pos}(s_\eta)} F^\rho$ (which is clearly a front of T^r) witness that $\eta \in A_1$, finishing the proof of the claim.

Now we construct inductively a condition q:

$\mathrm{root}(T^q) = \eta_0$, $T^q \subseteq T^p$,
if $\nu \in T^q$ and $\mathbf{nor}[t^p_\nu] < n+2$ then $t^q_\nu = t^p_\nu$, $\mathrm{succ}_{T^q}(\nu) = \mathrm{pos}(t^p_\nu)$,
if $\nu \in T^q$ and $\mathbf{nor}[t^p_\nu] \geq n+2$ then $t^q_\nu = s_\nu$, and $\mathrm{succ}_{T^q}(\nu) = \mathrm{pos}(s_\nu)$.
It should be clear that $q \in \mathbb{Q}^{\mathrm{tree}}_1(K, \Sigma)$ and

$$\{t^q_\nu : \nu \in T^q \ \&\ \mathbf{nor}[t^q_\nu] \leq n\} \subseteq \{t^p_\nu : \nu \in T^{p^{[\eta_0]}}\}.$$

CLAIM 2.3.6.3. $B \stackrel{\mathrm{def}}{=} \{\eta \in T^q : (\forall \nu \in T^q)(\eta \trianglelefteq \nu \quad \Rightarrow \quad \mathbf{nor}[t^q_\nu] \geq n+2)\} \subseteq A_1$.

Proof of the claim: Suppose that $\eta \in B$. Take a condition $r \in \mathbb{Q}^{\mathrm{tree}}_1(K, \Sigma)$, $q^{[\eta]} \leq r$ which decides $\dot{\tau}$. We may assume that $(\forall \nu \in T^r)(\mathbf{nor}[t^r_\nu] \geq n+2)$ (by 1.3.8(3)). Consequently $\emptyset = \{t^r_\nu : \nu \in T^r \ \&\ \mathbf{nor}[t^r_\nu] \leq n\} \subseteq \{t^p_\nu : \nu \in T^p\}$ and $\mathrm{root}(r) \in A_1$. But now we note that for each $\nu \in T^q$, if $\eta \trianglelefteq \nu \vartriangleleft \mathrm{root}(r)$ then $\mathbf{nor}[t^p_\nu] \geq n+2$ (as $\eta \in B$) and $t^q_\nu = s_\nu$. Thus we may apply 2.3.6.2 inductively to conclude that all these ν, including η, are in A_1, finishing the proof of the claim.

CLAIM 2.3.6.4. *If $\eta \in T^q$, $\eta \notin A_1$ then there is $\nu \in \mathrm{pos}(t^q_\eta)$ such that $\nu \notin A_1$.*

Proof of the claim: Should be clear.

CLAIM 2.3.6.5. $\eta_0 \in A_1$.

Proof of the claim: Assume not. Then we inductively choose a sequence

$$\eta_0 \vartriangleleft \eta_1 \vartriangleleft \eta_2 \vartriangleleft \ldots \in T^q$$

such that

$$(\forall i \in \omega)(\eta_i \notin A_1 \ \&\ \mathbf{nor}[t^q_{\eta_{i+1}}] < n+2).$$

For this suppose that we have defined $\eta_i \notin A_1$. Take $\eta^* \in \mathrm{pos}(t^q_{\eta_i}) \setminus A_1$ (possible by 2.3.6.4). By claim 2.3.6.3 we know that $\eta^* \notin B$ so there is $\nu \in (T^q)^{[\eta^*]}$ such that $\mathbf{nor}[t^q_\nu] < n+2$. Let η_{i+1} be the shortest such ν, i.e. η_{i+1} is such that

$$[\eta \in T^q \ \&\ \eta^* \trianglelefteq \eta \vartriangleleft \eta_{i+1}] \Rightarrow \mathbf{nor}[t^q_\eta] \geq n+2 \qquad \text{and } \mathbf{nor}[t^q_{\eta_{i+1}}] < n+2.$$

By repeating applications of 2.3.6.2 we conclude that $\eta_{i+1} \notin A_1$, as otherwise $\eta^* \in A_1$.
Now look at the branch through T^q determined by $\langle \eta_i : i < \omega \rangle$ – it contradicts $q \in \mathbb{Q}^{\mathrm{tree}}_1(K, \Sigma)$. This finishes the proof of the claim and the lemma. $\square$

THEOREM 2.3.7. *Suppose (K, Σ) is a tree–creating pair, $e < 3$, $p \in \mathbb{Q}^{\mathrm{tree}}_e(K, \Sigma)$. Further suppose that if $e = 2$ then Σ is bounded. Let $n < \omega$ and let $\dot{\tau}_k$ be $\mathbb{Q}^{\mathrm{tree}}_e(K, \Sigma)$-names for ordinals (for $k \in \omega$). Then:*

1. *There exist a condition $q \in \mathbb{Q}^{\mathrm{tree}}_e(K, \Sigma)$ and e-thick antichains $A_k \subseteq T^q$ of T^q such that for each $k \in \omega$:*
   (α) *$p \leq^e_n q$,*
   (β) *for every $\eta \in A_k$ the condition $q^{[\eta]}$ decides the value of $\dot{\tau}_k$,*
   (γ) *$(\forall \nu \in A_{k+1})(\exists \eta \in A_k)(\eta \vartriangleleft \nu)$.*
2. *If* either *$e = 0$ and (K, Σ) is t-omittory* or *$e = 1$ and (K, Σ) is 2-big then there are $q \in \mathbb{Q}^{\mathrm{tree}}_e(K, \Sigma)$ and fronts F_k of T^q such that for each $k \in \omega$ the conditions (α)–(γ) of (1) above are satisfied.*

PROOF. This is an inductive application of 2.3.6 and 2.3.1 (and 1.3.8 + 1.3.11). $\square$

LEMMA 2.3.8. *Let (K,Σ) be a tree–creating pair, $\dot\tau$ be a $\mathbb{Q}^{\rm tree}_3(K,\Sigma)$–name for an ordinal, $p\in\mathbb{Q}^{\rm tree}_3(K,\Sigma)$, $n<\omega$. Then there are a condition $q\in\mathbb{Q}^{\rm tree}_3(K,\Sigma)$ and a 3–thick antichain $A\subseteq T^p$ of T^p such that $q\geq p$ and*

(α) $(\forall\eta\in A)(\forall\nu\in T^p)(\nu\vartriangleleft\eta \quad\Rightarrow\quad \nu\in T^q\ \&\ t^q_\nu=t^p_\nu)$,
(β) *for every $\eta\in A$ the condition $q^{[\eta]}$ decides the value of $\dot\tau$,*
(γ) $(\forall\eta\in A)(\forall\nu\in T^q)(\eta\trianglelefteq\nu \quad\Rightarrow\quad \mathbf{nor}[t^q_\nu]>n)$.

PROOF. Look at the set

$$B\stackrel{\rm def}{=}\{\eta\in T^p:\ \ \text{there is a condition } q\geq p^{[\eta]} \text{ such that } \operatorname{root}(q)=\eta,\ (\forall\nu\in T^q)(\mathbf{nor}[t^q_\nu]>n) \text{ and } q \text{ decides the name } \dot\tau\}.$$

Easily, $B\cap T^r\neq\emptyset$ for every condition $r\geq p$. Hence the set

$$A\stackrel{\rm def}{=}\{\eta\in B:\neg(\exists\nu\in B)(\nu\vartriangleleft\eta)\}$$

is a 3–thick antichain of T^p. Now we finish in a standard way. □

THEOREM 2.3.9. *Let (K,Σ) be a tree–creating pair, $\dot\tau_k$ be $\mathbb{Q}^{\rm tree}_3(K,\Sigma)$–names for ordinals (for $k<\omega$) and $p\in\mathbb{Q}^{\rm tree}_3(K,\Sigma)$. Then there are a condition $q\in\mathbb{Q}^{\rm tree}_3(K,\Sigma)$ and 3–thick antichains A_k of T^p such that $q\geq p$ and for every $k\in\omega$:*

(α) $A_k\subseteq T^q$,
(β) *if $\eta\in A_k$ then $q^{[\eta]}$ decides $\dot\tau_k$,*
(γ) $(\forall\nu\in A_{k+1})(\exists\eta\in A_k)(\eta\vartriangleleft\nu)$.

PROOF. Build the condition q by induction using 2.3.8. □

COROLLARY 2.3.10. *If $e<4$, (K,Σ) is a tree creating pair which is bounded if $e=2$, and $\bigcup\limits_{i\in\omega}\mathbf{H}(i)$ is countable then the forcing notion $\mathbb{Q}^{\rm tree}_e(K,\Sigma)$ is proper (and even more).*

PROOF. By 2.3.7, 2.3.9 (or rather the proofs of them) and the definition of thick antichains (remember 1.3.9). □

LEMMA 2.3.11. *Assume that (K,Σ) is a 2–big tree–creating pair, $n<\omega$, and $p\in\mathbb{Q}^{\rm tree}_1(K,\Sigma)$. Then there are $q\in\mathbb{Q}^{\rm tree}_1(K,\Sigma)$ and a front F of T^q such that $p\leq^1_n q$ and*

$$(\forall\nu\in F)(\forall\eta\in T^q)(\nu\trianglelefteq\eta\quad\Rightarrow\quad\mathbf{nor}[t^q_\eta]>n+1).$$

PROOF. It is like 2.3.6(2). We consider the set

$$A^*_1\stackrel{\rm def}{=}\{\nu\in T^p:\ \ (\exists\eta\in F^1_n(p))(\eta\trianglelefteq\nu) \text{ and there is } q\geq_0 p^{[\nu]} \text{ such that } \{t^q_\rho:\rho\in T^q\ \&\ \mathbf{nor}[t^q_\rho]\leq n\}\subseteq\{t^p_\rho:\rho\in T^p\} \text{ and there is a front } F \text{ of } T^q \text{ with } (\forall\rho\in F)(\forall\eta\in T^q)(\rho\trianglelefteq\eta\quad\Rightarrow\quad\mathbf{nor}[t^q_\eta]>n+1)\}.$$

We proceed exactly as in 2.3.6(2) to show that $F^1_n(p)\subseteq A^*_1$. □

COROLLARY 2.3.12. *Suppose that (K,Σ) is a 2-big tree creating pair, $n<\omega$, $p\in\mathbb{Q}^{\rm tree}_1(K,\Sigma)$. Then there are $q\in\mathbb{Q}^{\rm tree}_1(K,\Sigma)$ and fronts F_m of T^q (for $m\in\omega$) such that*

$$p\leq^1_n q\quad\text{and}\quad(\forall\eta\in T^q)(\forall\nu\in F_m)(\nu\trianglelefteq\eta\quad\Rightarrow\quad\mathbf{nor}[t^q_\eta]\geq m).$$

Hence, in particular, $\mathbb{Q}^{\rm tree}_4(K,\Sigma)$ is dense in $\mathbb{Q}^{\rm tree}_1(K,\Sigma)$.

2.4. Examples

In this part we will give several examples of weak creating pairs, putting some of the known forcing notions into our setting. It seems that the main ingredient of any application of our technique is, next to an appropriate choice of the function **H**, the definition of the norm we use to measure possibilities. Often such a norm is an application of a particular type of pre–norms.

DEFINITION 2.4.1. A function $H : \mathcal{P}(A) \longrightarrow \mathbb{R}^{\geq 0}$ is *a pre-norm on the set A* (or rather $\mathcal{P}(A)$) if

(a) $B \subseteq C \subseteq A$ implies $H(B) \leq H(C)$
(b) $H(A) > 0$ and if $a \in A$ then $H(\{a\}) \leq 1$.

A pre-norm H on A is *nice* if additionally

(c) if $B \subseteq C \subseteq A$, $H(C) > 1$ then either $H(B) \geq H(C) - 1$ or $H(C \setminus B) \geq H(C) - 1$.

DEFINITION 2.4.2. 1. For a non-empty finite set A we let $\mathrm{dp}^0(A) = |A|$.
2. For a finite family $A \subseteq [\omega]^{<\omega}$ such that $(\forall a \in A)(|a| > 1)$ we define $\mathrm{dp}^1(A) \in \omega$ by the following induction

$$\begin{array}{ll} \mathrm{dp}^1(A) \geq 0 & \text{always,} \\ \mathrm{dp}^1(A) \geq 1 & \text{if } A \neq \emptyset, \\ \mathrm{dp}^1(A) \geq n+2 & \text{if for every set } X \subseteq \omega \text{ one of the following conditions holds:} \\ & \mathrm{dp}^1(\{a \in A : a \subseteq X\}) \geq n+1 \ \text{ or} \\ & \mathrm{dp}^1(\{a \in A : a \subseteq \omega \setminus X\}) \geq n+1. \end{array}$$

3. For a non-empty finite family A of non-empty subsets of ω we let

$$\mathrm{dp}^2(A) = \min\{|I| : (\forall a \in A)(a \cap I \neq \emptyset)\}.$$

4. For $n \in \omega$, $i < 3$ and A in the domain of dp^i we let

$$\mathrm{dp}^i_n = \log_{2+n}(\mathrm{dp}^i(A)).$$

PROPOSITION 2.4.3. 1. *Let $i < 3$, $n \in \omega$. Suppose that A is a finite set in the domain of* dp^i *such that* $\mathrm{dp}^i_n(A) > 0$. *Then* $\mathrm{dp}^i_n \restriction \mathcal{P}(A)$ *is a nice pre-norm on* $\mathcal{P}(A)$.
2. *If H is a nice pre-norm on $\mathcal{P}(A)$, $r < H(A)$ is a positive real number and $H^r : \mathcal{P}(A) \longrightarrow \mathbb{R}^{\geq 0}$ is defined by*

$$H^r(B) = \max\{0, H(B) - r\} \qquad \textit{for } B \subseteq A,$$

then H^r is a nice pre-norm on $\mathcal{P}(A)$.

PROOF. 1) Note that (in all cases), if $B, C \subseteq A$ then

$$\mathrm{dp}^i(B \cup C) \leq \mathrm{dp}^i(B) + \mathrm{dp}^i(C).$$

The only unclear instance here might be $i = 1$, but note that if $\bigcup A \subseteq [m_0, m_1)$, $B \subseteq A$ then

$\mathrm{dp}^1(B) \geq k+2$ if and only if
for every partition $\langle I_\ell : \ell < \ell_0 \rangle$ of $[m_0, m_1)$, $\ell_0 \leq 2^{k+1}$, there are $b \in B$ and $\ell < \ell_0$ such that $b \subseteq I_\ell$.

2) Check. □

REMARK 2.4.4. 2.4.3(2) will be of special importance when defining creating pairs with the Halving Property. Then we will use H^r for $r = \lfloor \frac{1}{2} H(A) \rfloor$ (see 2.4.6, 4.4.2 and 7.5.1).

Our first example of a creating pair recalls Blass–Shelah forcing notion applied in [**BsSh 242**] to show the consistency of the following statement:

if $\mathcal{D}_1, \mathcal{D}_2$ are non-principal ultrafilters on ω
then there is a finite-to-one function $f \in \omega^\omega$ such that $f(\mathcal{D}_1) = f(D_2)$.

(The suitable model was obtained there by a countable support iteration of forcing notions close to $\mathbb{Q}^*_{s\infty}(K^*_{2.4.5}, \Sigma^*_{2.4.5})$ over a model of CH.)

EXAMPLE 2.4.5. Let $\mathbf{H}(m) = 2$ for $m \in \omega$. We build a full, omittory and omittory–big (and smooth) creating pair $(K_{2.4.5}, \Sigma_{2.4.5})$ for $\mathbf{H}$.

CONSTRUCTION. A creature $t \in \mathrm{CR}[\mathbf{H}]$ is in $K_{2.4.5}$ if $m^t_{\mathrm{dn}} + 2 < m^t_{\mathrm{up}}$ and there is a sequence $\langle A^t_u : u \in \prod\limits_{i<m^t_{\mathrm{dn}}} \mathbf{H}(i)\rangle$ such that for every $u \in \prod\limits_{i<m^t_{\mathrm{dn}}} \mathbf{H}(i)$:

(α) A^t_u is a non-empty family of subsets of $[m^t_{\mathrm{dn}}, m^t_{\mathrm{up}})$, each member of A^t_u has at least 2 elements,

(β) $\langle u, v\rangle \in \mathbf{val}[t]$ if and only if
$$u \vartriangleleft v \in \prod_{i<m^t_{\mathrm{up}}} \mathbf{H}(i) \text{ and } \{i \in [m^t_{\mathrm{dn}}, m^t_{\mathrm{up}}) : v(i) = 1\} \in A^t_u \cup \{\emptyset\},$$

(γ) $\mathbf{nor}[t] = \min\{\mathrm{dp}^1_0(A^t_u) : u \in \prod\limits_{i<m^t_{\mathrm{dn}}} \mathbf{H}(i)\}$.

[Note that we do not specify here what are the $\mathbf{dis}[t]$ for $t \in K_{2.4.5}$. We have a total freedom in this, we may allow all possible values of $\mathbf{dis}[t]$ to appear.]

The composition operation $\Sigma_{2.4.5}$ on $K_{2.4.5}$ is defined as follows. Suppose that $t_0, \ldots, t_n \in K_{2.4.5}$ are such that $m^{t_i}_{\mathrm{up}} = m^{t_{i+1}}_{\mathrm{dn}}$ for $i < n$. Then

$s \in \Sigma_{2.4.5}(t_0, \ldots, t_n)$ if and only if
$s \in K_{2.4.5}$, $m^s_{\mathrm{dn}} = m^{t_0}_{\mathrm{dn}}$, $m^s_{\mathrm{up}} = m^{t_n}_{\mathrm{up}}$ and for every $\langle u, v\rangle \in \mathbf{val}[s]$
for each $i \leq n$ we have $\langle v{\restriction} m^{t_i}_{\mathrm{dn}}, v{\restriction}^{t_i}_{\mathrm{up}}\rangle \in \mathbf{val}[t_i]$.

It is an easy exercise to check that $(K_{2.4.5}, \Sigma_{2.4.5})$ is a full, omittory, smooth and omittory–big creating pair (for the last property use 2.4.3). Note that the forcing notion $\mathbb{Q}^*_{s\infty}(K_{2.4.5}, \Sigma_{2.4.5})$ is non-trivial as for each $m_0 < m_0 + 2^{n+1} < m_1$ there is $t \in K_{2.4.5} \cap \mathrm{CR}_{m_0,m_1}[\mathbf{H}]$ such that $\mathbf{nor}[t] = \log_2(n+1)$.

One may consider a modification of $(K_{2.4.5}, \Sigma_{2.4.5})$ making it forgetful. For this we let $K^*_{2.4.5} = \{t \in K_{2.4.5} : (\forall u_0, u_1 \in \prod\limits_{i<m^t_{\mathrm{dn}}} \mathbf{H}(i))(A^t_{u_0} = A^t_{u_1})\}$ and $\Sigma^*_{2.4.5}(t_0, \ldots, t_n) = \Sigma_{2.4.5}(t_0, \ldots, t_n) \cap K^*_{2.4.5}$ (for suitable $t_0, \ldots, t_1 \in K^*_{2.4.5}$). Check that $(K^*_{2.4.5}, \Sigma^*_{2.4.5})$ is a forgetful, omittory and omittory–big creating pair. $\square$

EXAMPLE 2.4.6. We define functions $\mathbf{H} : \omega \longrightarrow \omega$ and $f : \omega \times \omega \longrightarrow \omega$ and a creating pair $(K_{2.4.6}, \Sigma_{2.4.6})$ for $\mathbf{H}$ such that:

- f is $\mathbf{H}$–fast,
- $(K_{2.4.6}, \Sigma_{2.4.6})$ is $\bar{2}$–big, forgetful, simple and has the Halving Property,
- the forcing notion $\mathbb{Q}^*_f(K_{2.4.6}, \Sigma_{2.4.6})$ is non–trivial.

CONSTRUCTION. Let $F \in \omega^\omega$ be an increasing function. Define inductively functions $\mathbf{H} = \mathbf{H}^F$ and $f = f^F$ such that for each $n, k, \ell \in \omega$:

(i) $\mathbf{H}(n) = F(\varphi_{\mathbf{H}}(n)) \cdot 2^{f(n,n)}$ (remember $\varphi_{\mathbf{H}}(n) = \prod\limits_{i<n} |\mathbf{H}(i)|$, $\varphi_{\mathbf{H}}(0) = 1$),

(ii) $f(0,\ell) = \ell + 1$, $f(k+1,\ell) = 2^{\varphi_{\mathbf{H}}(\ell)} \cdot (f(k,\ell) + F(\varphi_{\mathbf{H}}(\ell)) \cdot \varphi_{\mathbf{H}}(\ell) + 2)$

(note that **(i)+(ii)** uniquely determine $\mathbf{H}$ and f and f is $\mathbf{H}$–fast).

A creature $t \in \mathrm{CR}[\mathbf{H}]$ belongs to $K_{2.4.6}$ if $m^t_{\mathrm{up}} = m^t_{\mathrm{dn}} + 1$ and

- $\mathbf{dis}[t] = \langle m^t_{\mathrm{dn}}, A_t, H_t \rangle$, where A_t is a subset of $\mathbf{H}(m^t_{\mathrm{dn}})$ and $H_t : \mathcal{P}(A_t) \longrightarrow \omega$ is a nice pre-norm,
- $\mathbf{val}[t] = \{\langle u, v \rangle \in \prod\limits_{i<m^t_{\mathrm{dn}}} \mathbf{H}(i) \times \prod\limits_{i \le m^t_{\mathrm{dn}}} \mathbf{H}(i) : u \lhd v \ \&\ v(m^t_{\mathrm{dn}}) \in A_t\}$,
- $\mathbf{nor}[t] = H_t(A_t)$.

For $t \in K_{2.4.6}$ let $\Sigma_{2.4.6}(t)$ consist of all $s \in K_{2.4.6}$ such that $m^s_{\mathrm{dn}} = m^t_{\mathrm{dn}}$, $A_s \subseteq A_t$ and $(\forall B \subseteq A_s)(H_s(B) \le H_t(B))$. (And if $\mathcal{S} \subseteq K_{2.4.6}$, $|\mathcal{S}| \ne 1$ then we let $\Sigma_{2.4.6}(\mathcal{S}) = \emptyset$.) Next, for $t \in K_{2.4.6}$ we define $\mathrm{half}(t) \in K_{2.4.6}$ as follows:

if $\mathbf{nor}[t] < 2$ then $\mathrm{half}(t) = t$,

if $\mathbf{nor}[t] \ge 2$ then $\mathrm{half}(t) \in K_{2.4.6}$ is (the unique creature) such that

$$m^{\mathrm{half}(t)}_{\mathrm{dn}} = m^t_{\mathrm{dn}}, \quad A_{\mathrm{half}(t)} = A_t, \quad \text{and} \quad H_{\mathrm{half}(t)} = (H_t)^r \quad \text{(see 2.4.3(2))},$$

where $r = \lfloor \frac{1}{2} \mathbf{nor}[t] \rfloor$.

It should be clear that $(K_{2.4.6}, \Sigma_{2.4.6})$ is a forgetful, simple and $\bar{2}$–big creating pair (for the last remember the definition of nice pre-norms). Moreover, the function half witnesses that $(K_{2.4.6}, \Sigma_{2.4.6})$ has the weak Halving Property (and so the Halving Property). [Why? Note that if $s \in \Sigma_{2.4.6}(\mathrm{half}(t))$, $\mathbf{nor}[s] > 0$, $\mathbf{nor}[t] \ge 2$ then $A_s \subseteq A_t$ and

$$1 \le H_s(A_s) \le H_{\mathrm{half}(t)}(A_s) = H_t(A_s) - \lfloor \frac{1}{2} \mathbf{nor}[t] \rfloor.$$

Thus $H_t(A_s) \ge \lfloor \frac{1}{2} \mathbf{nor}[t] \rfloor + 1 > \frac{1}{2} \mathbf{nor}[t] \ge 1$. Now look at a creature $t' \in K_{2.4.6}$ such that $\mathbf{dis}[t'] = \langle m^t_{\mathrm{dn}}, A_s, H_t \restriction \mathcal{P}(A_s) \rangle$.] Finally note that if $m < \omega$ and $t \in K_{2.4.6}$ is such that $\mathbf{dis}[t] = \langle m, \mathbf{H}(m), \mathrm{dp}^0_0 \restriction \mathcal{P}(\mathbf{H}(m)) \rangle$ then $\mathbf{nor}[t] > f(m,m)$. □

Note that the creating pair (K, Σ) described in the Prologue to represent the Silver forcing $\mathbb{Q}$ "below 2^n" is an example of a finitary creating pair which captures singletons.

The first serious application of tree–creating pairs appeared in [**Sh 326**]. The forcing notion $\mathcal{LT}^f_d$ constructed there was later modified in various ways and several variants of it found their applications (see e.g. [**BJSh 368**], [**FrSh 406**] and 2.4.8 below). This forcing notion is essentially the $\mathbb{Q}^{\mathrm{tree}}_1(K^0_{2.4.7}, \Sigma^0_{2.4.7})$. (One should note similarities with the forcing notion $\mathbb{Q}^*_{\mathrm{s}\infty}(K_{2.4.5}, \Sigma_{2.4.5})$.)

EXAMPLE 2.4.7. Let $f \in \omega^\omega$ be a strictly increasing function, $\mathbf{H}(m) = f(m)+1$ for $m \in \omega$. We construct finitary tree–creating pairs $(K^\ell_{2.4.7}, \Sigma^\ell_{2.4.7})$, $\ell < 4$, for $\mathbf{H}$ such that

1. $K^0_{2.4.7} = K^1_{2.4.7}$, $K^2_{2.4.7} = K^3_{2.4.7}$,
2. $(K^0_{2.4.7}, \Sigma^0_{2.4.7})$, $(K^1_{2.4.7}, \Sigma^1_{2.4.7})$ are 2–big local tree–creating pairs,
3. $(K^2_{2.4.7}, \Sigma^2_{2.4.7})$, $(K^3_{2.4.7}, \Sigma^3_{2.4.7})$ are 2–big t-omittory tree–creating pairs.

CONSTRUCTION. First we define $(K^0_{2.4.7}, \Sigma^0_{2.4.7})$. A tree creature $t \in \mathrm{TCR}_\eta[\mathbf{H}]$ (where $\eta \in \bigcup\limits_{n<\omega} \prod\limits_{i<n} \mathbf{H}(i)$) is in $K^0_{2.4.7}$ if

- $\mathbf{dis}[t] = \langle \eta, A_t, H_t \rangle$, where $A_t \subseteq \mathbf{H}(\ell g(\eta))$ and H_t is a nice pre-norm on $\mathcal{P}(A_t)$,
- $\mathbf{val}[t] = \{\langle \eta, \nu \rangle : \eta \vartriangleleft \nu \ \ \& \ \ \ell g(\nu) = \ell g(\eta) + 1 \ \ \& \ \ \nu(\ell g(\eta)) \in A_t\}$,
- $\mathbf{nor}[t] = H_t(A_t)$.

The operation $\Sigma^0_{2.4.7}$ is the trivial one and for $t \in \mathrm{TCR}_\eta[\mathbf{H}] \cap K^0_{2.4.7}$:

$$\Sigma^0_{2.4.7}(t) = \{s \in \mathrm{TCR}_\eta[\mathbf{H}] \cap K^0_{2.4.7} : A_s \subseteq A_t \ \ \& \ \ H_s = H_t {\restriction} \mathcal{P}(A_s)\}.$$

Easily, $(K^0_{2.4.7}, \Sigma^0_{2.4.7})$ is a finitary 2–big simple tree–creating pair.

To define $\Sigma^1_{2.4.7}$ on $K^1_{2.4.7} = K^0_{2.4.7}$ we let for $t \in \mathrm{TCR}_\eta[\mathbf{H}]$:

$$\Sigma^1_{2.4.7}(t) = \{t \in \mathrm{TCR}_\eta[\mathbf{H}] \cap K^1_{2.4.7} : A_s \subseteq A_t \ \ \& \ \ (\forall B \subseteq A_s)(H_s(B) \leq H_t(B))\}.$$

Plainly, $(K^1_{2.4.7}, \Sigma^1_{2.4.7})$ is a finitary 2–big simple tree–creating pair too.

To have t-omittory variants of the tree-creating pairs defined above we declare that a tree–creature $t \in \mathrm{TCR}_\eta[\mathbf{H}]$ is in $K^2_{2.4.7} = K^3_{2.4.7}$ if

- $\mathbf{dis}[t] = \langle \eta, \eta^*_t, A_t, H_t \rangle$, where $\eta \trianglelefteq \eta^*_t \in \bigcup\limits_{n<\omega} \prod\limits_{i<n} \mathbf{H}(i)$, $A_t \subseteq \mathbf{H}(\ell g(\eta^*_t))$ and H_t is a nice pre-norm on $\mathcal{P}(A_t)$,
- $\mathbf{val}[t] = \{\langle \eta, \nu \rangle : \eta^*_t \vartriangleleft \nu \ \ \& \ \ \ell g(\nu) = \ell g(\eta^*_t) + 1 \ \ \& \ \ \nu(\ell g(\eta^*_t)) \in A_t\}$,
- $\mathbf{nor}[t] = H_t(A_t)$.

The operations $\Sigma^2_{2.4.7}$, $\Sigma^3_{2.4.7}$ are such that if T is a well founded quasi tree, $\langle s_\nu : \nu \in \hat{T} \rangle$ is a system of tree–creatures from $K^2_{2.4.7}$ such that

$$(\forall \nu \in \hat{T})(s_\nu \in \mathrm{TCR}_\nu[\mathbf{H}] \ \& \ \mathrm{rng}(\mathbf{val}[s_\nu]) = \mathrm{succ}_T(\nu))$$

then $\Sigma^2_{2.4.7}(s_\nu : \nu \in \hat{T})$ consists of all tree-creatures $s \in K^2_{2.4.7} \cap \mathrm{TCR}_{\mathrm{root}(T)}[\mathbf{H}]$ such that for some $\nu_0 \in \hat{T}$ we have

$$\mathrm{rng}(\mathbf{val}[s]) \subseteq \mathrm{rng}(\mathbf{val}[s_{\nu_0}]) \subseteq \max(T) \quad \text{ and } \quad H_s = H_{s_{\nu_0}} {\restriction} \mathcal{P}(A_s),$$

and $\Sigma^3_{2.4.7}(s_\nu : \nu \in \hat{T})$ is defined in a similar manner but we replace the last demand (on H_s) by "$(\forall B \subseteq A_s)(H_s(B) \leq H_{s_{\nu_0}}(B))$". Now check that $\Sigma^2_{2.4.7}$, $\Sigma^3_{2.4.7}$ are t-omittory 2–big tree compositions on $K^2_{2.4.7} = K^3_{2.4.7}$. □

Let us finish our overview of "classical" examples recalling Fremlin–Shelah forcing notion. This forcing notion is essentially $\mathbb{Q}^{\mathrm{tree}}_1(K_{2.4.8}, \Sigma_{2.4.8})$, and it is a relative of $\mathbb{Q}^{\mathrm{tree}}_1(K^1_{2.4.7}, \Sigma^1_{2.4.7})$. It was applied in [**FrSh 406**] to construct a model in which there is a countable relatively pointwise compact set of Lebesgue measurable functions which is not stable.

EXAMPLE 2.4.8. We build a function $\mathbf{H}$ and a finitary, local 2–big tree–creating pair $(K_{2.4.8}, \Sigma_{2.4.8})$.

CONSTRUCTION. Choose inductively increasing sequences $\langle n_k : k < \omega \rangle$ and $\langle m_k : k < \omega \rangle$ such that $n_0 = m_0 = 4$, $m_{k+1} > m_k \cdot 2^{n_k}$ and

$$(n_{k+1})^{(m_{k+1})^{-(k+6)}} \cdot (m_{k+1})^{-(k+6)} > \log_2(n_{k+1}), \quad n_{k+1} > 2^{(m_{k+1})^{k+6}} \cdot (m_{k+1})^{k+6}.$$

For $i \in \omega$ let $\mathbf{H}(i) = \{a \subseteq n_i : \frac{|a|}{n_i} \geq 1 - \frac{1}{2^{i+2}}\}$.
A tree–creature $t \in \mathrm{TCR}_\eta[\mathbf{H}]$ is taken to $K_{2.4.8}$ if

- $\mathbf{dis}[t] = \langle \eta, A_t \rangle$, where $A_t \subseteq \mathbf{H}(\ell g(\eta))$ is non-empty,
- $\mathbf{val}[t] = \{\langle \eta, \nu \rangle : \eta \vartriangleleft \nu \ \ \& \ \ \ell g(\nu) = \ell g(\eta) + 1 \ \ \& \ \ \nu(\ell g(\eta)) \in A_t\}$,
- $\mathbf{nor}[t] = \frac{\ell g(\eta)+1}{\log_2(n_{\ell g(\eta)}) - (\ell g(\eta)+2)} \cdot \mathrm{dp}^2_0(A_t)$.

[Note that $\mathrm{dp}_0^2(\mathbf{H}(i)) \geq \log_2(n_i) - (i+2)$.] The operation $\Sigma_{2.4.8}$ is trivial and for $t \in \mathrm{TCR}_\eta[\mathbf{H}] \cap K_{2.4.8}$

$$\Sigma_{2.4.8}(t) = \{s \in K_{2.4.8} \cap \mathrm{TCR}_\eta[\mathbf{H}] : A_s \subseteq A_t\}.$$

Check that $(K_{2.4.8}, \Sigma_{2.4.8})$ is a local 2–big tree–creating pair. □

REMARK 2.4.9. 1. Note that if $\dot{W}$ is the $\mathbb{Q}_1^{\mathrm{tree}}(K_{2.4.8}, \Sigma_{2.4.8})$–name for the generic real (see 1.1.13), then

$$\Vdash_{\mathbb{Q}_1^{\mathrm{tree}}(K_{2.4.8},\Sigma_{2.4.8})} \text{“ } (\forall i \in \omega)(\dot{W}(i) \subseteq n_i \ \ \& \ \ \tfrac{|\dot{W}(i)|}{n_i} \geq 1 - \tfrac{1}{2^{i+2}}) \quad \text{and} \quad (\forall \eta \in \mathbf{V} \cap \prod_{i<\omega} n_i)(\forall^\infty i \in \omega)(\eta(i) \notin \dot{W}(i)) \text{ ”}.$$

Thus, after forcing with $\mathbb{Q}_1^{\mathrm{tree}}(K_{2.4.8}, \Sigma_{2.4.8})$, the ground model reals are of measure zero.

2. One can define a t-omittory variant of $(K_{2.4.8}, \Sigma_{2.4.8})$ (similarly to the definition of the pair $(K^2_{2.4.7}, \Sigma^2_{2.4.7})$; in forcing this would correspond to considering $\mathbb{Q}_0^{\mathrm{tree}}(K_{2.4.8}, \Sigma_{2.4.8})$).
3. In practical applications, forcing notions of the type $\mathbb{Q}_0^{\mathrm{tree}}(K, \Sigma)$ can be represented in an equivalent form as $\mathbb{Q}_1^{\mathrm{tree}}(K^*, \Sigma^*)$ for some t-omittory pair (K^*, Σ^*).

In the next two examples we want to show that the choice of the type of forcing notion or the norm condition may be very crucial. Even if we use the same or very similar weak creating pairs, different approaches may result in forcing notions with extremely different properties.

EXAMPLE 2.4.10. There exists a finitary, local and 2–big tree–creating pair $(K_{2.4.10}, \Sigma_{2.4.10})$ such that the forcing notion $\mathbb{Q}_1^{\mathrm{tree}}(K_{2.4.10}, \Sigma_{2.4.10})$ is ω^ω–bounding but the forcing notion $\mathbb{Q}_3^{\mathrm{tree}}(K_{2.4.10}, \Sigma_{2.4.10})$ adds an unbounded real.

CONSTRUCTION. Let $\mathbf{H}(i) = 2^{i+1}$ for $i \in \omega$. A tree–creature $t \in \mathrm{TCR}_\eta[\mathbf{H}]$ is taken to $K_{2.4.10}$ if

- $\mathbf{dis}[t] = \langle \eta, A_t \rangle$ for some $A_t \subseteq \mathbf{H}(\ell g(\eta))$,
- $\mathbf{val}[t] = \{\langle \eta, \nu \rangle : \eta \vartriangleleft \nu \ \& \ \ell g(\nu) = \ell g(\eta) + 1 \ \& \ \nu(\ell g(\eta)) \in A_t\}$,
- $\mathbf{nor}[t] = \mathrm{dp}_0^0(A_t)$.

The operation $\Sigma_{2.4.10}$ is trivial and $\Sigma_{2.4.10}(t) = \{s \in K_{2.4.10} : \mathbf{val}[s] \subseteq \mathbf{val}[t]\}$.

Plainly, $(K_{2.4.10}, \Sigma_{2.4.10})$ is a finitary, local and 2–big tree creating pair. By 2.3.7(2) we conclude that $\mathbb{Q}_1^{\mathrm{tree}}(K_{2.4.10}, \Sigma_{2.4.10})$ is ω^ω–bounding (compare 3.1.1). Note that, by 2.3.12, $\mathbb{Q}_4^{\mathrm{tree}}(K_{2.4.10}, \Sigma_{2.4.10})$ is dense in $\mathbb{Q}_1^{\mathrm{tree}}(K_{2.4.10}, \Sigma_{2.4.10})$.

Suppose now that $p \in \mathbb{Q}_3^{\mathrm{tree}}(K_{2.4.10}, \Sigma_{2.4.10})$. By induction on $i < \omega$ we build an increasing sequence $\langle n_i : i < \omega \rangle \subseteq \omega$, a condition $q \in \mathbb{Q}_3^{\mathrm{tree}}(K_{2.4.10}, \Sigma_{2.4.10})$ and a function $f : \{\eta \in T^q : (\exists i < \omega)(\ell g(\eta) = n_i)\} \longrightarrow \omega$ such that

(α) $q \geq p$ and $\mathrm{root}(q) = \mathrm{root}(p)$,
(β) $n_0 = \ell g(\mathrm{root}(q))$, $f(\mathrm{root}(q)) = 0$,
(γ) if $\nu, \eta \in T^q$, $\ell g(\eta) = n_i$, $\ell g(\nu) = n_{i+1}$ and $\eta \vartriangleleft \nu$ then $f(\nu) \in \{f(\eta), f(\eta)+1\}$,
(δ) for each $\eta \in T^q$ such that $\ell g(\eta) = n_i$ there is exactly one $\nu \in T^q$ such that $\eta \vartriangleleft \nu$, $\ell g(\nu) = n_{i+1}$ and $f(\nu) = f(\eta) + 1$,
(ε) if $\nu \in T^q$, $n_i \leq \ell g(\nu) < n_{i+1}$ then $\mathbf{nor}[t^q_\nu] = f(\nu{\restriction}n_i)$.

The construction is quite straightforward. Suppose we have defined n_i, $T^q \cap \omega^{\leq n_i} \subseteq T^p \cap \omega^{\leq n_i}$ and $f{\restriction} T^q \cap \omega^{\leq n_i}$ in such a way that

$$(\forall \eta \in T^q \cap \omega^{n_i})(\forall \nu \in T^p)(\eta \trianglelefteq \nu \quad \Rightarrow \quad \mathbf{nor}[t^p_\nu] \geq f(\eta)).$$

By the definition of $\mathbb{Q}^{\rm tree}_3(K_{2.4.10}, \Sigma_{2.4.10})$ we find $n_{i+1} > n_i$ such that for each $\eta \in T^q \cap \omega^{n_i}$ there is $\eta^* \in T^p \cap \omega^{n_{i+1}}$ such that $\eta \vartriangleleft \eta^*$ and

$$(\forall \nu \in T^p)(\eta^* \trianglelefteq \nu \quad \Rightarrow \quad \mathbf{nor}[t^p_\nu] \geq f(\eta) + 1)$$

(remember $T^q \cap \omega^{n_i}$ is finite). Now, for each $\eta \in T^q \cap \omega^{n_i}$ we continue building the condition q above η in such a manner that each t^q_ν (for $n_i \leq \ell g(\nu) < n_{i+1}$) has norm $f(\eta)$ and the η^* is taken to T^q. Declare $f(\eta^*) = f(\eta) + 1$ and $f(\eta') = f(\eta)$ for all $\eta' \in T^q \cap \omega^{n_{i+1}}$ extending η but different from η^*.
It is easy to check that q built in this manner is a condition in $\mathbb{Q}^{\rm tree}_3(K_{2.4.10}, \Sigma_{2.4.10})$ stronger than p.

Note that for each $m \in \omega$ the set $\{\eta \in T^q : \eta \in \mathrm{dom}(f)\ \&\ f(\eta) = m+1\}$ is the $B_m(q)$ (and thus it is a 3–thick antichain of T^q). We will be done if we show the following claim.

CLAIM 2.4.10.1. *If $q \leq r \in \mathbb{Q}^{\rm tree}_3(K_{2.4.10}, \Sigma_{2.4.10})$ then for some $m \in \omega$ the set $\{\eta \in T^r : \eta \in \mathrm{dom}(f)\ \&\ f(\eta) = m\}$ is infinite.*

Proof of the claim: Choose $\eta \in T^r$ such that

$$(\forall \nu \in T^r)(\eta \trianglelefteq \nu \quad \Rightarrow \quad \mathbf{nor}[t^r_\nu] \geq 2) \quad \text{and} \quad \ell g(\eta) = n_i \text{ (for some } i \in \omega).$$

Let $m = f(\eta) + 1$. Note that if $\nu \in T^r$, $\ell g(\nu) = n_j \geq n_i$, $\eta \trianglelefteq \nu$ and $f(\nu) = f(\eta)$ then:

1. $|\{\nu^* \in T^r : \nu \vartriangleleft \nu^*\ \&\ \ell g(\nu^*) = n_{j+1}\}| \geq 4$,
2. there is at most one $\nu^* \in T^r$ such that
$$\ell g(\nu^*) = n_{j+1}, \quad \nu \vartriangleleft \nu^*, \quad \text{and} \quad f(\nu^*) = m,$$
3. there are $j^* > j$ and $\nu^* \in T^r$ such that
$$\nu \vartriangleleft \nu^*, \quad \ell g(\nu^*) = n_{j^*} \quad \text{and} \quad f(\nu^*) = m.$$

Hence the set $\{\nu \in T^r : \eta \trianglelefteq \nu\ \&\ \nu \in \mathrm{dom}(f)\ \&\ f(\nu) = m\}$ is infinite. □

REMARK 2.4.11. Note that the proof that $\mathbb{Q}^{\rm tree}_3(K_{2.4.10}, \Sigma_{2.4.10})$ adds an unbounded real does not use the specific form of $(K_{2.4.10}, \Sigma_{2.4.10})$. With not much changes we may repeat it for any local tree creating pair (K, Σ) such that

1. if $\nu \in \mathrm{pos}(t)$, $t \in K$ and $m \leq \mathbf{nor}[t]$, $m \in \omega$
   then there is $s \in \Sigma(t)$ such that $\mathbf{nor}[s] = m$ and $\nu \in \mathrm{pos}(s)$, and
2. if $\mathbf{nor}[t] > 2$ then $|\mathrm{pos}(t)| > 2$.

In the last example of this section we try to show the difference between the use of tree creating pairs and that of creating pairs.

EXAMPLE 2.4.12. Let $\langle n_i : i < \omega \rangle \subseteq \omega$, $f : \omega \times \omega \longrightarrow \omega$ and $\mathbf{H} : \omega \longrightarrow [\omega]^{<\omega}$ be such that

1. $f(0, \ell) = \ell + 1$, $f(k+1, \ell) = 2^{\varphi_{\mathbf{H}}(\ell)+1} \cdot (f(k, \ell) + \varphi_{\mathbf{H}}(\ell) + 2)$,
2. $n_0 = 0$, $n_{i+1} \geq (n_i + 1) \cdot 2^{2^{2 \cdot f(i,i) \cdot \varphi_{\mathbf{H}}(i)^2 + 2}} + n_i$,
3. $\mathbf{H}(i) = [[n_i, n_{i+1})]^{n_i + 1}$

(so f is $\mathbf{H}$–fast).
We construct weak creating pairs $(K^0_{2.4.12}, \Sigma^0_{2.4.12})$ and $(K^1_{2.4.12}, \Sigma^1_{2.4.12})$ for $\mathbf{H}$ such that

(a) $(K^0_{2.4.12}, \Sigma^0_{2.4.12})$ is a 2–big, local and finitary tree creating pair,
(b) $(K^1_{2.4.12}, \Sigma^1_{2.4.12})$ is a simple, $\bar{2}$–big, finitary and forgetful creating pair with the Halving Property,
(c) if $\mathbb{P}$ is either $\mathbb{Q}^{\rm tree}_1(K^0_{2.4.12}, \Sigma^0_{2.4.12})$ or $\mathbb{Q}^*_f(K^1_{2.4.12}, \Sigma^1_{2.4.12})$ and $\dot{W}$ is the corresponding name for the generic real (see 1.1.13) interpreted as an infinite subset of ω, then

$$\Vdash_{\mathbb{P}} \text{“ } (\forall i \in \omega)(|\dot{W} \cap [n_i, n_{i+1})| = n_i + 1) \text{ and} \\ (\forall X \in [\omega]^\omega \cap \mathbf{V})(\forall^\infty i \in \omega)(\dot{W} \cap [n_i, n_{i+1}) \subseteq X \text{ or } \dot{W} \cap [n_i, n_{i+1}) \cap X = \emptyset)\text{”}.$$

CONSTRUCTION. We try to define minimal forcing notions adding a set $\dot{W} \subseteq [\omega]^\omega$ with the property stated in the clause **(c)**. The most natural way is to use weak creatures giving approximations to $\dot{W}$ with norms related to dp^1.

Defining $(K^0_{2.4.12}, \Sigma^0_{2.4.12})$ we may follow the simplest possible pattern presented already in 2.4.8 and 2.4.10. So a tree–creature $t \in \mathrm{TCR}_\eta[\mathbf{H}]$ is in $K^0_{2.4.12}$ if

- $\mathbf{dis}[t] = \langle \eta, A_t \rangle$, where $A_t \subseteq \mathbf{H}(\ell g(\eta))$,
- $\mathbf{val}[t] = \{\langle \eta, \nu \rangle : \eta \lhd \nu \ \&\ \ell g(\nu) = \ell g(\eta) + 1 \ \&\ \nu(\ell g(\eta)) \in A_t\}$,
- $\mathbf{nor}[t] = \frac{\mathrm{dp}^1_0(A_t)}{2 \cdot \varphi_{\mathbf{H}}(\ell g(\eta))^2}$.

The operation $\Sigma^0_{2.4.12}$ is trivial (and so $s \in \Sigma^0_{2.4.12}(t)$ if and only if $\mathbf{val}[s] \subseteq \mathbf{val}[t]$). One easily checks that $(K^0_{2.4.12}, \Sigma^0_{2.4.12})$ is a local 2–big and finitary tree creating pair. Note that (see the proof of 2.4.3)

$$\mathrm{dp}^1(\mathbf{H}(i)) \geq 2^{2 \cdot \varphi_{\mathbf{H}}(i)^2 \cdot f(i,i)} + 1 \quad \text{and thus} \quad \mathrm{dp}^1_0(\mathbf{H}(i)) > f(i,i) \cdot 2 \cdot \varphi_{\mathbf{H}}(i)^2.$$

Consequently, $\mathbb{Q}^{\rm tree}_1(K^0_{2.4.12}, \Sigma^0_{2.4.12})$ is a non-trivial forcing notion. Checking that it satisfies the demand **(c)** is easy if you remember the definition of dp^1.

Now we want to define a creating pair $(K^1_{2.4.12}, \Sigma^1_{2.4.12})$ in a similar way as $(K^0_{2.4.12}, \Sigma^0_{2.4.12})$. However, we cannot just copy the previous case (making suitable adjustments) as we have to get a new quality: the Halving Property. But we use 2.4.3(2) for this. Thus a creature $t \in \mathrm{CR}[\mathbf{H}]$ is taken to $K^1_{2.4.12}$ if $m^t_{\rm up} = m^t_{\rm dn} + 1$ and

- $\mathbf{dis}[t] = \langle m^t_{\rm dn}, B_t, r_t \rangle$, where $B_t \subseteq \mathbf{H}(m^t_{\rm dn})$ and r_t is a non-negative real,
- $\mathbf{val}[t] = \{(u, v) \in \prod\limits_{i < m^t_{\rm dn}} \mathbf{H}(i) \times \prod\limits_{i \leq m^t_{\rm dn}} : u \lhd v \ \&\ v(m^t_{\rm dn}) \in B_t\}$,
- $\mathbf{nor}[t] = \max\{0, \frac{\mathrm{dp}^1_0(B_t)}{2 \cdot \varphi_{\mathbf{H}}(m^t_{\rm dn})^2} - r_t\}$.

The operation $\Sigma^1_{2.4.12}$ is defined by:

$$\Sigma^1_{2.4.12}(t) = \{s \in K^1_{2.4.12} : m^t_{\rm dn} = m^s_{\rm dn} \ \&\ B_s \subseteq B_t \ \&\ r_s \geq r_t\}.$$

It is not difficult to verify that $(K^1_{2.4.12}, \Sigma^1_{2.4.12})$ is a finitary, forgetful, simple and $\bar{2}$–big creating pair (remember 2.4.3(2)). Let $\mathrm{half} : K^1_{2.4.12} \longrightarrow K^1_{2.4.12}$ be such that $s = \mathrm{half}(t)$ if and only if $m^s_{\rm dn} = m^t_{\rm dn}$, $B_s = B_t$ and $r_s = r_t + \frac{1}{2}\mathbf{nor}[t]$.
We claim that the function half witnesses the Halving Property for $(K^1_{2.4.12}, \Sigma^1_{2.4.12})$.

Clearly $\mathrm{half}(t)\in\Sigma(t)$ and

$$\begin{aligned}&\mathbf{nor}[\mathrm{half}(t)]=\max\{0,\frac{\mathrm{dp}^1_0(B_t)}{2\cdot\varphi_{\mathbf{H}}(m^t_{\mathrm{dn}})^2}-r_t-\tfrac12\mathbf{nor}[t]\}=\\&\max\{0,\frac{\mathrm{dp}^1_0(B_t)}{2\cdot\varphi_{\mathbf{H}}(m^t_{\mathrm{dn}})^2}-r_t-\frac12\cdot\frac{\mathrm{dp}^1_0(B_t)}{2\cdot\varphi_{\mathbf{H}}(m^t_{\mathrm{dn}})^2}+\tfrac12 r_t\}=\\&\tfrac12\max\{0,\frac{\mathrm{dp}^1_0(B_t)}{2\cdot\varphi_{\mathbf{H}}(m^t_{\mathrm{dn}})^2}-r_t\}=\tfrac12\mathbf{nor}[t].\end{aligned}$$

Suppose now that $t_0\in K^1_{2.4.12}$, $\mathbf{nor}[t_0]\geq 2$ and $t\in\Sigma^1_{2.4.12}(\mathrm{half}(t_0))$ is such that $\mathbf{nor}[t]>0$. Then $m^t_{\mathrm{dn}}=m^{t_0}_{\mathrm{dn}}$, $B_t\subseteq B_{t_0}$ and $r_t\geq r_{t_0}+\frac12\mathbf{nor}[t_0]$. Let $s\in\Sigma^1_{2.4.12}(t_0)$ be such that $B_s=B_t$ and $r_s=r_{t_0}$. Clearly $\mathbf{val}[s]=\mathbf{val}[t]$ and $\mathbf{nor}[s]=\max\{0,\frac{\mathrm{dp}^1_0(B_s)}{2\cdot\varphi_{\mathbf{H}}(m^s_{\mathrm{dn}})^2}-r_{t_0}\}$. But we know that

$$\begin{aligned}&0<\mathbf{nor}[t]=\max\{0,\frac{\mathrm{dp}^1_0(B_t)}{2\cdot\varphi_{\mathbf{H}}(m^t_{\mathrm{dn}})^2}-r_t\}=\\&\frac{\mathrm{dp}^1_0(B_s)}{2\cdot\varphi_{\mathbf{H}}(m^s_{\mathrm{dn}})^2}-r_t\leq\frac{\mathrm{dp}^1_0(B_s)}{2\cdot\varphi_{\mathbf{H}}(m^s_{\mathrm{dn}})^2}-r_{t_0}-\tfrac12\mathbf{nor}[t_0]\end{aligned}$$

and hence $\frac12\mathbf{nor}[t_0]\leq\mathbf{nor}[s]$.
Moreover, by standard arguments, the forcing notion $\mathbb{Q}^*_f(K^1_{2.4.12},\Sigma^1_{2.4.12})$ is not trivial and satisfies the demand **(c)**.

Let us try to show what may distinguish the two forcing notions. We do not have a clear property of the extensions, but we will present a technical hint that they may work differently. Let us start with noting the following property of the pre-norm dp^1.

CLAIM 2.4.12.1. *Suppose that $A_0,\dots,A_{k-1}\subseteq[\omega]^{<\omega}$ are finite families of sets with at least 2 elements, $m>\frac{k(k+3)}{2}$ and $\mathrm{dp}^1(A_i)\geq m$ for each $i<k$. Then there is a set $X\subseteq\omega$ such that for each $i<k$*

$$\begin{aligned}&\textit{both}\quad \mathrm{dp}^1(\{a\in A_i: a\subseteq X\})\geq m-\tfrac{k(k+3)}{2}\\&\textit{and}\quad \mathrm{dp}^1(\{a\in A_i: a\cap X=\emptyset\})\geq m-\tfrac{k(k+3)}{2}.\end{aligned}$$

Proof of the claim: We prove the claim by induction on k.
STEP $k=1$.
We have $A_0\subseteq\mathcal{P}([m_0,m_1))$ such that $\mathrm{dp}^1(A_0)\geq m>2$. Take a set $X\subseteq[m_0,m_1)$ of the smallest possible size such that $\mathrm{dp}^1(\{a\in A_0: a\subseteq X\})\geq m-1$. Pick any point $n\in X$ and let $Y=X\setminus\{n\}$. Then

$$\mathrm{dp}^1(\{a\in A_0: a\subseteq Y\})<m-1\quad\text{and}\quad \mathrm{dp}^1(\{a\in A_0: a\subseteq\{n\}\})=0.$$

Hence we get $\mathrm{dp}^1(\{a\in A_0: a\cap X=\emptyset\})\geq m-2$ (remember the characterization of dp^1 from the proof of 2.4.3(1)). Consequently, the set X is as required.

STEP $k+1$.
Suppose $A_0,\dots,A_{k-1},A_k\subseteq\mathcal{P}([m_0,m_1))$ are such that $\mathrm{dp}^1(A_i)\geq m>\frac{(k+1)(k+4)}{2}$. Let $X_0\subseteq[m_0,m_1)$ be such that for each $i<k$

$$\begin{aligned}&\mathrm{dp}^1(\{a\in A_i: a\subseteq X_0\})\geq m-\tfrac{k(k+3)}{2}\quad\text{and}\\&\mathrm{dp}^1(\{a\in A_i: a\cap X_0=\emptyset\})\geq m-\tfrac{k(k+3)}{2}\end{aligned}$$

(exists by the inductive hypothesis). Since $\mathrm{dp}^1(A_k)\geq m$, one of the following holds:

$$\mathrm{dp}^1(\{a\in A_k: a\subseteq X_0\})\geq m-1\quad\text{or}\quad \mathrm{dp}^1(\{a\in A_k: a\cap X_0=\emptyset\})\geq m-1.$$

We may assume that the first takes place. Take $X_1\subseteq X_0$ such that both

$$\mathrm{dp}^1(\{a\in A_k: a\subseteq X_1\})\geq m-3\ \text{ and }\ \mathrm{dp}^1(\{a\in A_k: a\subseteq X_0\setminus X_1\})\geq m-3.$$

Let $I_0 = \{i < k : \mathrm{dp}^1(\{a \in A_i : a \subseteq X_0 \setminus X_1\}) \geq m - \frac{k(k+3)}{2} - 1\}$. If $I_0 = k$ then we finish this procedure, otherwise we fix $i_0 \in k \setminus I_0$ and we choose a set $X_2 \subseteq X_1$ such that

$$\begin{aligned}
&\mathrm{dp}^1(\{a \in A_k : a \subseteq X_2\}) \geq m - 5,\\
&\mathrm{dp}^1(\{a \in A_k : a \subseteq X_1 \setminus X_2\}) \geq m - 5, \quad \text{and}\\
&\mathrm{dp}^1(\{a \in A_{i_0} : a \subseteq X_1 \setminus X_2\}) \geq m - \tfrac{k(k+3)}{2} - 2.
\end{aligned}$$

We let $I_1 = \{i < k : i \in I_0 \text{ or } \mathrm{dp}^1(\{a \in A_i : a \subseteq X_1 \setminus X_2\}) \geq m - \frac{k(k+3)}{2} - 2\}$. Note that $I_0 \subsetneq I_1$. We continue in this fashion till we get $I_\ell = k$. Note that this has to happen for some $\ell \leq k$. Look at the set $X_{\ell+1}$ constructed at this stage. It has the property that

$$\begin{aligned}
&\mathrm{dp}^1(\{a \in A_k : a \subseteq X_{\ell+1}\}) \geq m - (2k+1),\\
&\mathrm{dp}^1(\{a \in A_k : a \subseteq X_0 \setminus X_{\ell+1}\}) \geq m - (2k+1), \quad \text{and for } i < k\\
&\mathrm{dp}^1(\{a \in A_i : a \subseteq X_0 \setminus X_{\ell+1}\}) \geq m - \tfrac{k(k+3)}{2} - (k+1) \geq m - \tfrac{(k+1)(k+4)}{2}.
\end{aligned}$$

So let $X = X_0 \setminus X_{\ell+1}$ and check that it is as required for $A_0, \ldots, A_k$ (and $k+1$).

Now we may show an extra property of $\dot{W}$ which we may get in the case of $\mathbb{Q}_1^{\mathrm{tree}}(K^0_{2.4.12}, \Sigma^0_{2.4.12})$.

CLAIM 2.4.12.2. *The following holds in* $\mathbf{V}^{\mathbb{Q}_1^{\mathrm{tree}}(K^0_{2.4.12}, \Sigma^0_{2.4.12})}$:
there are sequences $\langle i_k : k < \omega\rangle$, $\langle X_i : i < \omega\rangle$ *from* $\mathbf{V}$ *such that*

1. $i_0 < i_1 < \ldots < \omega$, $X_i \subseteq [n_i, n_{i+1})$,
2. *for each* $k \in \omega$, *for exactly one* $i \in [i_k, i_{k+1})$ *we have* $\dot{W} \cap [n_i, n_{i+1}) \subseteq X_i$,
3. *the set* $\{i \in \omega : \dot{W} \cap [n_i, n_{i+1}) \subseteq X_i\}$ *is* not *in* $\mathbf{V}$.

[The last demand is to avoid a triviality like $X_i \in \{\emptyset, [n_i, n_{i+1})\}$.*]*

Proof of the claim: Let $p \in \mathbb{Q}_1^{\mathrm{tree}}(K^0_{2.4.12}, \Sigma^0_{2.4.12})$. We may assume that $(\forall \nu \in T^p)(\mathbf{nor}[t^p_\nu] > 4)$. Let $i \geq \ell g(\mathrm{root}(p))$ and let $\eta \in T^p$, $\ell g(\eta) = i$. Then

$$\mathrm{dp}^1(A_{t^p_\eta}) > 2^{8\cdot\varphi_{\mathbf{H}}(i)^2} > \frac{\varphi_{\mathbf{H}}(i)(\varphi_{\mathbf{H}}(i)+3)}{2}.$$

Since $|\{\eta \in T^p : \ell g(\eta) = i\}| \leq \varphi_{\mathbf{H}}(i)$, we may use 2.4.12.1 to find a set $X_i \subseteq [n_i, n_{i+1})$ such that for every $\eta \in T^p$ with $\ell g(\eta) = i$ we have both

$$\begin{aligned}
&\mathrm{dp}^1(\{a \in A_{t^p_\eta} : a \subseteq X_i\}) \geq \mathrm{dp}^1(A_{t^p_\eta}) - \tfrac{\varphi_{\mathbf{H}}(i)(\varphi_{\mathbf{H}}(i)+3)}{2} \quad \text{and}\\
&\mathrm{dp}^1(\{a \in A_{t^p_\eta} : a \cap X_i = \emptyset\}) \geq \mathrm{dp}^1(A_{t^p_\eta}) - \tfrac{\varphi_{\mathbf{H}}(i)(\varphi_{\mathbf{H}}(i)+3)}{2}.
\end{aligned}$$

Note that

$$\frac{\log_2(\mathrm{dp}^1(A_{t^p_\eta}) - \frac{\varphi_{\mathbf{H}}(i)(\varphi_{\mathbf{H}}(i)+3)}{2})}{2\cdot\varphi_{\mathbf{H}}(i)^2} \geq \mathbf{nor}[t^p_\eta] - 1.$$

Therefore we may inductively build a condition $q \in \mathbb{Q}_1^{\mathrm{tree}}(K^0_{2.4.12}, \Sigma^0_{2.4.12})$ and a sequence $\langle i_k : k < \omega\rangle$ such that

1. $p \leq q$, $\mathrm{root}(q) = \mathrm{root}(p)$, $\ell g(\mathrm{root}(q)) = i_0 < i_1 < i_2 < \ldots < \omega$,
2. for each $\eta \in T^q$, if $\ell g(\eta) = i$ then either

$$A_{t^q_\eta} = \{a \in A_{t^p_\eta} : a \cap X_i = \emptyset\} \quad \text{or} \quad A_{t^q_\eta} = \{a \in A_{t^p_\eta} : a \subseteq X_i\}$$

   (and so $\mathbf{nor}[t^q_\eta] \geq \mathbf{nor}[t^p_\eta] - 1$),
3. for each $\eta \in T^q$ with $\ell g(\eta) = i_k$ there is exactly one $i = i(\eta) \in [i_k, i_{k+1})$ such that

if $\eta \lhd \nu \in T^q$, $\ell g(\nu) = i_{k+1}$ and $\nu_0 = \nu \restriction i$
then $A_{t^q_{\nu_0}} = \{a \in A_{t^p_{\nu_0}} : a \subseteq X_i\}$,
and for distinct η as above the values of $i(\eta)$ are distinct.

Now check that the condition q forces that $\langle X_i : i < \omega \rangle$ and $\langle i_k : k < \omega \rangle$ are as required.

Finally look at 2.4.12.2 in the context of the *forgetful* creating pair $(K^1_{2.4.12}, \Sigma^1_{2.4.12})$. $\square$

CHAPTER 3

More properties

While the properness is the first property we usually ask for when building a forcing notion, the next request is preserving some properties of ground model reals. In this chapter we start investigations in this direction dealing with three properties of this kind. We formulate conditions on weak creating pairs which imply that the corresponding forcing notions: do not add unbounded reals, preserve non–null sets or preserve non–meager sets. Applying the methods developed here we answer Bartoszyński's request (see [**Ba94**, Problem 5]), building a proper forcing notion $\mathbb{P}$ which

1. preserves non–meager sets, and
2. preserves non–null sets, and
3. is ω^ω-bounding, and
4. does not have the Sacks property.

A forcing notion with these properties is associated with the cofinality of the null ideal (see [**BaJu95**]). The construction is done in 3.5.1, 3.5.2 and it fulfills promise of [**BaJu95**, 7.3A].

3.1. Old reals are dominating

Recall that a forcing notion $\mathbb{P}$ is *ω^ω–bounding* if it does not add unbounded reals, i.e.

$$\Vdash_{\mathbb{P}} (\forall x \in \omega^\omega)(\exists y \in \omega^\omega \cap \mathbf{V})(\forall^\infty n)(x(n) < y(n)).$$

Any countable support iteration of proper ω^ω–bounding forcing notions is ω^ω–bounding (see [**Sh:f**, Ch VI, 2.8A–C, 2.3]).

Conclusion 3.1.1. Suppose that (K,Σ) is a finitary tree-creating pair.

1. If (K,Σ) is 2-big then the forcing notion $\mathbb{Q}^{\rm tree}_1(K,\Sigma)$ is ω^ω-bounding.
2. If (K,Σ) is t-omittory then the forcing notion $\mathbb{Q}^{\rm tree}_0(K,\Sigma)$ is ω^ω-bounding.

Proof. By 2.3.7(2) and 1.3.8(5). ☐

Conclusion 3.1.2. Let (K,Σ) be a finitary creating pair for $\mathbf{H}$, and let $f : \omega\times\omega \longrightarrow \omega$ be an $\mathbf{H}$-fast function. Suppose that (K,Σ) is $\bar{2}$-big, has the Halving Property and is either simple or gluing. Then the forcing notion $\mathbb{Q}^*_f(K,\Sigma)$ is ω^ω-bounding.

Proof. By 2.2.11. ☐

Conclusion 3.1.3. Let (K,Σ) be a finitary creating pair which captures singletons. Then the forcing notion $\mathbb{Q}^*_{\rm w\infty}(K,\Sigma)$ is ω^ω-bounding.

Proof. By 2.1.12. ☐

3.2. Preserving non-meager sets

An important question concerning forcing notions is if "large" sets of reals from the ground model remain "large" after the forcing. Here we interpret "large" as "non-meager". Preserving this property in countable support iteration is relatively easy. Any countable support iteration of proper ω^ω–bounding forcing notions which preserve non-meager sets is of the same type (see [**BaJu95**, 6.3.21, 6.3.22]). If we omit "ω^ω–bounding" then we may consider a condition slightly stronger than "preserving non-meager sets":

DEFINITION 3.2.1. Let $\mathbb{P}$ be a proper forcing notion. We say that $\mathbb{P}$ is *Cohen–preserving* if

$\otimes^{\rm meager}_{\mathbb{P}}$ for every countable elementary submodel N of $(\mathcal{H}(\chi), \in, <^*_\chi)$, a condition $p \in \mathbb{P}$ and a real $x \in 2^\omega$ such that $p, \mathbb{P}, \ldots \in N$ and x is a Cohen real over N, *there is* an $(N, \mathbb{P})$–generic condition $q \in \mathbb{P}$ stronger than p such that

$$q \Vdash_{\mathbb{P}} \text{"}x \text{ is a Cohen real over } N[\Gamma_{\mathbb{P}}]\text{"}.$$

In practice, forcing notions preserving non-meagerness of sets from the ground model are Cohen–preserving. Now, to deal with iterations we may use [**Sh:f**, Ch XVIII, 3.10] (considering $(\bar{R}, S, \mathbf{g})$ as there with $\mathbf{g}_a$ being a Cohen real over a).

THEOREM 3.2.2. *Suppose that* $\bigcup\limits_{i\in\omega} \mathbf{H}(i)$ *is countable and* (K, Σ) *is a t-omittory tree–creating pair. Then the forcing notion* $\mathbb{Q}^{\rm tree}_0(K, \Sigma)$ *is Cohen–preserving.*

PROOF. Suppose that $N \prec (\mathcal{H}(\chi), \in, <^*_\chi)$ is countable, $\mathbf{H}, K, \Sigma, p, \ldots \in N$, $p \in \mathbb{Q}^{\rm tree}_0(K, \Sigma)$ and $x \in 2^\omega$ is a Cohen real over N. Let $\langle \dot{\tau}_n : n < \omega\rangle$, $\langle\langle \dot{k}^n_i : i < \omega\rangle : n < \omega\rangle$, $\langle\langle \dot{\sigma}^n_i : i < \omega\rangle : n < \omega\rangle$ list all $\mathbb{Q}^{\rm tree}_0(K, \Sigma)$–names from N for ordinals and sequences of integers and sequences of finite functions, respectively, such that for each $n < \omega$:

(a) $\Vdash_{\mathbb{Q}^{\rm tree}_0(K,\Sigma)}$ "the sequence $\langle \dot{k}^n_i : i < \omega\rangle$ is strictly increasing",
(b) $\Vdash_{\mathbb{Q}^{\rm tree}_0(K,\Sigma)}$ "$(\forall i < \omega)(\dot{\sigma}^n_i : [\dot{k}^n_i, \dot{k}^n_{i+1}) \longrightarrow 2)$".

Thus each $\langle \dot{k}^n_i, \dot{\sigma}^n_i : i < \omega\rangle$ is essentially a name for a canonical co-meager set

$$\{y \in 2^\omega : (\exists^\infty i)(y{\restriction}[\dot{k}^n_i, \dot{k}^n_{i+1}) = \dot{\sigma}^n_i)\}.$$

Of course, the enumerations are not in N, but their initial segments are there.

CLAIM 3.2.2.1. *Suppose that* $q \in \mathbb{Q}^{\rm tree}_0(K, \Sigma) \cap N$, $\nu \in T^q$, $n \in \omega$. *Then there exists a condition* $q^n_\nu \in \mathbb{Q}^{\rm tree}_0(K, \Sigma) \cap N$ *such that*

1. $q^{[\nu]} \leq^0_0 q^n_\nu$,
2. $\mathbf{nor}[t^{q^n_\nu}_\nu] > n$,
3. q^n_ν *decides the values of* $\dot{\tau}_m$ *for* $m \leq n$,
4. *for every* $m \leq n$:

$$q^n_\nu \Vdash (\exists i_0 < \ldots < i_n < \omega)\big(x{\restriction}[\dot{k}^m_{i_0}, \dot{k}^m_{i_0+1}) = \dot{\sigma}^m_{i_0} \ \& \ \ldots \ \& \ x{\restriction}[\dot{k}^m_{i_n}, \dot{k}^m_{i_n+1}) = \dot{\sigma}^m_{i_n}\big).$$

Proof of the claim: First, using 2.3.7(2), choose a condition $q^* \in \mathbb{Q}^{\rm tree}_0(K, \Sigma) \cap N$ such that $q^{[\nu]} \leq q^*$, q^* decides the values of $\dot{\tau}_m$ for $m \leq n$ and for some fronts F^*_j of T^{q^*} (for $j < \omega$) we have

(α) $(\forall j \in \omega)(\forall \eta_0 \in F^*_{j+1})(\exists \eta_1 \in F^*_j)(\eta_1 \vartriangleleft \eta_0)$,

(β) for each $m \leq n$, $j \in \omega$, $\eta \in F^*_j$ and $i \leq j$, the condition $(q^*)^{[\eta]}$ decides the values of $\dot{k}^m_i$, $\dot{\sigma}^m_i$,

(γ) $\langle F^*_j : j < \omega \rangle \in N$

(remember that N is an elementary submodel of $(\mathcal{H}(\chi), \in, <^*_\chi)$). Now take an infinite branch $\rho \in N \cap \lim(T^{q^*})$ through the quasi tree T^{q^*}. Take an increasing sequence $\langle \ell_j : j < \omega \rangle \in N$ of integers such that $\rho{\restriction}\ell_j \in F^*_j$. Then, by ($\beta$), we have two sequences $\langle k^m_j : m \leq n, j < \omega \rangle$ and $\langle \sigma^m_j : m \leq n, j < \omega \rangle$, both in N, such that for $m \leq n$, $j < \omega$:

$$(q^*)^{[\rho\restriction\ell_j]} \Vdash_{\mathbb{Q}^{\rm tree}_0(K,\Sigma)} \text{“}\dot{k}^m_j = k^m_j \ \ \& \ \ \dot{\sigma}^m_j = \sigma^m_j\text{”}.$$

Look at the set

$$\{y \in 2^\omega : (\forall m \leq n)(\exists^\infty j)(y{\restriction}[k^m_j, k^m_{j+1}) = \sigma^m_j)\}.$$

It is a dense Π^0_2–set (coded) in N and therefore x belongs to it. Hence we find $j^* < \omega$ such that

$$(\forall m \leq n)(|\{j < j^* : x{\restriction}[k^m_j, k^m_{j+1}) = \sigma^m_j\}| > n).$$

Now take $\eta^* \in T^{(q^*)^{[\rho\restriction\ell_{j^*}]}}$ such that $\mathbf{nor}[t^{q^*}_{\eta^*}] > n+1$. Since (K,Σ) is t-omittory we find a condition $q^n_\nu \in \mathbb{Q}^{\rm tree}_0(K,\Sigma) \cap N$ such that

$$q^{[\nu]} \leq^0_0 q^n_\nu, \quad \mathrm{pos}(t^{q^n_\nu}_\nu) \subseteq \mathrm{pos}(t^{q^*}_{\eta^*}), \quad \mathbf{nor}[t^{q^n_\nu}_\nu] > n, \quad \text{and}$$

$t^{q^n_\nu}_\eta = t^{q^*}_\eta$ for all $\eta \in T^{q^n_\nu}$, $\nu \lhd \eta$ (compare 2.3.5). Now one easily checks that q^n_ν is as required in the claim.

We inductively build a sequence $\langle q_n : n < \omega \rangle$, a condition $q \in \mathbb{Q}^{\rm tree}_0(K,\Sigma)$ and an enumeration $\langle \nu_n : n < \omega \rangle$ of T^q such that for all $m, n \in \omega$:

1. $\nu_n \lhd \nu_m \quad \Rightarrow \quad n < m$,
2. $\langle \nu_n : n < \omega \rangle \subseteq T^p$,
3. $q_n \in \mathbb{Q}^{\rm tree}_0(K,\Sigma) \cap N$, $p^{[\nu_n]} \leq^0_0 q_n$ and if $\nu_m \lhd \nu_n$ (so $m < n$) and $\nu_n \in \mathrm{pos}(t^{q_m}_{\nu_m})$ then $(q_m)^{[\nu_n]} \leq^0_0 q_n$,
4. $\mathbf{nor}[t^{q_n}_{\nu_n}] > n$,
5. the condition q_n decides the values of $\dot{\tau}_m$ for $m \leq n$,
6. for every $m \leq n$:

$$q_n \Vdash (\exists i_0 < \ldots < i_n < \omega)\big(x{\restriction}[\dot{k}^m_{i_0}, \dot{k}^m_{i_0+1}) = \dot{\sigma}^m_{i_0} \ \& \ \ldots \ \& \ x{\restriction}[\dot{k}^m_{i_n}, \dot{k}^m_{i_n+1}) = \dot{\sigma}^m_{i_n}\big),$$

7. $t^q_{\nu_n} = t^{q_n}_{\nu_n}$.

The construction is actually described by the conditions above: with a suitable bookkeeping we build sequences $\langle \nu_n : n \in \omega \rangle$ and $\langle A_n : n < \omega \rangle \subseteq \mathcal{P}(\omega)$. Arriving at the stage n of the construction we know ν_m for $m \leq n$ and q_m for $m < n$. Applying 3.2.2.1 we find $q_n \in \mathbb{Q}^{\rm tree}_0(K,\Sigma) \cap N$ such that the requirements (3)–(6) above are satisfied. Next we choose $A_n \subseteq \omega \setminus \bigcup\limits_{m<n} A_m$ of size $|\mathrm{pos}(t^{q_n}_{\nu_n})|$ and we assign numbers from A_n to elements of $\mathrm{pos}(t^{q_n}_{\nu_n})$ in such a way that $\mathrm{pos}(t^{q_n}_{\nu_n}) = \{\nu_k : k \in A_n\}$, the set $\omega \setminus \bigcup\limits_{m\leq n} A_m$ is infinite and $\min(\omega \setminus \bigcup\limits_{m<n} A_m) \in A_n$.

The q constructed above is a condition in $\mathbb{Q}^{\rm tree}_0(K,\Sigma)$ due to (4), and it is stronger than p by (3) and (7). Clearly q is not in N, but as every finite step of the construction takes place in N, the condition q is $(N,\mathbb{Q}^{\rm tree}_0(K,\Sigma))$–generic (by (5)). Moreover, by (6),

$$q\Vdash_{\mathbb{Q}^{\rm tree}_0(K,\Sigma)} (\forall m<\omega)(\exists^\infty i)(x{\restriction}[\dot{k}^m_i,\dot{k}^m_{i+1})=\dot{\sigma}^m_i),$$

what implies that

$$q\Vdash_{\mathbb{Q}^{\rm tree}_0(K,\Sigma)} \text{“}x \text{ is a Cohen real over } N[\Gamma_{\mathbb{Q}^{\rm tree}_0(K,\Sigma)}]\text{”}.$$

This finishes the proof. □

The definition 3.2.3 below was inspired by 3.2.2 and its proof. We distinguish here the two cases: “(K,Σ) is a creating pair” and “(K,Σ) is a tree–creating pair”, but in both of them the flavor of being of the **NMP**–type is the same: it generalizes somehow the notion of t-omittory tree–creating pairs. (Note that if (K,Σ) is a t-omittory tree–creating pair then it is of the **NMP**–type.) One could formulate a uniform condition here, but that would result in unnecessary complications in formulation.

DEFINITION 3.2.3. 1. A finitary creating pair (K,Σ) is *of the* **NMP**–*type* if the following condition is satisfied:

$(\circledast)_{\mathbf{NMP}}$ Suppose that $(w,t_0,t_1,\dots)\in\mathbb{Q}^*_\emptyset(K,\Sigma)$ is such that

$$(\forall k\in\omega)(\mathbf{nor}[t_k]>\varphi_{\mathbf{H}}(m^{t_k}_{\rm dn}))$$

and let $n_0<n_1<n_2<\ldots<\omega$. Further, assume that

$$g:\bigcup_{i\in\omega}{\rm pos}(w,t_0,\dots,t_{n_i-1})\longrightarrow\bigcup_{i\in\omega}{\rm pos}(w,t_0,\dots,t_{n_{i+1}-1})$$

is such that $g(v)\in{\rm pos}(v,t_{n_i},\dots,t_{n_{i+1}})$ for $v\in{\rm pos}(w,t_0,\dots,t_{n_i-1})$ (so $v\vartriangleleft g(v)$). *Then* there are $0<i<\omega$ and a creature $s\in\Sigma(t_{n_0},\dots,t_{n_i-1})$ such that

(α) $\mathbf{nor}[s]\geq\min\{\mathbf{nor}[t_m]-\varphi_{\mathbf{H}}(m^{t_m}_{\rm dn}):n_0\leq m<n_i\}$,

(β) for each $v\in{\rm pos}(w,t_0,\dots,t_{n_0-1},s)$ there is $j<i$ such that

$$v{\restriction}m^{t_{n_{j+1}-1}}_{\rm up}=g(v{\restriction}m^{t_{n_j-1}}_{\rm up}).$$

2. A tree–creating pair (K,Σ) is *of the* **NMP**–*type* if the following condition is satisfied:

$(\circledast)^{\rm tree}_{\mathbf{NMP}}$ Suppose that $\langle t_\eta:\eta\in T\rangle\in\mathbb{Q}^{\rm tree}_\emptyset(K,\Sigma)$ (see 1.3.5(4)) is such that

$$(\forall\eta\in T)(\mathbf{nor}[t_\eta]>1)$$

and let $F_0,F_1,F_2,\dots$ be fronts of the quasi tree T such that for $i<\omega$:

$$(\forall\nu\in F_{i+1})(\exists\nu'\in F_i)(\nu'\vartriangleleft\nu).$$

Assume that a function $g:\bigcup\limits_{i\in\omega}F_i\longrightarrow\bigcup\limits_{i\in\omega}F_{i+1}$ is such that $\nu\vartriangleleft g(\nu)\in F_{i+1}$ provided $\nu\in F_i$. *Then* there are $0<i<\omega$ and a tree–creature $s\in\Sigma(t_\eta:(\exists\nu\in F_i)(\eta\vartriangleleft\nu))$ such that

$(\alpha)^{\rm tree}$ $\mathbf{nor}[s]\geq\inf\{\mathbf{nor}[t_\eta]-1:\eta\in T\}$,

$(\beta)^{\rm tree}$ for each $\nu\in{\rm pos}(s)$ there are $j<i$ and $k<\ell g(\nu)$ such that

$$\nu{\restriction}k\in F_j\quad\text{and}\quad g(\nu{\restriction}k)\trianglelefteq\nu.$$

THEOREM 3.2.4. *Assume* (K,Σ) *is a finitary, gluing and* $\bar{2}$*–big creating pair. Suppose that* (K,Σ) *is of the* **NMP**–*type and has the Halving Property. Let* $f:\omega\times\omega\longrightarrow\omega$ *be an* **H**-*fast function. Then the forcing notion* $\mathbb{Q}^*_f(K,\Sigma)$ *is Cohen–preserving.*

PROOF. By 3.1.2 we know that the forcing notion $\mathbb{Q}^*_f(K,\Sigma)$ is ω^ω-bounding. Consequently it is enough to show that if $A\subseteq 2^\omega$ is a non-meager set then

$$\Vdash_{\mathbb{Q}^*_f(K,\Sigma)} \text{“}A \text{ is not meager”}$$

(see [**BaJu95**, 6.3.21]). So suppose that $A\subseteq 2^\omega$ is not meager but some condition in $\mathbb{Q}^*_f(K,\Sigma)$ forces that this set is meager. Thus we find a condition $p_0\in\mathbb{Q}^*_f(K,\Sigma)$ and $\mathbb{Q}^*_f(K,\Sigma)$–names $\dot{k}_n,\dot{\sigma}_n$ such that

$$p_0 \Vdash_{\mathbb{Q}^*_f(K,\Sigma)} \text{“ } \dot{k}_0<\dot{k}_1<\ldots<\omega \text{ and } \dot{\sigma}_n:[\dot{k}_n,\dot{k}_{n+1})\longrightarrow 2 \text{ (for } n<\omega) \quad\text{and}\quad (\forall x\in A)(\forall^\infty n)(x{\restriction}[\dot{k}_n,\dot{k}_{n+1})\neq\dot{\sigma}_n)\text{”}.$$

As $\mathbb{Q}^*_f(K,\Sigma)$ is ω^ω-bounding, we find a condition $p_1\geq p_0$, a sequence $0=k_0<k_1<k_2<\ldots<\omega$ and names $\dot{\rho}_n$ such that

$$p_1 \Vdash_{\mathbb{Q}^*_f(K,\Sigma)} \text{“}\dot{\rho}_n:[k_n,k_{n+1})\longrightarrow 2 \text{ (for } n<\omega) \text{ and } (\forall x\in A)(\forall^\infty n)(x{\restriction}[k_n,k_{n+1})\neq\dot{\rho}_n)\text{”}.$$

Further, applying 2.2.11 we find $p_2\geq p_1$ such that p_2 essentially decides all the names $\dot{\rho}_n$. Clearly we may assume that $\mathbf{nor}[t^{p_2}_n]>f(2,m^{t^{p_2}_n}_{\rm dn})$ for all $n<\omega$ (and thus $\mathbf{nor}[t^{p_2}_n]>\varphi_{\mathbf{H}}(m^{t^{p_2}_n}_{\rm dn})$). Choose $0=n_0<n_1<\ldots<\omega$ and $\ell_0<\ell_1<\ell_2<\ldots<\omega$ such that

$$(\forall m\in\omega)(\varphi_{\mathbf{H}}(m^{t^{p_2}_{n_m}}_{\rm dn})<\ell_{m+1}-\ell_m)$$

and for each $m<\omega$ and every sequence $w\in{\rm pos}(w^{p_2},t^{p_2}_0,\ldots,t^{p_2}_{n_{m+1}-1})$ the condition $(w,t^{p_2}_{n_{m+1}},t^{p_2}_{n_{m+1}+1},\ldots)$ decides all the names $\dot{\rho}_j$ for $j\in[\ell_m,\ell_{m+1})$ (remember the choice of p_2). Let

$$g:\bigcup_{m\in\omega}{\rm pos}(w^{p_2},t^{p_2}_0,\ldots,t^{p_2}_{n_m-1})\longrightarrow\bigcup_{m\in\omega}{\rm pos}(w^{p_2},t^{p_2}_0,\ldots,t^{p_2}_{n_{m+1}-1})$$

be such that $v\vartriangleleft g(v)\in{\rm pos}(w^{p_2},t^{p_2}_0,\ldots,t^{p_2}_{n_{m+1}-1})$ for $v\in{\rm pos}(w^{p_2},t^{p_2}_0,\ldots,t^{p_2}_{n_m-1})$. Next, for each $v\in{\rm pos}(w^{p_2},t^{p_2}_0,\ldots,t^{p_2}_{n_m-1})$, $m<\omega$ fix $\ell(v)\in[\ell_m,\ell_{m+1})$ and $\rho(v)$ such that

- there are no repetitions in $\langle\ell(v): v\in{\rm pos}(w^{p_2},t^{p_2}_0,\ldots,t^{p_2}_{n_m-1})\rangle$,
- $\rho(v):[k_{\ell(v)},k_{\ell(v)+1})\longrightarrow 2$ is such that

$$(g(v),t^{p_2}_{n_{m+1}},t^{p_2}_{n_{m+1}+1},\ldots)\Vdash_{\mathbb{Q}^*_f(K,\Sigma)}\text{“ }\dot{\rho}_{\ell(v)}=\rho(v)\text{ ”}.$$

Now we apply successively 3.2.3(1) to the condition $(w^{p_2},t^{p_2}_0,t^{p_2}_1,\ldots)$, the sequence $\langle n_i:i<\omega\rangle$ and the mapping g. As a result we construct an increasing sequence $0=i_0<i_1<i_2<\ldots<\omega$ of integers and creatures $s_j\in\Sigma(t^{p_2}_{n_{i_j}},\ldots,t^{p_2}_{n_{i_{j+1}}-1})$ such that for all $j<\omega$:

1. $\mathbf{nor}[s_j]\geq\min\{\mathbf{nor}[t^{p_2}_{n_{i_j}}]-\varphi_{\mathbf{H}}(m^{t^{p_2}_{n_{i_j}}}_{\rm dn}),\ldots,\mathbf{nor}[t^{p_2}_{n_{i_{j+1}}-1}]-\varphi_{\mathbf{H}}(m^{t^{p_2}_{n_{i_{j+1}}-1}}_{\rm dn})\}$,
2. for each $v\in{\rm pos}(w^{p_2},t^{p_2}_0,\ldots,t^{p_2}_{n_{i_j}-1},s_j)$ there is $i^*\in[i_j,i_{j+1})$ such that $v{\restriction}m^{t^{p_2}_{n_{i^*+1}}}_{\rm dn}=g(v{\restriction}m^{t^{p_2}_{n_{i^*}}}_{\rm dn})$.

It should be clear that $(w^{p_2}, s_0, s_1, \dots) \in \mathbb{Q}^*_f(K,\Sigma)$ (note that if $\mathbf{nor}[t^{p_2}_n] > f(k+1, m^{t^{p_2}_n}_{\rm dn})$ for all $n \in [n_{i_j}, n_{i_{j+1}})$ then $\mathbf{nor}[s_j] > f(k, m^{s_j}_{\rm dn})$). Look at the set

$$A^* \stackrel{\rm def}{=} \{x \in 2^\omega : \text{ for infinitely many } j \in \omega, \text{ for every } i^* \in [i_j, i_{j+1}) \\ (\forall v \in \mathrm{pos}(w^{p_2}, t^{p_2}_0, \dots, t^{p_2}_{n_{i^*}-1}))(x{\restriction}[k_{\ell(v)}, k_{\ell(v)+1}) = \rho(v))\}.$$

It is a dense Π^0_2–subset of 2^ω and hence $A^* \cap A \neq \emptyset$. Take $x \in A^* \cap A$. Note that the choice of the s_j's (see clause 2. above) implies that for each $j < \omega$ and $v \in \mathrm{pos}(w^{p_2}, s_0, \dots, s_j)$ there is $i^* \in [i_j, i_{j+1})$ such that letting $v' = v{\restriction}m^{t^{p_2}_{n_{i^*}}}_{\rm dn}$ we have

$$(v, s_{j+1}, s_{j+2}, \dots) \Vdash_{\mathbb{Q}^*_f(K,\Sigma)} \text{“ } \dot{\rho}_{\ell(v')} = \rho(v') \text{ ”}.$$

Hence $(w^{p_2}, s_0, s_1, \dots) \Vdash_{\mathbb{Q}^*_f(K,\Sigma)}$ “ $(\exists^\infty n \in \omega)(x{\restriction}[k_n, k_{n+1}) = \dot{\rho}_n)$ ”, a contradiction. □

THEOREM 3.2.5. *Assume that (K,Σ) is a finitary 2-big tree–creating pair of the* **NMP**–*type. Then the forcing notion $\mathbb{Q}^{\rm tree}_1(K,\Sigma)$ is Cohen-preserving.*

PROOF. Like 3.2.4 but using 3.1.1, 2.3.7(2), 2.3.12, and 3.2.3(2): we choose p_0, $\dot{k}_n$, $\dot{\sigma}_n$, p_1, k_n and $\dot{\rho}_n$ as there. Further we inductively build an increasing sequence $\langle \ell_m : m \in \omega\rangle$ of integers, a condition $p_2 \geq p_1$ and fronts F_m of T^{p_2} (for $m < \omega$) such that $|F_m| < \ell_{m+1} - \ell_m$ and for every $\eta \in F_{m+1}$

$$\eta \trianglelefteq \nu \in T^{p_2} \Rightarrow \mathbf{nor}[t^{p_2}_\nu] \geq m+1 \quad \text{and} \quad p^{[\eta]}_2 \text{ decides all } \dot{\rho}_j \text{ for } j \in [\ell_m, \ell_{m+1}),)$$

and the front F_{m+1} is above F_m. For $\nu \in F_m$ we define $g(\nu) \in F_{m+1}$, $\ell(\nu) \in [\ell_m, \ell_{m+1})$ and $\rho(\nu) : [k_{\ell(\nu)}, k_{\ell(\nu)+1}) \longrightarrow 2$ in a manner parallel to that in the proof of 3.2.4. Next we build a condition $q \geq p_2$ and an increasing sequence $\langle m_i : i \in \omega\rangle$ such that each $F_{m_i} \cap T^q$ is a front of T^q and

if $\eta \in F_{m_i} \cap T^q$, $i \in \omega$ then $\mathrm{pos}(t^q_\eta) \subseteq \bigcup\{F_m : m_i < m \leq m_{i+1}\}$ and for every $\nu \in \mathrm{pos}(t^q_\eta)$ there are j and $k < \ell g(\nu)$ such that $\nu{\restriction}k \in F_j$ and $g(\nu{\restriction}k) \trianglelefteq \nu$.

Finally we let

$$A^* \stackrel{\rm def}{=} \{x \in 2^\omega : (\exists^\infty i \in \omega)(\forall m \in [m_i, m_{i+1}))(\forall \nu \in F_m)(x{\restriction}[k_{\ell(\nu)}, k_{\ell(\nu)+}) = \rho(\nu))\}$$

and we finish as in 3.2.4. □

THEOREM 3.2.6. *Suppose that (K,Σ) is a finitary creating pair which captures singletons. Then the forcing notion $\mathbb{Q}^*_{\rm w\infty}(K,\Sigma)$ is Cohen-preserving.*

PROOF. Like 3.2.4, but using 3.1.3, 2.1.12 and 2.1.10. Note that the last implies that the pair (K,Σ) has the following property:

(⊚) *If $(t_0, \dots, t_N) \in \mathrm{PFC}(K,\Sigma)$, $\varphi_{\mathbf{H}}(m^{t_0}_{\rm dn}) \leq k$, $0 = n_0 < n_1 < \dots < n_k < N$ and $u \in \mathrm{pos}(u{\restriction}m^{t_0}_{\rm dn}, t_0, \dots, t_N)$ then* there are $s_0, \dots, s_\ell \in K$ such that $\mathrm{pos}(u{\restriction}m^{s_0}_{\rm dn}, s_0, \dots, s_\ell) = \{u\}$, $m^{s_0}_{\rm dn} = m^{t_0}_{\rm dn}$, $m^{s_\ell}_{\rm up} = m^{t_N}_{\rm up}$ and $\langle t_0, \dots, t_N\rangle \leq \langle s_0, \dots, s_\ell\rangle$. *Consequently,* choosing an enumeration $\{w_j : j < \varphi_{\mathbf{H}}(m^{t_0}_{\rm dn})\}$ of $\mathbf{basis}(t_0)$ and letting $u_j = w_j {}^\frown u{\restriction}[m^{t_0}_{\rm dn}, m^{t_{n_{j+1}-1}}_{\rm up})$ we will have (remember (K,Σ) is forgetful)

(α) $u_j \in \mathrm{pos}(w_j, t_0, \dots, t_{n_{j+1}-1})$,
(β) $(t_0, \dots, t_N) \leq (s_0, \dots, s_\ell)$,

(γ) for each $w \in \mathbf{basis}(s_0)$ and $v \in \mathrm{pos}(w, s_0, \ldots, s_\ell)$ there is $j < \varphi_{\mathbf{H}}(m^{t_0}_{\mathrm{dn}})$ such that $v{\restriction}m^{t_{n_{j+1}-1}}_{\mathrm{up}} = u_j$.

Thus we may repeat the proof of 3.2.4 with not many changes. □

Let us note that it is not an accident that we have the results on preserving of non-meager sets only for forcing notions which are very much like the ones coming from t-omittory tree–creating pairs. If we look at the opposite pole: local weak creating pairs, then we notice that they easily produce forcing notions making the ground reals meager.

DEFINITION 3.2.7. Let $\mathbf{H}$ be of countable character. We say that a weak creating pair (K, Σ) for $\mathbf{H}$ is *trivially meagering* if for every $t \in K$ with $\mathbf{nor}[t] > 1$ and each $u \in \mathbf{basis}(t)$ and $v \in \mathrm{pos}(u, t)$ there is $s \in \Sigma(t)$ such that $\mathbf{nor}[s] \geq \mathbf{nor}[t] - 1$ and $v \notin \mathrm{pos}(u, s)$.

PROPOSITION 3.2.8. 1. *If (K, Σ) is a local trivially meagering tree–creating pair for $\mathbf{H}$, $\mathbf{H}$ is of countable character and $e = 1, 3$ then*

$$\Vdash_{\mathbb{Q}^{\mathrm{tree}}_e(K,\Sigma)} \text{“ } \omega^\omega \cap \mathbf{V} \text{ is meager ”.}$$

2. *If (K, Σ) is a simple finitary and trivially meagering creating pair for $\mathbf{H}$ and $f : \omega \times \omega \longrightarrow \omega$ is $\mathbf{H}$–fast then*

$$\Vdash_{\mathbb{Q}^*_f(K,\Sigma)} \text{“ } \omega^\omega \cap \mathbf{V} \text{ is meager ”.}$$

PROOF. In the first case remember that if $p \in \mathbb{Q}^{\mathrm{tree}}_e(K, \Sigma)$, $e = 1, 3$ then for some $\nu \in T^p$ we have $(\forall \eta \in T^p)(\nu \trianglelefteq \eta \Rightarrow \mathbf{nor}[t^p_\eta] \geq 2)$. Hence, as (K, Σ) is local and trivially meagering we easily get

$$\Vdash_{\mathbb{Q}^{\mathrm{tree}}_e(K,\Sigma)} \text{“ } (\forall x \in \prod_{m\in\omega} \mathbf{H}(m) \cap \mathbf{V})(\forall^\infty m \in \omega)(\dot{W}(m) \neq x(m)) \text{ ”.}$$

For the second case suppose that $p \in \mathbb{Q}^*_f(K, \Sigma)$. Since $f(n+1, \ell) > \varphi_{\mathbf{H}}(\ell) + f(n, \ell)$ using the assumptions that (K, Σ) is simple and trivially meagering we immediately see that

$$p \Vdash_{\mathbb{Q}^*_f(K,\Sigma)} \text{“ } (\forall x \in \prod_{m\in\omega} \mathbf{H}(m) \cap \mathbf{V})(\forall^\infty n \in \omega)(\dot{W}{\restriction}[m^{t^p_n}_{\mathrm{dn}}, m^{t^p_n}_{\mathrm{up}}) \neq x{\restriction}[m^{t^p_n}_{\mathrm{dn}}, m^{t^p_n}_{\mathrm{up}})) \text{ ”,}$$

finishing the proof. □

Later, in 4.1.3, we will see that the forcing notions $\mathbb{Q}^*_{\mathrm{s}\infty}(K, \Sigma)$ may make the ground model reals meager too. This suggests that if one wants to build a forcing notion preserving non-meagerness then the most natural approach is $\mathbb{Q}^{\mathrm{tree}}_e$ for $e = 0, 2$ or $\mathbb{Q}^*_{\mathrm{w}\infty}$.

3.3. Preserving non-null sets

In this section we introduce a property of tree–creating pairs which implies that forcing notions $\mathbb{Q}^{\mathrm{tree}}_e(K, \Sigma)$ preserve non-null sets. Though preserving non-null sets alone is not enough to use the preservation theorem of [**Sh:f**, Ch. XVIII, §3], one may apply the methods of [**Sh 630**], [**Sh 669**] when dealing with countable support iterations of forcing notions of the type presented here.

DEFINITION 3.3.1. We say that a weak creating pair (K, Σ) is of the **NNP**–type if the following condition is satisfied:

$(\circledast)_{\mathbf{NNP}}$ there are increasing sequences $\bar{a} = \langle a_n : n < \omega \rangle \subseteq (0,1)$ and $\bar{k} = \langle k_n : n \in \omega \rangle \subseteq \omega$ such that $\lim_{n\to\infty} a_n < 1$ and:
if $t \in K$, $\mathbf{nor}[t] > 2$, $v \in \mathbf{basis}(t)$, $\ell g(v) \geq k_n$, $N < \omega$ and a function

$$g : \mathrm{pos}(v,t) \longrightarrow \mathcal{P}(N)$$

is such that $(\forall u \in \mathrm{pos}(v,t))(a_{n+1} \leq \frac{|g(u)|}{N})$ *then* the set

$$\{n < N : \text{ there is } s \in \Sigma(t) \text{ such that } \mathbf{nor}[s] \geq \mathbf{nor}[t] - 1 \text{ and } \mathbf{basis}(t) \subseteq \mathbf{basis}(s) \text{ and } (\forall u \in \mathrm{pos}(v,s))(n \in g(u))\}$$

has not less than $N \cdot a_n$ elements.

(The sequences $\bar{a}, \bar{k}$ from $(\circledast)_{\mathbf{NNP}}$ are said to witness **NNP**.)

DEFINITION 3.3.2. We say that a tree–creating pair (K,Σ) for $\mathbf{H}$ is *gluing* (respectively: *weakly gluing*) if for each well founded quasi tree $T \subseteq \bigcup_{n<\omega} \prod_{i<n} \mathbf{H}(i)$ and a system $\langle s_\nu : \nu \in \hat{T} \rangle \subseteq K$ such that

$$(\forall \nu \in \hat{T})(s_\nu \in \mathrm{TCR}_\nu[\mathbf{H}] \quad \& \quad \mathrm{pos}(s_\nu) = \mathrm{succ}_T(\nu))$$

there is $s \in \Sigma(s_\nu : \nu \in \hat{T})$ such that

$$\mathbf{nor}[s] \geq \sup\{\mathbf{nor}[s_\nu] - 1 : \nu \in \hat{T}\}$$

$(\mathbf{nor}[s] \geq \inf\{\mathbf{nor}[s_\nu] - 1 : \nu \in \hat{T}\}$, respectively).

REMARK 3.3.3. The above definition, though different from 2.1.7(2) (for creating pairs), has actually the same meaning: we may glue together creatures without loosing too much on norms.

DEFINITION 3.3.4. We say that a weak creating pair (K,Σ) is *strongly finitary* if K is finitary (see 1.1.3(2)) and $\Sigma(\mathcal{S})$ is finite for each $\mathcal{S} \subseteq K$. If $\sim_\Sigma$ (see 1.1.4(3)) is an equivalence relation on K and Σ depends on $\sim_\Sigma$–equivalence classes only, then what we actually require is that $\Sigma(\mathcal{S})/\sim_\Sigma$ is finite.

THEOREM 3.3.5. *Suppose (K,Σ) is a strongly finitary tree–creating pair of the* **NNP***–type. Further suppose that:*
either (K,Σ) is t-omittory and $e = 0$
or (K,Σ) is 2-big and weakly gluing and $e = 1$.
If $A \subseteq 2^\omega$ *is a non-null set* then $\Vdash_{\mathbb{Q}^{\mathrm{tree}}_e(K,\Sigma)}$ *"A is not null".*

PROOF. In both cases (i.e. $e = 0$ and $e = 1$) the proof is actually the same so let us deal with the case $e = 1$ only (and thus we assume that (K,Σ) is 2-big and weakly gluing). Suppose that $A \subseteq 2^\omega$ is a set which is not null but for some $p_0 \in \mathbb{Q}^{\mathrm{tree}}_1(K,\Sigma)$

$$p_0 \Vdash_{\mathbb{Q}^{\mathrm{tree}}_1(K,\Sigma)} \text{"}A \text{ is null"}.$$

We may assume that A is of outer measure 1 – just consider the set

$$\{x \in 2^\omega : (\exists y \in A)(\forall^\infty n)(x(n) = y(n))\}.$$

Let $\bar{a} = \langle a_n : n < \omega \rangle$, $\bar{k} = \langle k_n : n < \omega \rangle$ witness that (K,Σ) is of the **NNP**–type. Let $\dot{T}$ be a $\mathbb{Q}^{\mathrm{tree}}_1(K,\Sigma)$-name for a subtree of $2^{<\omega}$ such that

$$p_0 \Vdash_{\mathbb{Q}^{\mathrm{tree}}_1(K,\Sigma)} \text{"}\mu([\dot{T}]) > \lim_{n\to\infty} a_n \quad \& \quad [\dot{T}] \cap A = \emptyset\text{"}.$$

By 2.3.7(2) we find a condition $p_1 \geq p_0$ and fronts F_n of T^{p_1} such that for each $n < \omega$:

$$(\forall \eta \in F_n)(\text{the condition } p_1^{[\eta]} \text{ decides } \dot{T} \cap 2^n).$$

Clearly we may assume that

$$F_0 = \{\text{root}(p_1)\} \quad \text{and} \quad (\forall n \in \omega)(\forall \nu \in F_{n+1})(\exists \nu' \in F_n)(\nu' \lhd \nu \ \& \ k_{n+1} \leq \ell g(\nu))$$

and $(\forall n \in \omega)(\forall \nu \in F_n)(\forall \eta \in T^{p_1})(\nu \trianglelefteq \eta \Rightarrow \mathbf{nor}[t^{p_1}_\eta] > 4 + n)$ (remember 2.3.12). As (K, Σ) is weakly gluing and finitary we find a condition $p_2 \geq p_1$ such that

$$T^{p_2} = \bigcup_{n \in \omega} F_n \quad \text{and} \quad (\forall \eta \in T^{p_2})(\mathbf{nor}[t^{p_2}_\eta] > 3).$$

For $\eta \in F_n$, $n < \omega$ let $g_\eta : \text{pos}(t^{p_2}_\eta) \longrightarrow \mathcal{P}(2^{n+1})$ be a function such that

$$(\forall \nu \in \text{pos}(t^{p_2}_\eta))(p_2^{[\nu]} \Vdash_{\mathbb{Q}_1^{\text{tree}}(K,\Sigma)} \dot{T} \cap 2^{n+1} = g_\eta(\nu)).$$

Clearly, if $\eta \in F_n$, $n < \omega$ and $\nu \in \text{pos}(t^{p_2}_\eta)$ then $a_{n+1} \leq \frac{|g_\eta(\nu)|}{2^{n+1}}$. Let (for $\eta \in F_n$, $n < \omega$)

$$X^n_\eta \stackrel{\text{def}}{=} \{\sigma \in 2^{n+1} : \ \text{there is } s \in \Sigma(t^{p_2}_\eta) \text{ such that } \mathbf{nor}[s] \geq \mathbf{nor}[t^{p_2}_\eta] - 1, \text{ and } (\forall \nu \in \text{pos}(s))(\sigma \in g_\eta(\nu))\}.$$

Due to $(\circledast)_{\mathbf{NNP}}$ we know that $2^{n+1} \cdot a_n \leq |X^n_\eta|$ (for all $\eta \in F_n$, $n \in \omega$). Fix $n < \omega$. By downward induction on $m < n$ we define sets X^n_η for $\eta \in F_m$:
if $\eta \in F_m$, $m < n$ then

$$X^n_\eta = \{\sigma \in 2^{n+1} : \ \text{there is } s \in \Sigma(t^{p_2}_\eta) \text{ such that } \mathbf{nor}[s] \geq \mathbf{nor}[t^{p_2}_\eta] - 1 \text{ and } (\forall \nu \in \text{pos}(t^{p_2}_\eta))(\sigma \in X^n_\nu))\}.$$

Now we may apply the choice of $\bar{a}, \bar{k}$ (remembering that $\ell g(\eta) \geq k_m$ for each $\eta \in F_m$) to conclude (by the downward induction on $m \leq n$) that

$$(\forall m \leq n)(\forall \eta \in F_m)(|X^n_\eta| \geq 2^{n+1} \cdot a_m).$$

Hence, in particular, $2^{n+1} \cdot a_0 \leq |X^n_{\text{root}(p_2)}|$ for each $n < \omega$. So look at the set

$$F \stackrel{\text{def}}{=} \{x \in 2^\omega : (\exists^\infty n)(x{\restriction}(n+1) \in X^n_{\text{root}(p_2)})\}.$$

Necessarily $\mu(F) \geq a_0 > 0$ and thus we may take $x \in F \cap A$. For each $n < \omega$ such that $x{\restriction}(n+1) \in X^n_{\text{root}(p_2)}$ we may choose a well founded quasi tree S_n and a system $\langle s^n_\nu : \nu \in \hat{S}_n \rangle$ of creatures from K such that:

$$\hat{S}_n \subseteq \bigcup_{m \leq n} F_m, \ \max(S_n) \subseteq F_{n+1}, \ \text{root}(S_n) = \text{root}(p_2) \ \text{ and for all } \nu \in \hat{S}_n \text{ we have}$$

$$\text{pos}(s^n_\nu) = \text{succ}_{S_n}(\nu), \quad s^n_\nu \in \Sigma(t^{p_2}_\nu), \quad \mathbf{nor}[s^n_\nu] \geq \mathbf{nor}[t^{p_2}_\nu] - 1 \quad \text{and } x{\restriction}(n+1) \in X^n_\nu,$$

and if $\nu \in S_n \cap F_n$, $\nu^* \in \text{pos}(s^n_\nu)$ then $x{\restriction}(n+1) \in g_\nu(\nu^*)$. Note that if one constructs a condition q_n such that

$$S_n \subseteq T^{q_n} \subseteq \bigcup_{m < \omega} F_m,$$
$$(\forall m \leq n)(\forall \nu \in T^{q_n} \cap F_m)(\nu \in \hat{S}_n \ \& \ t^{q_n}_\nu = s^n_\nu) \quad \text{and}$$
$$(\forall m > n)(\forall \nu \in T^{q_n} \cap F_m)(t^{q_n}_\nu = t^{p_2}_\nu)$$

then $q_n \Vdash_{\mathbb{Q}_1^{\text{tree}}(K,\Sigma)} x{\restriction}(n+1) \in \dot{T}$. Hence, in particular,

($\oplus$) *if* n is such that $x{\restriction}(n+1) \in X^n_{\mathrm{root}(p_2)}$, $m \leq n+1$ and $\nu \in S_n \cap F_m$ *then* $p_2^{[\nu]} \Vdash x{\restriction}m \in \dot{T}$.

Applying the König Lemma (remember that (K, Σ) is strongly finitary!) we find a quasi tree $S \subseteq \bigcup\limits_{n\in\omega} F_n$ and a system $\langle s_\nu : \nu \in S\rangle$ such that

1. $\max(S) = \emptyset$,
2. for all $\nu \in S$:

$$\mathrm{succ}_S(\nu) = \mathrm{pos}(s_\nu) \quad \text{and} \quad \mathbf{nor}[s_\nu] \geq \mathbf{nor}[t^{p_2}_\nu] - 1 \quad \text{and} \quad s_\nu \in \Sigma(t^{p_2}_\nu),$$

3. for some increasing sequence $0 < n_0 < n_1 < n_2 < \ldots < \omega$ we have:

$$(\forall i \in \omega)(x{\restriction}(n_i+1) \in X^{n_i}_{\mathrm{root}(p_2)}) \quad \text{and}$$
$$(\forall i \in \omega)(\forall \nu \in S \cap F_i)(\forall j \geq i)(\nu \in S_{n_j} \ \&\ s_\nu = s^{n_j}_\nu).$$

The quasi tree S determines a condition $q \in \mathbb{Q}^{\mathrm{tree}}_1(K, \Sigma)$ stronger than p_2. We claim that $q \Vdash_{\mathbb{Q}^{\mathrm{tree}}_1(K,\Sigma)} (\forall i < \omega)(x{\restriction}i \in \dot{T})$. Why? Let $i \in \omega$ and $\nu \in S \cap F_i$. Then

$$s_\nu = s^{n_i}_\nu, \text{ and } \nu \in S_{n_i} \cap F_i,\ i \leq n_i \text{ and } x{\restriction}(n_i+1) \in X^{n_i}_{\mathrm{root}(p_2)}.$$

But now we may use ($\oplus$) to conclude that for each such ν

$$p_2^{[\nu]} \Vdash_{\mathbb{Q}^{\mathrm{tree}}_1(K,\Sigma)} x{\restriction}i \in \dot{T},$$

and hence $q \Vdash x{\restriction}i \in \dot{T}$. Consequently $q \Vdash x \in [\dot{T}]$, contradicting $q \geq p_0$. □

We may get a variant of 3.3.5 for tree creating pairs which are not gluing (e.g. for local (K, Σ), see 1.4.3). Then, however, we have to require more from the witnesses for the **NNP**–type.

THEOREM 3.3.6. *Suppose that (K, Σ) is a strongly finitary 2–big tree creating pair of the* **NNP**–*type with witnesses $\bar{a}, \bar{k}$ such that $k_n = n$. Then*

$$\Vdash_{\mathbb{Q}^{\mathrm{tree}}_1(K,\Sigma)} \text{“ } A \text{ is non-null ”}$$

whenever $A \subseteq 2^\omega$ is a set of positive outer measure.

PROOF. It is similar to 3.3.5. We start exactly like there choosing A, $\dot{T}$, p_1, fronts F_n of T^{p_1} and functions $g_n : F_n \longrightarrow \mathcal{P}(2^n)$ such that $p_1^{[\eta]} \Vdash \dot{T} \cap 2^n = g_n(\eta)$ for each $n \in \omega$ and $\eta \in F_n$. Next we define $g_n(\nu)$ for $\nu \in T^{p_1}$ below F_n by downward induction, in such a way that:

if $\nu \in T^{p_1}$ and there is $\eta_0 \in F_n$ such that $\nu \vartriangleleft \eta_0$ and $g_n(\eta)$ has been defined already for all $\eta \in \mathrm{pos}(t^{p_1}_\nu)$ and $(\forall \eta \in \mathrm{pos}(t^{p_1}_\nu))(a_{\ell g(\eta)} \leq \frac{|g_n(\eta)|}{2^n})$ then

$$g_n(\nu) = \{\sigma \in 2^n\colon \ \text{there is } s \in \Sigma(t^{p_1}_\nu) \text{ such that } \mathbf{nor}[s] \geq \mathbf{nor}[t^{p_1}_\nu] - 1, \text{ and } (\forall \eta \in \mathrm{pos}(s))(\sigma \in g_n(\eta))\}.$$

We continue as in 3.3.5 getting suitable S_n for $n \in \omega$ and applying the König lemma we get $S \subseteq T^{p_1}$ with the corresponding properties.
Note that the main difference is that, in the above construction, we may keep the demand $\frac{|g_n(\nu)|}{2^n} \geq a_{\ell g(\nu)}$ (and this is the replacement for “weakly gluing”). □

3.4. (No) Sacks Property

Recall that a forcing notion $\mathbb{P}$ has *the Sacks property* is for every $\mathbb{P}$–name $\dot{x}$ for a real in ω^{ω} we have

$$\Vdash_{\mathbb{P}} (\exists F \in \prod_{n\in\omega} [\omega]^{n+1} \cap \mathbf{V})(\forall n \in \omega)(\dot{x}(n) \in F(n)).$$

The Sacks property is equivalent to preserving the basis of the null ideal: every Lebesgue null set in the extension may be covered by a null set (coded) in the ground model. Here we are interested in refusing this property, i.e. getting forcing notions which do not preserve the basis of the null ideal.

DEFINITION 3.4.1. We say that a weak–creating pair (K, Σ) *strongly violates the Sacks property* if

$(\otimes)_{3.4.1}$ for some nondecreasing unbounded function $f \in \omega^{\omega}$ we have
if $t \in K$, $\mathbf{nor}[t] > 1$
then for each $u \in \mathbf{basis}(t)$ there is $n \geq \ell g(u)$ such that

$$f(n) < |\{w(n) : w \in \mathrm{pos}(u,t) \ \&\ n < \ell g(w)\}|.$$

THEOREM 3.4.2. *Let* $\mathbf{H}$ *be of countable character and let* (K, Σ) *be a weak creating pair for* $\mathbf{H}$ *which strongly violates the Sacks property. Assume that*
either (K, Σ) *is a creating pair and* $\mathbb{P}$ *is one of* $\mathbb{Q}^*_{\mathrm{s}\infty}(K, \Sigma)$, $\mathbb{Q}^*_{\infty}(K, \Sigma)$, $\mathbb{Q}^*_{\mathrm{w}\infty}(K, \Sigma)$, $\mathbb{Q}^*_{f}(K, \Sigma)$
or (K, Σ) *is a tree-creating pair and then* $\mathbb{P}$ *is one of* $\mathbb{Q}^{\mathrm{tree}}_{e}(K, \Sigma)$ $(e < 5)$.
Then the forcing notion $\mathbb{P}$ *fails the Sacks property.*

PROOF. For simplicity we may assume that $\mathbf{H}(i) = \omega$ (for all $i \in \omega$). Take an increasing sequence $\langle n_k : k < \omega\rangle$ of positive integers such that $k + 1 < f(n_k)$ for all $k \in \omega$. Let $\dot{W}$ be the $\mathbb{P}$–name for the generic real (see 1.1.13) and let $\dot{x}$ be the $\mathbb{P}$-name for an element of ω^{ω} such that

$$\Vdash_{\mathbb{P}} (\forall k \in \omega)(\dot{x}(k) = \pi(\dot{W}{\restriction}n_k)).$$

where $\pi : \omega^{<\omega} \longrightarrow \omega$ is the canonical bijection. Now we claim that

$$\Vdash_{\mathbb{P}} (\forall F \in \prod_{n\in\omega} [\omega]^{n+1} \cap \mathbf{V})(\exists^{\infty} n)(\dot{x}(n) \notin F(n)).$$

Why? Suppose that $p \in \mathbb{P}$, $F \in \prod\limits_{n\in\omega} [\omega]^{n}$, $N \in \omega$. By 3.4.1, in all relevant cases, we find $k > N$ and $n \in [n_k, n_{k+1})$ such that p "allows" more than $f(n)$ values for $\dot{W}(n)$. But $k + 2 \leq f(n_k) \leq f(n)$ and thus the condition p "allows" more than $k + 1$ values for $\dot{x}(k + 1)$. □

3.5. Examples

EXAMPLE 3.5.1. We build a tree–creating pair $(K_{3.5.1}, \Sigma_{3.5.1})$ which is: strongly finitary, 2-big, t-omittory, gluing, of the **NNP**–type, and which strongly violates the Sacks property.

CONSTRUCTION. Before we define $(K_{3.5.1}, \Sigma_{3.5.1})$ let us note some basic properties of the nice pre-norm dp^0_k defined in 2.4.2(1),(4).

CLAIM 3.5.1.1. *Let* $M < \omega$, $k < \omega$.

1. *If* $A \subseteq M$ *then either* $\mathrm{dp}^0_k(A) \geq \mathrm{dp}^0_k(M) - \log_{2+k}(2)$ *or* $\mathrm{dp}^0_k(M \setminus A) \geq \mathrm{dp}^0_k(M) - \log_{2+k}(2)$.
2. *Suppose that* $a \in (0,1)$, $N < \omega$ *and a function* $g : M \longrightarrow \mathcal{P}(N)$ *is such that*

$$(\forall m < M)(N \cdot \frac{a(k+1)+1}{k+2} \leq |g(m)|).$$

Then $a \cdot N \leq |\{n < N : \mathrm{dp}^0_k(\{m < M : n \in g(m)\}) \geq \mathrm{dp}^0_k(M) - 1\}|$.

Proof of the claim: *2)* Let $u(n) = \{m < M : n \in g(m)\}$ and

$$X = \{n < N : \mathrm{dp}^0_k(u(n)) \geq \mathrm{dp}^0_k(M) - 1\}.$$

Look at the set $Y \stackrel{\mathrm{def}}{=} \{(m,n) \in M \times N : n \in g(m)\}$ and note that

$$|Y| = \sum_{m<M} |g(m)| \geq N \cdot \frac{a(k+1)+1}{k+2} \cdot M.$$

On the other hand, noticing that $n \in X \quad \Leftrightarrow \quad |u(n)| \geq \frac{M}{k+2}$, we have:

$$|Y| = \sum_{n<N} |u(n)| \leq \sum_{n \in X} |u(n)| + \sum_{n \notin X} \frac{M}{k+2} \leq |X| \cdot M + (N - |X|) \cdot \frac{M}{k+2}.$$

Consequently $\frac{a(k+1)+1}{k+2} \cdot N \leq |X| + \frac{N-|X|}{k+2}$ and hence $a \cdot N \leq |X|$, proving the claim.

Let $\mathbf{H}(n) = (n+2)^n$.

Let $K_{3.5.1}$ be the collection of all tree–creatures t for $\mathbf{H}$ such that

1. $\mathbf{dis}[t]$ is a pair $(d_0(t), d_1(t))$ such that $d_0(t) \subseteq \bigcup\limits_{n \in \omega} \prod\limits_{m<n} \mathbf{H}(m)$ is a finite tree, $|d_0(t)| > 2$ and $d_1(t) \trianglelefteq \mathrm{root}(d_0(t))$,
2. $\mathbf{nor}[t] = \min\{\mathrm{dp}^0_k(\mathrm{succ}_{d_0(t)}(\eta)) : \eta \in \mathrm{split}(d_0(t)) \ \& \ \ell g(\eta) = k\}$,
3. $\mathbf{val}[t] = \{\langle d_1(t), \nu \rangle : \nu \in \max(d_0(t))\}$.

For a well founded quasi tree T and a system $\langle s_\nu : \nu \in \hat{T} \rangle$ of tree–creatures from $K_{3.5.1}$ such that the requirement **(a)** of 1.3.3 is satisfied we let

$$\Sigma_{3.5.1}(s_\nu : \nu \in \hat{T}) = \{s \in K_{3.5.1} : d_1(s) = \mathrm{root}(T) \ \& \ \mathrm{rng}(\mathbf{val}[s]) \subseteq \max(T)\}.$$

It should be clear that $\Sigma_{3.5.1}$ is a tree-composition on $K_{3.5.1}$. Now, the tree–creating pair $(K_{3.5.1}, \Sigma_{3.5.1})$ is strongly finitary, t-omittory and gluing. For the last two properties we apply the procedure similar to the one below.

Note that if $t \in K_{3.5.1}$ and $\nu \in d_0(t)$ is a splitting point of $d_0(t)$ then choosing $\eta_\rho \in \max(d_0(t))$ (for $\rho \in \mathrm{succ}_{d_0(t)}(\nu)$) such that $\rho \trianglelefteq \eta_\rho$ we may build a tree creature $s \in \Sigma_{3.5.1}(t)$ such that $\mathrm{pos}(s) = \{\eta_\rho : \rho \in \mathrm{succ}_{d_0(t)}(\nu)\}$. Then we will have $\mathbf{nor}[s] \geq \mathbf{nor}[t]$. If additionally $\nu \in d_0(t)$ is a splitting point such that $\mathbf{nor}[t] = \mathrm{dp}^0_{\ell g(\nu)}(\mathrm{succ}_{d_0(t)}(\nu))$ then $\mathbf{nor}[s] = \mathbf{nor}[t]$ (see the definition of $K_{3.5.1}$). In this case, let us call the respective tree–creature $s(t)$ (here we just fix one such).

Considering suitable $s(t)$'s and using 3.5.1.1(1) one can easily show that the creating pair $(K_{3.5.1}, \Sigma_{3.5.1})$ is 2-big.

Let $k_n = 2^{n+4} - 2$, $a_0 = \frac{1}{2}$, $a_{n+1} = \frac{a_n \cdot (k_n+1)+1}{k_n+2} = a_n - \frac{a_n}{2^{n+4}} + \frac{1}{2^{n+4}}$. We are going to show that the sequences $\bar{k} = \langle k_n : n < \omega \rangle$, $\bar{a} = \langle a_n : n < \omega \rangle$ witness that $(K_{3.5.1}, \Sigma_{3.5.1})$ is of the **NNP**–type.

First note that $\bar{a}, \bar{k}$ are strictly increasing, $\bar{a} \subseteq (0,1)$ and $\lim\limits_{n\to\infty} a_n \leq \frac{5}{8}$. Now suppose that $t \in K_{3.5.1}$, $k_n \leq m^t_{\rm dn}$, $N < \omega$ and $g : {\rm pos}(t) \longrightarrow \mathcal{P}(N)$ is such that

$$(\forall \nu \in {\rm pos}(t))(a_{n+1} \leq \frac{|g(\nu)|}{N}).$$

Take $s(t)$ (defined above) and look at $h = g{\restriction}{\rm pos}(s(t))$. We may think that actually $h : {\rm succ}_{d_0(s(t))}(\nu) \longrightarrow \mathcal{P}(N)$, where ν is the unique splitting point of $s(t)$ (note that $\ell g(\nu) \geq m^{s(t)}_{\rm dn} \geq k_n$). Applying claim 3.5.1.1(2) we get

$$a_n \cdot N \leq |\{n < N : {\rm dp}^0_{k_n}(\{m : n \in h(m)\}) \geq {\rm dp}^0_{k_n}({\rm succ}_{d_0(s(t))}(\nu)) - 1\}|.$$

For each $n < N$ from the set on the right hand side of the inequality above choose $s_n \in \Sigma_{3.5.1}(s(t))$ such that ${\rm pos}(s_n) = \{\eta \in {\rm pos}(s(t)) : n \in h(\eta(\ell g(\nu)))\}$. By the definition of ${\rm dp}^0_k$, ${\rm dp}^0_k(w_0) \geq {\rm dp}^0_k(w_1) - 1$ implies that ${\rm dp}^0_{k+1}(w_0) \geq {\rm dp}^0_{k+1}(w_1) - 1$, and therefore ${\rm dp}^0_{k_n}(\{m : n \in h(m)\}) \geq {\rm dp}^0_{k_n}({\rm succ}_{d_0(s(t))}(\nu)) - 1$ implies that $\mathbf{nor}[s_n] \geq \mathbf{nor}[s(t)] - 1$. Now we may conclude that the set

$$\{n < N : \text{ there is } s \in \Sigma_{3.5.1}(t) \text{ such that } \mathbf{nor}[s] \geq \mathbf{nor}[t] - 1 \text{ and } (\forall \nu \in {\rm pos}(s))(n \in g(\nu))\}.$$

has not less than $a_n \cdot N$ elements.

Finally note that $(K_{3.5.1}, \Sigma_{3.5.1})$ satisfies the condition $(\oplus_{3.4.1})$ for $f(n) = n$ (so it strongly violates the Sacks property). □

CONCLUSION 3.5.2. The forcing notions $\mathbb{Q}^{\rm tree}_e(K_{3.5.1}, \Sigma_{3.5.1})$ for $e<5$ are equivalent. They are proper, preserve the outer measure, preserve non-meager sets, are ω^ω-bounding, but do not have the Sacks property.

EXAMPLE 3.5.3. We construct a finitary, 2–big and local tree–creating pair $(K_{3.5.3}, \Sigma_{3.5.3})$ which is trivially meagering and of the **NNP**–type with the sequence $\bar{k} = \langle n : n < \omega\rangle$ witnessing it.

CONSTRUCTION. This is similar to 3.5.1. We define $\bar{k}^*$, $\bar{k}$ and $\bar{a}$ letting $k^*_n = 2^{n+4} - 2$, $k_n = n$, $a_0 = \frac{1}{2}$, $a_{n+1} = \frac{a_n(k^*_n+1)+1}{k^*_n+2}$. Let $\mathbf{H}(n) = 2^{(n+4)^2}$.

The family $K_{3.5.3}$ consists of these $t \in {\rm TCR}[\mathbf{H}]$ that:

- $\mathbf{dis}[t] = \langle \eta, A_t\rangle$, where η is such that $t \in {\rm TCR}_\eta[\mathbf{H}]$ and $A_t \subseteq \mathbf{H}(\ell g(\eta))$,
- $\mathbf{val}[t] = \{\langle \eta, \nu\rangle : \eta \vartriangleleft \nu \ \&\ \ell g(\nu) = \ell g(\eta) + 1 \ \&\ \nu(\ell g(\eta)) \in A_t\}$,
- $\mathbf{nor}[t] = {\rm dp}^0_{k^*_{\ell g(\eta)}}(A_t)$.

The operation $\Sigma_{3.5.3}$ is trivial and $s \in \Sigma_{3.5.3}(t)$ if and only if $\mathbf{val}[s] \subseteq \mathbf{val}[t]$.

Plainly, $(K_{3.5.3}, \Sigma_{3.5.3})$ is a 2–big local and trivially meagering tree–creating pair. Checking that it is of the **NNP**–type (with witnesses $\bar{a}$ and $\bar{k}$) is exactly like in 3.5.1 (just apply 3.5.1.1). □

CONCLUSION 3.5.4. $\mathbb{Q}^{\rm tree}_1(K_{3.5.3}, \Sigma_{3.5.3})$ is a proper ω^ω–bounding forcing notion which preserves outer measure but makes the ground model reals meager.

EXAMPLE 3.5.5. We define functions $\mathbf{H}$, f and a creating pair $(K_{3.5.5}, \Sigma_{3.5.5})$ such that

1. $\mathbf{H}$ is finitary, $f : \omega \times \omega \longrightarrow \omega$ is $\mathbf{H}$–fast,
2. $(K_{3.5.5}, \Sigma_{3.5.5})$ is gluing, forgetful, $\bar{2}$–big, has the Halving Property and is of the **NMP**–type,
3. the forcing notion $\mathbb{Q}^*_f(K_{3.5.5}, \Sigma_{3.5.5})$ is not trivial.

CONSTRUCTION. Define inductively $\mathbf{H}$ and f such that for all $n, k, \ell \in \omega$:

(i) $\mathbf{H}(0) = 8$, $\mathbf{H}(n) = 2^{\varphi_{\mathbf{H}}(n)+f(n,n)}$,

(ii) $f(0,\ell) = \ell + 1$, $f(k+1,\ell) = 2^{\varphi_{\mathbf{H}}(\ell)+1} \cdot (f(k,\ell) + \varphi_{\mathbf{H}}(\ell) + 2)$

(compare with 2.4.6; note that the above conditions uniquely determine $\mathbf{H}$ and f). By their definition $\mathbf{H}$ is finitary and f is $\mathbf{H}$–fast.

A creature $t \in \mathrm{CR}_{m_0,m_1}[\mathbf{H}]$ belongs to $K_{3.5.5}$ if:

- $\mathbf{dis}[t] = \langle m_0, m_1, X_t, H_t, \langle A^k_t, H^k_t, u^k_t : k \in X_t \rangle \rangle$, where $X_t \subseteq [m_0, m_1)$, H_t is a nice pre-norm on $\mathcal{P}(X_t)$, and for each $k \in X_t$: $A^k_t \subseteq \mathbf{H}(k)$, H^k_t is a nice pre-norm on $\mathcal{P}(A^k_t)$ and $u^k_t \in \prod\limits_{i \in [m_0,m_1)\setminus\{k\}} \mathbf{H}(i)$,
- $\mathbf{val}[t] = \{\langle u, v \rangle \in \prod\limits_{i<m_0} \mathbf{H}(i) \times \prod\limits_{i<m_1} \mathbf{H}(i) : u \lhd v \ \&\ (\exists k \in X_t)(u^k_t \subseteq v \ \&\ v(k) \in A^k_t)\}$,
- $\mathbf{nor}[t] = \min\{H_t(X_t), H^k_t(A^k_t) : k \in X_t\}$.

Now we describe the operation $\Sigma_{3.5.5}$. Suppose that $t_0, \ldots, t_n \in K_{3.5.5}$ are such that $m^{t_i}_{\mathrm{up}} = m^{t_{i+1}}_{\mathrm{dn}}$ for $i < n$. Let $\Sigma_{3.5.5}(t_0, \ldots, t_n)$ consist of all creatures $s \in K_{3.5.5}$ such that $m^s_{\mathrm{dn}} = m^{t_0}_{\mathrm{dn}}$, $m^s_{\mathrm{up}} = m^{t_n}_{\mathrm{up}}$ and for some $i \leq n$

(α) $X_s \subseteq X_{t_i}$, $(\forall B \subseteq X_s)(H_s(B) \leq H_{t_i}(B))$ and for every $k \in X_s$:

(β) $A^k_s \subseteq A^k_{t_i}$ and $(\forall A \subseteq A^k_s)(H^k_s(A) \leq H^k_{t_i}(A))$, and

(γ) $u^k_{t_i} \subseteq u^k_s$ and for every $j \in (n+1) \setminus \{i\}$ for some $\ell \in X_{t_j}$ we have

$$u^\ell_{t_j} \subseteq u^k_s \quad \text{and} \quad u^k_s(\ell) \in A^\ell_{t_j}.$$

It should be clear that $(K_{3.5.5}, \Sigma_{3.5.5})$ is a gluing and forgetful creating pair for $\mathbf{H}$. It is $\bar{2}$–big as for each $t \in K_{3.5.5}$ both H_t and H^k_t (for $k \in X_t$) are nice pre-norms. We may use similar arguments as in 3.2.6 to show that $(K_{3.5.5}, \Sigma_{3.5.5})$ is of the **NMP**–type. Now define function half : $K_{3.5.5} \longrightarrow K_{3.5.5}$ by

half$(t) = s$ if and only if

$$m^s_{\mathrm{dn}} = m^t_{\mathrm{dn}},\ m^s_{\mathrm{up}} = m^t_{\mathrm{up}},\ X_s = X_t,\ A^k_s = A^k_t \text{ for } k \in X_s \text{ and}$$

$$H_s = (H_t)^{\frac{1}{2}\mathbf{nor}[t]}, \quad H^k_s = (H^k_t)^{\frac{1}{2}\mathbf{nor}[t]} \quad \text{for } k \in X_s$$

(see 2.4.3(2)).

It is not difficult to check that the function half witnesses that $(K_{3.5.5}, \Sigma_{3.5.5})$ has the Halving Property.

Note that $(K_{3.5.5}, \Sigma_{3.5.5})$ resembles an omittory creating pair. □

CHAPTER 4

Omittory with Halving

In 2.2.11, 3.1.2 we saw how the Halving Property and the bigness apply to forcing notions $\mathbb{Q}^*_f(K,\Sigma)$. In this chapter we will look at another combination: omittory creating pairs with the weak Halving Property. Since an omittory creating pair cannot be big, it is natural that we consider in this context the (s∞) norm condition. The first example of a forcing notion of the type $\mathbb{Q}^*_{\mathrm{s}\infty}(K,\Sigma)$ for an omittory creating pair (K,Σ) with the weak Halving Property appeared in [**Sh 207**] (but in the real application there a different norm condition was used). A direct application of a forcing notion of this type was presented in [**RoSh 501**]. In the last part of this chapter we will develop the example from that paper. Before, in the first section, we show that the forcing notions $\mathbb{Q}^*_{\mathrm{s}\infty}(K,\Sigma)$ with (K,Σ) omittory tend to add Cohen reals and make ground reals meager. Next we introduce some general operations on creating pairs and, in the third section, we explain how the weak Halving Property may prevent them from adding dominating reals.

4.1. What omittory may easily do

Natural examples of omittory creating pairs with the weak Halving Property are meagering and anti-big (see 4.1.2 below). We will show how these properties cause that forcing notions $\mathbb{Q}^*_{\mathrm{s}\infty}(K,\Sigma)$ do some harm to the old reals. Examples and applications are presented in the last part of this chapter.

First note the following easy observation.

PROPOSITION 4.1.1. *If (K,Σ) is an omittory creating pair such that for each $t\in K$, $u\in\mathbf{basis}(t)$*

$$\mathbf{nor}[t]>0\quad\Rightarrow\quad|\mathrm{pos}(u,t)|>2$$

then $\Vdash_{\mathbb{Q}^*_{\mathrm{s}\infty}(K,\Sigma)}$ "there is an unbounded real over $\mathbf{V}$".

DEFINITION 4.1.2. Let (K,Σ) be a creating pair.

1. We say that (K,Σ) is *meagering* if for every $(t_0,\dots,t_{n-1})\in\mathrm{PFC}(K,\Sigma)$, $t\in\Sigma(t_0,\dots,t_{n-1})$ and $\langle k_i:i<n\rangle$ such that for each $i<n$:

   $$\mathbf{nor}[t_i]>2\quad\text{and}\quad m^{t_i}_{\mathrm{dn}}\le k_i<m^{t_i}_{\mathrm{up}}\quad\text{and}\quad\mathbf{nor}[t]>2$$

   there is $s\in\Sigma(t)$ satisfying

   $\mathbf{nor}[s]\ge\mathbf{nor}[t]-1$ and
   $(\forall u\in\mathbf{basis}(t_0))(\exists v\in\mathrm{pos}(u,s))(\exists k\in[\ell g(u),\ell g(v)))(v(k)\ne0)$ and
   $(\forall u\in\mathbf{basis}(t_0))(\forall v\in\mathrm{pos}(u,s))(\forall i<n)(v(k_i)=0)$.
2. The creating pair (K,Σ) is called *anti-big* if there are colourings

   $$c_t:\bigcup_{u\in\mathbf{basis}(t)}\mathrm{pos}(u,t)\longrightarrow3\qquad\text{for }t\in K$$

such that: if $(t_0,\ldots,t_{n-1}) \in \mathrm{PFC}(K,\Sigma)$, $\mathbf{nor}[t_i] > 1$ (for $i < n$) and $t \in \Sigma(t_0,\ldots,t_{n-1})$, $\mathbf{nor}[t] > 1$ then for each $u \in \mathbf{basis}(t_0)$ there are $v_0, v_1 \in \mathrm{pos}(u,t)$ and $\ell < n$ satisfying

$$v_0{\restriction}m^{t_\ell}_{\mathrm{dn}} = v_1{\restriction}m^{t_\ell}_{\mathrm{dn}},\quad c_{t_\ell}(v_0{\restriction}m^{t_\ell}_{\mathrm{up}}) = 0,\quad c_{t_\ell}(v_1{\restriction}m^{t_\ell}_{\mathrm{up}}) = 1,\qquad \text{and}$$
$$(\forall i \in n\setminus\{\ell\})(c_{t_i}(v_0{\restriction}m^{t_i}_{\mathrm{up}}) = c_{t_i}(v_1{\restriction}m^{t_i}_{\mathrm{up}}) = 2).$$

THEOREM 4.1.3. *Let (K,Σ) be a growing creating pair.*

1. *If (K,Σ) is meagering then*

$$\Vdash_{\mathbb{Q}^*_{\mathrm{s}\infty}(K,\Sigma)} \text{“}\omega^\omega \cap \mathbf{V} \textit{ is meager”}.$$

2. *If (K,Σ) is anti-big then*

$$\Vdash_{\mathbb{Q}^*_{\mathrm{s}\infty}(K,\Sigma)} \text{“\textit{there is a Cohen real over} } \mathbf{V}\text{”}.$$

PROOF. 1) Let $p = (w^p, t^p_0, t^p_1, \ldots) \in \mathbb{Q}^*_{\mathrm{s}\infty}(K,\Sigma)$ and for $n \in \omega$ let $f(n) = \mathcal{P}([m^{t^p_n}_{\mathrm{dn}}, m^{t^p_n}_{\mathrm{up}}))$. The space $\prod\limits_{n\in\omega} f(n)$ equipped with the product topology (of the discrete $f(n)$'s) is a perfect Polish space. Thus it is enough to show that

$$p \Vdash_{\mathbb{Q}^*_{\mathrm{s}\infty}(K,\Sigma)} \text{“} \prod_{n\in\omega} f(n) \cap \mathbf{V} \text{ is a meager subset of } \prod_{n\in\omega} f(n) \text{”}.$$

Note that if $X \in \prod\limits_{n\in\omega} f(n)$ is such that $(\exists^\infty n \in \omega)(X(n) \neq \emptyset)$ then the set

$$\{Y \in \prod_{n\in\omega} f(n) : (\forall^\infty n \in \omega)(Y(n) = \emptyset \text{ or } Y(n) \neq X(n))\}$$

is meager in $\prod\limits_{n\in\omega} f(n)$. Let $\dot{X}$ be a $\mathbb{Q}^*_{\mathrm{s}\infty}(K,\Sigma)$–name for an element of $\prod\limits_{n\in\omega} f(n)$ such that

$$p \Vdash_{\mathbb{Q}^*_{\mathrm{s}\infty}(K,\Sigma)} (\forall n \in \omega)(\dot{X}(n) = \{k \in [m^{t^p_n}_{\mathrm{dn}}, m^{t^p_n}_{\mathrm{up}}) : \dot{W}(k) \neq 0\}),$$

where $\dot{W}$ is the $\mathbb{Q}^*_{\mathrm{s}\infty}(K,\Sigma)$–name for the generic real (see 1.1.13). It follows from the remarks above that it is enough to show that

(α) $p \Vdash_{\mathbb{Q}^*_{\mathrm{s}\infty}(K,\Sigma)} (\exists^\infty n \in \omega)(\dot{X}(n) \neq \emptyset)$ and

(β) $p \Vdash_{\mathbb{Q}^*_{\mathrm{s}\infty}(K,\Sigma)} (\forall Y \in \prod\limits_{n\in\omega} f(n) \cap \mathbf{V})(\forall^\infty n \in \omega)(Y(n) = \emptyset \text{ or } Y(n) \neq \dot{X}(n))$.

To this end suppose that $p \le q = (w^q, t^q_0, t^q_1, \ldots) \in \mathbb{Q}^*_{\mathrm{s}\infty}(K,\Sigma)$ is such that $\ell g(w^q) > 1$ and $\mathbf{nor}[t^q_i] > m^{t^q_i}_{\mathrm{dn}} + 2$ for $i \in \omega$ and let $Y \in \prod\limits_{n\in\omega} f(n)$. For each $n \in \omega$ choose $k_n \in [m^{t^p_n}_{\mathrm{dn}}, m^{t^p_n}_{\mathrm{up}})$ such that $Y(n) \neq \emptyset \Rightarrow k_n \in Y(n)$. Let $0 \le n_0 < n_1 < n_2 < \ldots < \omega$ be such that $w^q \in \mathrm{pos}(w^p, t^p_0, \ldots, t^p_{n_0-1})$ and $t^q_i \in \Sigma(t^p_{n_i}, \ldots, t^p_{n_{i+1}-1})$ for $i \in \omega$. Note that necessarily $m^{t^p_{n_0}}_{\mathrm{dn}} \ge 2$ and thus $\mathbf{nor}[t^p_n] > 2$ for each $n \ge n_0$. Applying 4.1.2(1) we find $s_i \in \Sigma(t^q_i)$ such that for each $i \in \omega$:

$(*)^1_i$ $\mathbf{nor}[s_i] \ge \mathbf{nor}[t^q_i] - 1 > m^{t^q_i}_{\mathrm{dn}}$, and

$(*)^2_i$ $(\forall u \in \mathrm{pos}(w^p, t^p_0, \ldots, t^p_{n_i-1}))(\exists v \in \mathrm{pos}(u, s_i))(\exists k \in [\ell g(u), \ell g(v)))(v(k) \neq 0)$,

$(*)^3_i$ $(\forall u \in \mathrm{pos}(w^p, t^p_0, \ldots, t^p_{n_i-1}))(\forall v \in \mathrm{pos}(u, s_i))(\forall n \in [n_i, n_{i+1}))(v(k_n) = 0)$.

Look at $r \stackrel{\mathrm{def}}{=} (w^q, s_0, s_1, s_2, \ldots)$. Clearly $r \in \mathbb{Q}^*_{\mathrm{s}\infty}(K,\Sigma)$ is a condition stronger than q. Moreover, by the choice of the s_i's we have

$$r \Vdash_{\mathbb{Q}^*_{\mathrm{s}\infty}(K,\Sigma)} (\forall n \ge n_0)(\dot{W}(k_n) = 0)$$

and therefore, by the choice of the k_n's, we get

$$r \Vdash_{\mathbb{Q}^*_{s\infty}(K,\Sigma)} (\forall n \geq n_0)(Y(n) \neq \emptyset \quad \Rightarrow \quad Y(n) \neq \dot{X}(n)).$$

Further note that if $v \in \mathrm{pos}(w^p, s_0)$ is given by $(*)^2_0$ above for w^q then

$$(v, s_1, s_2, \dots) \Vdash (\exists n \in [n_0, n_1))(\dot{X}(n) \neq \emptyset).$$

We finish by density arguments.

2) Let $p = (w^p, t^p_0, t^p_1, \dots) \in \mathbb{Q}^*_{s\infty}(K,\Sigma)$ and let $\dot{Z}$ be a $\mathbb{Q}^*_{s\infty}(K,\Sigma)$–name for a subset of ω such that

$$p \Vdash_{\mathbb{Q}^*_{s\infty}(K,\Sigma)} \dot{Z} = \{n \in \omega : c_{t^p_n}(\dot{W} {\restriction} m^{t^p_n}_{\mathrm{up}}) < 2\}.$$

Note that $p \Vdash$"$\dot{Z}$ is infinite". Why? Suppose $p \leq q \in \mathbb{Q}^*_{s\infty}(K,\Sigma)$, $\ell g(w^q) > 1$. Let $n_0 < n_1 < \omega$ be such that $w^q \in \mathrm{pos}(w^p, t^p_0, \dots, t^p_{n_0-1})$, $t^q_0 \in \Sigma(t^p_{n_0}, \dots, t^p_{n_1-1})$. Necessarily $\mathbf{nor}[t^p_i] > 1$ for $i \in [n_0, n_1)$ and $\mathbf{nor}[t^q_0] > 1$. So we find $\ell \in [n_0, n_1)$ and $v_0, v_1 \in \mathrm{pos}(w^q, t^q_0)$ as in 4.1.2(2). Now look at the condition $r = (v_0, t^q_1, t^q_2, \dots) \in \mathbb{Q}^*_{s\infty}(K,\Sigma)$. It is stronger than q and forces that $\ell \in \dot{Z}$.

Now let $\dot{c}$ be a $\mathbb{Q}^*_{s\infty}(K,\Sigma)$–name for a real in 2^ω such that

$$p \Vdash_{\mathbb{Q}^*_{s\infty}(K,\Sigma)} \text{" if } k \in \omega \text{ and } n \text{ is the } k^{\mathrm{th}} \text{ member of } \dot{Z} \text{ then } \dot{c}(k) = c_{t^p_n}(\dot{W} {\restriction} m^{t^p_n}_{\mathrm{up}}) \text{ "}.$$

We claim that $p \Vdash$ "$\dot{c}$ is a Cohen real over $\mathbf{V}$". So suppose that $p \leq q \in \mathbb{Q}^*_{s\infty}(K,\Sigma)$, $\ell g(w^q) > 1$ and $\mathcal{U} \subseteq 2^\omega$ is an open dense set. Let $0 \leq n_0 < n_1 < \ldots < \omega$ be such that

$$w^q \in \mathrm{pos}(w^p, t^p_0, \dots, t^p_{n_0-1}) \quad \text{and} \quad t^q_i \in \Sigma(t^p_{n_i}, \dots, t^p_{n_{i+1}-1}) \quad \text{for } i \in \omega.$$

Let $m = |\{n < n_0 : c_{t^p_n}(w^q {\restriction} m^{t^p_n}_{\mathrm{up}}) < 2\}|$ and let $\nu \in 2^m$ be such that $\nu(k) = c_{t^p_n}(w^q {\restriction} m^{t^p_n}_{\mathrm{up}})$ if $k < m$ and $n < n_0$ is the kth member of the set $\{n < n_0 : c_{t^p_n}(w^q {\restriction} m^{t^p_n}_{\mathrm{up}}) < 2\}$. Choose $\eta \in 2^{<\omega}$ such that $\nu \lhd \eta$ and

$$(\forall x \in 2^\omega)(\eta \lhd x \ \Rightarrow \ x \in \mathcal{U}).$$

Let $j = \ell g(\eta) - m$. Use 4.1.2(2) to define inductively $u \in \mathrm{pos}(w^q, t^q_0, \dots, t^q_{j-1})$ such that for each $i < j$, for some $\ell \in [n_i, n_{i+1})$ we have

$$c_{t^p_\ell}(u {\restriction} m^{t^p_\ell}_{\mathrm{up}}) = \eta(\ell) \quad \text{and} \quad (\forall k \in [n_i, n_{i+1}) \setminus \{\ell\})(c_{t^p_k}(u {\restriction} m^{t^p_k}_{\mathrm{up}}) = 2).$$

Look at the condition $r = (u, t^q_j, t^q_{j+1}, \dots) \in \mathbb{Q}^*_{s\infty}(K,\Sigma)$: it is stronger than q and it forces that $\eta \lhd \dot{c}$. We finish by density argument. $\square$

4.2. More operations on weak creatures

Below we define some operations on creatures and tree–creatures which provide for (some) systems of weak creatures a new weak creature (of the same type). These operations may be used to define sub–composition operations.

DEFINITION 4.2.1. Suppose $0 < m < \omega$ and for $i < m$ we have $t_i \in \mathrm{CR}[\mathbf{H}]$ such that $m^{t_i}_{\mathrm{up}} \leq m^{t_{i+1}}_{\mathrm{dn}}$. Then we define *the sum of the creatures* t_i as a creature $t = \Sigma^{\mathrm{sum}}(t_i : i < m)$ such that (if well defined then):

(a) $m^t_{\mathrm{dn}} = m^{t_0}_{\mathrm{dn}}$, $m^t_{\mathrm{up}} = m^{t_{m-1}}_{\mathrm{up}}$,
(b) $\mathbf{val}[t]$ is the set of all pairs $\langle h_1, h_2 \rangle$ such that:

$\ell g(h_1) = m^t_{\rm dn}$, $\ell g(h_2) = m^t_{\rm up}$, $h_1 \lhd h_2$,
and $\langle h_2{\restriction}m^{t_i}_{\rm dn}, h_2{\restriction}m^{t_i}_{\rm up}\rangle \in \mathbf{val}[t_i]$ for $i < m$,
and $h_2{\restriction}[m^{t_i}_{\rm up}, m^{t_{i+1}}_{\rm dn})$ is identically zero for $i < m-1$,

(c) $\mathbf{nor}[t] = \min\{\mathbf{nor}[t_i] : i < m\}$,

(d) $\mathbf{dis}[t] = \langle 0\rangle^\frown\langle \mathbf{dis}[t_i] : i < m\rangle$.

If for all $i < m-1$ we have $m^{t_i}_{\rm up} = m^{t_{i+1}}_{\rm dn}$ then we call the sum *tight*.

REMARK 4.2.2. Note that the sum $\Sigma^{\rm sum}(t_i : i < m)$ is defined only for these sequences $\langle t_i : i < m\rangle \subseteq \mathrm{CR}[\mathbf{H}]$ for which $m^{t_i}_{\rm up} \leq m^{t_{i+1}}_{\rm dn}$ and part (b) of the definition gives a nonempty value of $\mathbf{val}[t]$.

DEFINITION 4.2.3. If $m < \omega$, $u \subseteq m$, $d \in \mathcal{H}(\chi)$ is a function such that $\mathrm{dom}(d) \supseteq (\mathbb{R}^{\geq 0})^u$ and $\mathrm{rng}(d) \subseteq \mathbb{R}^{\geq 0}$, and for $i < m$ creatures $t_i \in \mathrm{CR}[\mathbf{H}]$ are such that $m^{t_i}_{\rm up} \leq m^{t_{i+1}}_{\rm dn}$ then we define *the (d,u)-sum $t = \Sigma^{\rm sum}_{d,u}(t_i : i < m)$ of the t_i's* by:

(a) $m^t_{\rm dn} = m^{t_0}_{\rm dn}$, $m^t_{\rm up} = m^{t_{m-1}}_{\rm up}$,

(b) $\mathbf{val}[t]$ is the set of pairs $\langle h_1, h_2\rangle$ such that:
$\ell g(h_1) = m^t_{\rm dn}$, $\ell g(h_2) = m^t_{\rm up}$, $h_1 \lhd h_2$ and
$\langle h_2{\restriction}m^{t_i}_{\rm dn}, h_2{\restriction}m^{t_i}_{\rm up}\rangle \in \mathbf{val}[t_i]$ for $i \in u$,
$h_2{\restriction}[m^{t_i}_{\rm dn}, m^{t_i}_{\rm up})$ is identically zero for $i \notin u$ and
$h_2{\restriction}[m^{t_i}_{\rm up}, m^{t_{i+1}}_{\rm dn})$ is identically zero for $i < m-1$.

(c) $\mathbf{nor}[t] = d(\langle \mathbf{nor}[t_i] : i \in u\rangle)$,

(d) $\mathbf{dis}[t] = \langle 1, d, u\rangle^\frown\langle \mathbf{dis}[t_i] : i < m\rangle$.

[Note: the (d,u)-sum is defined only if clause (b) gives a nonempty value for $\mathbf{val}[t]$.]

DEFINITION 4.2.4. 1. For a pre-norm H on ω (see 2.4.1) let D_H be the family of all functions d such that for some finite set $u_d \subseteq \omega$, $H(u_d) > 0$ and

$$d : (\mathbb{R}^{\geq 0})^{u_d} \longrightarrow \mathbb{R}^{\geq 0} : \langle r_i : i \in u_d\rangle \mapsto \min\{H(u_d), r_i : i \in u_d\}.$$

2. We say that a creating pair (K,Σ) is *saturated with respect to a pre-norm H on ω* if for each $d \in D_H$ and $(t_i : m_0 \leq i < m_1) \in \mathrm{PFC}(K,\Sigma)$ such that $u_d \subseteq [m_0, m_1)$ and $\mathbf{nor}[t_i] > 0$ for $i \in u_d$:

$$\Sigma^{\rm sum}_{d,u_d}(t_i : m_0 \leq i < m_1) \text{ is well defined and belongs to } \Sigma(t_i : m_0 \leq i < m_1),$$

and if $t \in \Sigma(\Sigma^{\rm sum}_{d,u_d}(t_i : m_0 \leq i < m_1))$, $\mathbf{nor}[t] > 0$ then for some $d^* \in D_H$ and $s_i \in \Sigma(t_i)$ (for $m_0 \leq i < m_1$) we have

$$u_{d^*} \subseteq u_d, \quad \mathbf{val}[\Sigma^{\rm sum}_{d^*,u_{d^*}}(s_i : m_0 \leq i < m_1)] \subseteq \mathbf{val}[t], \quad \text{and } \mathbf{nor}[s_i] > 0 \text{ for } i \in u_{d^*}.$$

We say that (K,Σ) is *saturated with respect to (nice) pre–norms* if for each (nice) pre-norm H on ω, (K,Σ) is saturated with respect to H. Similarly for other classes of pre–norms.

REMARK 4.2.5. Note that in practical realizations of 4.2.4(2) the additional parameter **dis** may play a crucial role. Looking at a creature t we may immediately recognize if it comes from the operation $\Sigma^{\rm sum}_{d,u_d}$ and we do not have to worry that the last demand gives a contradiction. It may happen that for distinct d's from D_H we get (as a result of $\Sigma^{\rm sum}_{d,u_d}$) creatures with the same values of **val**, **nor**, however they are distinguished by **dis**. Moreover, the same effect appears for distinct pre–norms H: we can read from **dis** the function d and consequently the function H restricted to subsets of u_d. In applications we may redefine $\mathbf{dis}[\Sigma^{\rm sum}_{d,u_d}(t_i : i < m)]$, but we should keep this coding property.

DEFINITION 4.2.6. Let $T \subseteq \bigcup\limits_{n<\omega} \prod\limits_{i<n} \mathbf{H}(i)$ be a well founded quasi tree and let $\langle s_\nu : \nu \in \hat{T}\rangle \subseteq \mathrm{TCR}[\mathbf{H}]$ be a system of tree creatures such that for each $\nu \in \hat{T}$:

$$\mathrm{dom}(\mathbf{val}[s_\nu]) = \{\nu\} \quad \text{ and } \quad \mathrm{pos}(s_\nu) = \mathrm{succ}_T(\nu).$$

1. *The tree-sum* $t = \Sigma^{\rm tsum}(s_\nu : \nu \in \hat{T})$ *of tree creatures* s_ν (for $\nu \in \hat{T}$) is defined by:
   (α) $\mathbf{val}[t] = \{\langle \mathrm{root}(T), \eta\rangle : \eta \in \max(T)\}$,
   (β) $\mathbf{nor}[t] = \inf\{\mathbf{nor}[s_\nu] : \nu \in \hat{T}\ \&\ \mathbf{nor}[s_\nu] \geq 1\}$, if $\mathbf{nor}[s_\nu] < 1$ for all $\nu \in \hat{T}$ then we let $\mathbf{nor}[t] = 0$,
   (γ) $\mathbf{dis}[t] = \langle 2\rangle^\frown\langle s_\nu : \nu \in \hat{T}\rangle$.
2. For a function $g \in \omega^\omega$, *the special tree-sum* $t = \Sigma^{\rm tsum}_g(s_\nu : \nu \in \hat{T})$ *of tree creatures* s_ν *(for* $\nu \in \hat{T}$*) with respect to* g is defined in a similar manner as $\Sigma^{\rm tsum}$ but the conditions (β), (γ) introducing the norm and **dis** are replaced by
   $(\beta)^*_g$ $\mathbf{nor}[t] = \max\{k < \omega : (\forall \eta \in \max(T))(|\{\ell < \ell g(\eta) : \eta{\restriction}\ell \in \hat{T}$ and $\mathbf{nor}[s_{\eta\restriction\ell}] \geq k\}| \geq g(k))\}$,
   $(\gamma)^*_g$ $\mathbf{dis}[t] = \langle 3, g\rangle^\frown\langle \mathbf{dis}[s_\nu] : \nu \in \hat{T}\rangle$.

4.3. Old reals are unbounded

Recall that a forcing notion $\mathbb{P}$ is *almost ω^ω-bounding* if for every $\mathbb{P}$–name $\dot{f}$ for an element of ω^ω and any $p \in \mathbb{P}$ we have

$$(\exists g \in \omega^\omega)(\forall A \in [\omega]^\omega)(\exists q \geq p)(q \Vdash_{\mathbb{P}} \text{“}(\exists^\infty n \in A)(\dot{f}(n) < g(n))\text{”}).$$

Almost ω^ω–bounding forcing notions do not add dominating reals (i.e. they force that “$(\forall x \in \omega^\omega)(\exists y \in \omega^\omega \cap \mathbf{V})(\exists^\infty n)(x(n) < y(n))$”). If $\mathbb{Q}$ is a forcing notion not adding dominating reals and

$$\Vdash_{\mathbb{Q}} \text{“}\dot{\mathbb{P}} \text{ is almost } \omega^\omega\text{–bounding”}$$

then the composition $\mathbb{Q} * \dot{\mathbb{P}}$ does not add a dominating real (see [**Sh:f**, Ch VI, 3.6]). Thus the notion of “being almost ω^ω–bounding” is very useful from the point of view of iterations: in a countable support iteration of proper forcing notions no dominating reals are added at limit stages (see [**Sh:f**, Ch VI, 3.17]). (Note that “not adding dominating reals” is not preserved by compositions.)

In the definition 4.3.1(1) below one can think about the following situation (explaining the name “decision function”). Suppose that (K, Σ) is a creating pair, $\dot{\tau}$ is a $\mathbb{Q}^*_{\rm s\infty}(K,\Sigma)$–name for an ordinal and $p \in \mathbb{Q}^*_{\rm s\infty}(K,\Sigma)$, $N_0 < \omega$ are such that p approximates $\dot{\tau}$ at each $n \geq N_0$ (see 1.2.9, remember 2.1.4). Let us define a function

$$z : \mathrm{pos}(w^p, t^p_0, \ldots, t^p_{N_0-1}) \times \mathrm{PC}(K,\Sigma) \longrightarrow \bigcup_{k \geq N_0} \mathrm{pos}(w^p, t^p_0, \ldots, t^p_{k-1})$$

such that for every $v \in \mathrm{pos}(w^p, t^p_0, \ldots, t^p_{N_0-1})$ and $\langle t'_0, t'_1, \ldots\rangle \in \mathrm{PC}(K,\Sigma)$ satisfying $\langle t^p_{N_0}, t^p_{N_0+1}, \ldots\rangle \leq \langle t'_0, t'_1, \ldots\rangle$:

1. If $(v, t'_0, t'_1, \ldots) \in \mathbb{Q}^*_{\rm s\infty}(K,\Sigma)$ then $z(v, \langle t'_0, t'_1, \ldots\rangle)$ is the first (in a fixed ordering of $\bigcup\limits_{n<\omega} \prod\limits_{m<n} \mathbf{H}(m)$) of the shortest v^* such that $v^* \in \mathrm{pos}(v, t'_0, \ldots, t'_{k-1})$

(for some $k < \omega$) and $(v^*, t^p_m, t^p_{m+1}, \dots)$ decides the value of $\dot{\tau}$ (m is just suitable: $m^{t^p_m}_{\rm dn} = m^{t'_{k-1}}_{\rm up}$).

2. If $(v, t'_0, t'_1, \dots) \notin \mathbb{Q}^*_{\rm s\infty}(K, \Sigma)$ then we take the first k such that $\mathbf{nor}[t'_k] \leq m^{t'_k}_{\rm dn}$ and we ask if there is $\langle t''_0, t''_1, \dots\rangle \in {\rm PC}(K, \Sigma)$ such that

$$(v, t^p_{N_0}, t^p_{N_0+1}, \dots) \leq (v, t''_0, t''_1, \dots) \in \mathbb{Q}^*_{\rm s\infty}(K, \Sigma), \qquad \text{and for some } \ell \leq k$$

$$z(v, \langle t''_0, t''_1, \dots\rangle) \in {\rm pos}(v, t''_0, \dots, t''_{\ell-1}) \quad \text{and} \quad \langle t'_0, \dots, t'_{\ell-1}\rangle = \langle t''_0, \dots, t''_{\ell-1}\rangle.$$

If the answer is "yes" then we choose such a sequence $\langle t''_0, t''_1, \dots\rangle$ and we let

$$z(v, \langle t'_0, t'_1, \dots\rangle) = z(v, \langle t''_0, t''_1, \dots\rangle)$$

(note that this does not depend on the choice of the particular $\langle t''_0, t''_1, \dots\rangle$; see the previous case). If the answer is "no" then $z(v, \langle t'_0, t'_1, \dots\rangle)$ is the first element of ${\rm pos}(v, t'_0, \dots, t'_k)$.

This z is a canonical example of a decision function for p, N_0, (K, Σ); we will call it $z(p, N_0, \dot{\tau})$ (assuming that (K, Σ) is understood).

DEFINITION 4.3.1. Let (K, Σ) be a creating pair.

1. Let $p \in \mathbb{Q}^*_\emptyset(K, \Sigma)$, $N_0 \in \omega$. We say that a function

$$z : {\rm pos}(w^p, t^p_0, \dots, t^p_{N_0-1}) \times {\rm PC}(K, \Sigma) \longrightarrow \bigcup_{k \geq N_0} {\rm pos}(w^p, t^p_0, \dots, t^p_{k-1})$$

is *a decision function for p, N_0, (K, Σ)* if:

$(*)_{4.3.1}$ for every $v \in {\rm pos}(w^p, t^p_0, \dots, t^p_{N_0-1})$ and $\langle t'_0, t'_1, \dots\rangle \in {\rm PC}(K, \Sigma)$ such that $\langle t^p_{N_0}, t^p_{N_0+1}, \dots\rangle \leq \langle t'_0, t'_1, \dots\rangle$, there is $k \in \omega$ such that:

$$z(v, \langle t'_0, t'_1, \dots\rangle) \in {\rm pos}(v, t'_0, \dots, t'_{k-1})$$

and if $\langle t^p_{N_0}, t^p_{N_0+1}, \dots\rangle \leq \langle t''_0, t''_1, \dots\rangle \in {\rm PC}(K, \Sigma)$ is such that $t''_i \sim_\Sigma t'_i$ for all $i < k$, then

$$z(v, \langle t''_0, t''_1, \dots\rangle) = z(v, \langle t'_0, t'_1, \dots\rangle).$$

2. We say that (K, Σ) is *of the* **AB**-*type* whenever the following two conditions are satisfied:

$(\circledast)^0_{\mathbf{AB}}$ *if* $(t_0, \dots, t_{n-1}) \in {\rm PFC}(K, \Sigma)$, $k < n$ *then* there is $t \in \Sigma(t_0, \dots, t_{n-1})$ such that

$$\mathbf{nor}[t] \geq \min\{\mathbf{nor}[t_\ell] : \ell < n\}$$

and if $(w, t_0, \dots, t_{n-1}) \in {\rm FC}(K, \Sigma)$, $t' \in \Sigma(t)$, $\mathbf{nor}[t'] > 0$, then there is $t'' \in \Sigma(t_k)$ such that $\mathbf{nor}[t''] > 0$ and

$$(\exists u' \in {\rm pos}(w, t_0, \dots, t_{k-1}))(\forall u'' \in {\rm pos}(u, t''))(\exists v \in {\rm pos}(w, t'))(u'' \trianglelefteq v);$$

$(\circledast)^1_{\mathbf{AB}}$ *if* $p \in \mathbb{Q}^*_\emptyset(K, \Sigma)$, $N_0 \in \omega$, $\mathbf{nor}[t^p_i] > 2$ for $i \geq N_0$ and

$$z : {\rm pos}(w^p, t^p_0, \dots, t^p_{N_0-1}) \times {\rm PC}(K, \Sigma) \longrightarrow \bigcup_{k \geq N_0} {\rm pos}(w^p, t^p_0, \dots, t^p_{k-1})$$

is a decision function for $p, N_0, (K, \Sigma)$

then there are $N_1 > N_0$ and $t^* \in \Sigma(t^p_{N_0}, \dots, t^p_{N_1-1})$ such that

$$\mathbf{nor}[t^*] \geq \frac{1}{2}\min\{\mathbf{nor}[t^p_{N_0}], \dots, \mathbf{nor}[t^p_{N_1-1}]\} \qquad \text{and}$$

for each $v \in \mathrm{pos}(w^p, t_0^p, \dots, t_{N_0-1}^p)$ and $t \in \Sigma(t^*)$ with $\mathbf{nor}[t] > 0$ there is $\langle t_0', t_1', \dots\rangle \in \mathrm{PC}(K, \Sigma)$ such that $\langle t_{N_0}^p, t_{N_0+1}^p, \dots\rangle \leq \langle t_0', t_1', \dots\rangle$ and

(α) if $t_i' \in \Sigma(t_{N_0+k}^p, \dots, t_{N_0+k+\ell}^p)$ $(i, k, \ell < \omega)$ then $\mathbf{nor}[t_i'] \geq \frac{1}{2}\min\{\mathbf{nor}[t_{N_0+k}^p], \dots, \mathbf{nor}[t_{N_0+k+\ell}^p]\}$, and

(β) $(\exists w \in \mathrm{pos}(v, t))(z(v, \langle t_0', t_1', t_2', \dots\rangle) \trianglelefteq w)$.

3. We say that (K, Σ) is *condensed* if for every $(w, t_0, \dots, t_{n-1}) \in \mathrm{FC}(K, \Sigma)$ with $\mathbf{nor}[t_i] > 0$ for $i < n$, and $t \in \Sigma(t_0, \dots, t_{n-1})$, $\mathbf{nor}[t] > 0$, there exist $k < n$, a creature $s \in \Sigma(t_k)$ and $v \in \mathrm{pos}(w, t_0, \dots, t_{k-1})$ such that

$$\mathbf{nor}[s] > 0 \quad \text{and} \quad (\forall u \in \mathrm{pos}(v, s))(\exists u^* \in \mathrm{pos}(w, t))(u \trianglelefteq u^*).$$

REMARK 4.3.2. Note that the condition $(\circledast)^0_{\mathbf{AB}}$ is easy to satisfy: e.g. if (K, Σ) is omittory and has the property that for every $t \in K$:

$$\text{if } m_0 \leq m^t_{\mathrm{dn}} < m^t_{\mathrm{up}} \leq m_1 \text{ then } \Sigma(t \restriction [m_0, m_1)) = \{s \restriction [m_0, m_1) : s \in \Sigma(t)\}$$

then it satisfies this requirement (the $(\circledast)^0_{\mathbf{AB}}$ for $(t_0, \dots, t_{n-1})$, $k < n$ is witnessed by $t_k \restriction [m^{t_0}_{\mathrm{dn}}, m^{t_{n-1}}_{\mathrm{up}})$).

THEOREM 4.3.3. *Suppose that (K, Σ) is a finitary, growing and condensed creating pair of the* **AB***-type. Then the forcing notion $\mathbb{Q}^*_{\mathrm{s}\infty}(K, \Sigma)$ is almost ω^ω-bounding.*

PROOF. Let us start with the following claim which will be used later too.

CLAIM 4.3.3.1. *Let (K, Σ) be as in the assumptions of 4.3.3. Suppose that $\dot\tau$ is a $\mathbb{Q}^*_{\mathrm{s}\infty}(K, \Sigma)$-name for a function in ω^ω, $q \in \mathbb{Q}^*_{\mathrm{s}\infty}(K, \Sigma)$ and $n \in \omega$. Then there are a condition $p = (w^p, t_0^p, t_1^p, \dots) \in \mathbb{Q}^*_{\mathrm{s}\infty}(K, \Sigma)$ and a strictly increasing function $g \in \omega^\omega$ such that $q \leq^{\mathrm{s}\infty}_n p$ and for every $\ell \in \omega$*

($\boxplus^*_\ell$) *for each $v \in \mathrm{pos}(w^p, t_0^p, \dots, t_{n-1}^p, \dots, t_{n+\ell-1}^p)$ and $t \in \Sigma(t_{n+\ell}^p)$ with $\mathbf{nor}[t] > 0$ there is $w \in \mathrm{pos}(v, t)$ such that the condition $(w, t_{n+\ell+1}^p, t_{n+\ell+2}^p, \dots)$ decides the value of $\dot\tau(g(\ell))$ and the decision is smaller than $g(\ell+1)$.*

Proof of the claim: We define inductively conditions $p_\ell \in \mathbb{Q}^*_{\mathrm{s}\infty}(K, \Sigma)$ and the values $g(\ell)$ for $\ell \in \omega$ such that $g(0) = 0$, $q = p_0 \leq^{\mathrm{s}\infty}_n p_1 \leq^{\mathrm{s}\infty}_{n+1} \cdots \leq^{\mathrm{s}\infty}_{n+\ell} p_{n+\ell} \leq^{\mathrm{s}\infty}_{n+\ell+1} \cdots$ and $p_{\ell+1}$, $g(\ell)$, $g(\ell+1)$ have the property stated in $(\boxplus^*_\ell)$. It should be clear that then the limit condition $p = \lim\limits_\ell p_\ell$ (see 1.2.13(3)) is as required in the claim.

Suppose we have defined $p_\ell, g(\ell)$. Using 2.1.3 and 2.1.4 we find a condition $p^*_\ell \in \mathbb{Q}^*_{\mathrm{s}\infty}(K, \Sigma)$ such that

$p_\ell \leq_{n+\ell} p^*_\ell$, $\mathbf{nor}[t_k^{p^*_\ell}] > 2 \cdot m^{t_k^{p^*_\ell}} + 2$ for all $k \geq n+\ell$ and

p^*_ℓ approximates $\dot\tau(g(\ell))$ at each $t_k^{p^*_\ell}$ for $k \geq n+\ell$.

Let $z_\ell = z(p^*_\ell, n+\ell, \dot\tau(g(\ell)))$ be the canonical decision function as defined before 4.3.1 (remember the choice of p^*_ℓ). Thus

$$z_\ell : \mathrm{pos}(w^q, t_0^{p^*_\ell}, \dots, t_{n+\ell-1}^{p^*_\ell}) \times \mathrm{PC}(K, \Sigma) \longrightarrow \bigcup_{m \geq n+\ell} \mathrm{pos}(w^q, t_0^{p^*_\ell}, \dots, t_{m-1}^{p^*_\ell})$$

is a decision function such that

if $w \in \mathrm{pos}(w^q, t_0^{p^*_\ell}, \dots, t_{n+\ell-1}^{p^*_\ell})$, $(w, t_0', t_1', \dots) \in \mathbb{Q}^*_{\mathrm{s}\infty}(K, \Sigma)$,
$(w, t_{n+\ell}^{p^*_\ell}, t_{n+\ell+1}^{p^*_\ell}, \dots) \leq (w, t_0', t_1', \dots)$
then $z_\ell(w, \langle t_0', t_1', \dots\rangle) \in \mathrm{pos}(w, t_0', t_1', \dots, t_{m-1}')$ (for some $m \in \omega$)

is such that the condition $(z_\ell(w,\langle t'_0,t'_1,\ldots\rangle), t^{p^*_\ell}_M, t^{p^*_\ell}_{M+1},\ldots)$ gives a value to $\dot\tau(g(\ell))$, where M is such that $m_{\rm up}^{t'_{m-1}} = m_{\rm up}^{t^{p^*_\ell}_{M-1}}$.

Now apply $(\circledast)^1_{\mathbf{AB}}$ to p^*_ℓ, $n+\ell$ and z_ℓ to find $N>n+\ell$ and $t^*\in\Sigma(t^{p^*_\ell}_{n+\ell},\ldots,t^{p^*_\ell}_{N-1})$ such that

$$\mathbf{nor}[t^*]\geq\frac{1}{2}\min\{\mathbf{nor}[t^{p^*_\ell}_{n+\ell}],\ldots,\mathbf{nor}[t^{p^*_\ell}_{N-1}]\}>m_{\rm dn}^{t^{p^*_{n+\ell}}}+1$$

and for each $v\in{\rm pos}(w^q,t^{p^*_\ell}_0,\ldots,t^{p^*_\ell}_{n+\ell-1})$ and $t\in\Sigma(t^*)$ with $\mathbf{nor}[t]>0$ there is $w\in{\rm pos}(v,t)$ for which $(w,t^{p^*_\ell}_N,t^{p^*_\ell}_{N+1},\ldots)$ decides $\dot\tau(g(\ell))$. For this note that if $\langle t'_0,t'_1,\ldots\rangle\in{\rm PC}(K,\Sigma)$ is given by $(\circledast)^1_{\mathbf{AB}}$ for z_ℓ,t,v then, as for some $k_i\leq\ell_i<\omega$

$$\begin{array}{ll} t'_i\in\Sigma(t^{p^*_\ell}_{n+\ell+k_i},\ldots,t^{p^*_\ell}_{n+\ell+\ell_i}), & \text{and}\\ \mathbf{nor}[t'_i]\geq\frac{1}{2}\min\{\mathbf{nor}[t^{p^*_\ell}_{n+\ell+k_i}],\ldots,\mathbf{nor}[t^{p^*_\ell}_{n+\ell+\ell_i}]\}>m_{\rm dn}^{t'_i}, & \end{array}$$

the condition $(v,t'_0,t'_1,\ldots)\in\mathbb{Q}^*_{\rm s\infty}(K,\Sigma)$ is stronger than $(v,t^{p^*_\ell}_{n+\ell},t^{p^*_\ell}_{n+\ell+1},\ldots)$. Thus our requirements on z_k apply. Finally we define

$$p^*_{\ell+1}=(w^q,t^{p^*_\ell}_0,\ldots,t^{p^*_\ell}_{n+\ell-1},t^*,t^{p^*_\ell}_N,t^{p^*_\ell}_{N+1},\ldots)\qquad\text{and}$$

$$\begin{array}{l} g(\ell+1)=1+g(\ell)+\\ +\max\{i<\omega:(\exists v\in{\rm pos}(w^q,t^{p^*_\ell}_0,\ldots t^{p^*_\ell}_{n+\ell-1},t^*))((v,t^{p^*_\ell}_N,t^{p^*_\ell}_{N+1},\ldots)\Vdash\dot\tau(g(\ell))=i)\}. \end{array}$$

Clearly they are as required. This finishes the inductive construction and the proof of the claim.

Now we are going to show that $\mathbb{Q}^*_{\rm s\infty}(K,\Sigma)$ is almost ω^ω–bounding. For this suppose that $\dot\tau$ is a name for a strictly increasing function in ω^ω and $q\in\mathbb{Q}^*_{\rm s\infty}(K,\Sigma)$. Applying claim 4.3.3.1 to $\dot\tau,q$ and $n=0$ we get a condition $p\geq q$ and an increasing function $g\in\omega^\omega$ as there (so they satisfy $(\boxplus^*_\ell)$ for $\ell\in\omega$). Note that, as $\dot\tau$ is (forced to be) increasing, for every $\ell\in\omega$ we have

if $v\in{\rm pos}(w^p,t^p_0,\ldots,t^p_{\ell-1})$ and $t\in\Sigma(t^p_\ell)$ is such that $\mathbf{nor}[t]>0$
then $(w,t^p_{\ell+1},t^p_{\ell+2},\ldots)\Vdash$"$\dot\tau(\ell)<g(\ell+1)$", for some $w\in{\rm pos}(v,t)$.

We will be done when we show the following claim.

Claim 4.3.3.2. *For each $A\in[\omega]^\omega$ there is $p'\geq p$ such that*

$$p'\Vdash_{\mathbb{Q}^*_{\rm s\infty}(K,\Sigma)}(\exists^\infty k\in A)(\dot\tau(k)<g(k+1)).$$

Proof of the claim: Let $A\in[\omega]^\omega$. Choose $0=n_0<n_1<\ldots<\omega$ and creatures $t_i\in\Sigma(t^p_{n_i},\ldots,t^p_{n_{i+1}-1})$ and $k_i\in A$ such that

1. $n_i\leq k_i<n_{i+1}$, $\mathbf{nor}[t_i]\geq\min\{\mathbf{nor}[t^p_k]:n_i\leq k<n_{i+1}\}$,
2. if $w\in{\rm pos}(w^p,t^p_0,\ldots,t^p_{n_i-1})$, $t'\in\Sigma(t_i)$, $\mathbf{nor}[t']>0$ then there is $t''\in\Sigma(t^p_{k_i})$ such that $\mathbf{nor}[t'']>0$ and

$$(\exists u'\in{\rm pos}(w,t^p_{n_i},\ldots,t^p_{k_i-1}))(\forall u''\in{\rm pos}(u',t''))(\exists v\in{\rm pos}(w,t'))(u''\trianglelefteq v)$$

(possible by $(\circledast)^0_{\mathbf{AB}}$). Now let $p'=(w^p,t_0,t_1,\ldots)$. Plainly, $p'\in\mathbb{Q}^*_{\rm s\infty}(K,\Sigma)$, $p\leq p'$. We want to show that

$$p'\Vdash_{\mathbb{Q}^*_{\rm s\infty}(K,\Sigma)}(\exists^\infty k\in A)(\dot\tau(k)<g(k+1)).$$

So assume not. Thus we find a condition $p^+\geq p'$ and $k^*\in\omega$ such that

$$p^+\Vdash_{\mathbb{Q}^*_{\rm s\infty}(K,\Sigma)}(\forall k\in A)(k^*\leq k\quad\Rightarrow\quad\dot\tau(k)\geq g(k+1)).$$

Let $i \in \omega$ be such that $w^{p^+} \in \mathrm{pos}(w^p, t_0, t_1, \dots, t_{i-1})$. As we may pass to an extension of p^+ we may assume that $k^* < k_i$ and $\ell g(w^{p^+}) > 1$. Let $j > i$ be such that $t_0^{p^+} \in \Sigma(t_i, \dots, t_{j-1})$. Since (K, Σ) is condensed we find $\ell \in [i, j)$, $s \in \Sigma(t_\ell)$ and $v \in \mathrm{pos}(w^{p^+}, t_i, \dots, t_{\ell-1})$ such that

$$\mathbf{nor}[s] > 0 \quad \text{and} \quad (\forall u \in \mathrm{pos}(v, s))(\exists u^* \in \mathrm{pos}(w^{p^+}, t_0^{p^+}))(u \trianglelefteq u^*).$$

Now look at our choice of t_ℓ: since $v \in \mathrm{pos}(w^p, t_0^p, \dots, t_{n_\ell - 1}^p)$ and $s \in \Sigma(t_\ell)$, $\mathbf{nor}[s] > 0$, therefore we find $t'' \in \Sigma(t_{k_\ell}^p)$ and $u' \in \mathrm{pos}(v, t_{n_\ell}^p, \dots, t_{k_\ell - 1}^p)$ such that $\mathbf{nor}[t''] > 0$ and

$$(\forall u'' \in \mathrm{pos}(u', t''))(\exists u^+ \in \mathrm{pos}(v, s))(u'' \trianglelefteq u^+).$$

But now look at the choice of p: since $u' \in \mathrm{pos}(w^p, t_0^p, \dots, t_{k_\ell-1}^p)$, $t'' \in \Sigma(t_{k_\ell}^p)$ and $\mathbf{nor}[t''] > 0$, we find $w \in \mathrm{pos}(u', t'')$ such that the condition $(w, t_{k_\ell+1}^p, t_{k_\ell+2}^p, \dots)$ forces that $\dot\tau(k_\ell) < g(k_\ell + 1)$. But now, going back, we know that there is $u^+ \in \mathrm{pos}(v, s)$ such that $w \trianglelefteq u^+$. Further we find $u^* \in \mathrm{pos}(w^{p^+}, t_0^{p^+})$ such that $u^+ \trianglelefteq u^*$. So look at the condition $(u^*, t_1^{p^+}, t_2^{p^+}, \dots)$. It is stronger than p^+ and it forces that $\dot\tau(k_\ell) < g(k_\ell + 1)$, contradicting the choice of p^+ and $k^* < k_i \le k_\ell$. This finishes the proof of the claim and the theorem. □

LEMMA 4.3.4. *Suppose that (K, Σ) is a strongly finitary and omittory creating pair with the weak Halving Property which is saturated with respect to nice pre-norms with values in ω (see 4.2.4(2)). Further suppose that for each $t \in K$*

($\otimes_1$) *$(\forall s \in \Sigma(t))(\mathbf{val}[s] \subseteq \mathbf{val}[t])$ and*

($\otimes_2$) *if $m_0 \le m_{\mathrm{dn}}^t < m_{\mathrm{up}}^t \le m_1$, $s \in \Sigma(\mathrm{half}(t))$ then $s \restriction [m_0, m_1) \in \Sigma(\mathrm{half}(t \restriction [m_0, m_1)))$.*

Assume that $p \in \mathbb{Q}^*_\emptyset(K, \Sigma)$, $N_0 \in \omega$, $\mathbf{nor}[t_i^p] > 2$ for $i \ge N_0$, $m \ge 1$ and

$$z : \mathrm{pos}(w^p, t_0^p, \dots, t_{N_0-1}^p) \times \mathrm{PC}(K, \Sigma) \longrightarrow \bigcup_{n \ge N} \mathrm{pos}(w^p, t_0^p, \dots, t_{n-1}^p)$$

is a decision function for $p, N_0, (K, \Sigma)$.
Then *there are a nice pre-norm $H : [\omega]^{<\omega} \longrightarrow \omega$ (so (K, Σ) is saturated with respect to H) and $d \in D_H$ (see 4.2.4) such that $u_d = [N_0, N_1)$, $H([N_0, N_1)) \ge m$ and*

if $t \in \Sigma(\Sigma^{\mathrm{sum}}_{d, u_d}(\mathrm{half}(t_{N_0}^p), \dots, \mathrm{half}(t_{N_1-1}^p)))$, $v \in \mathrm{pos}(w^p, t_0^p, \dots, t_{N_0-1}^p)$, *and* $\mathbf{nor}[t] > 0$

then *there is $\langle t_0', t_1', \dots \rangle \in \mathrm{PC}(K, \Sigma)$ such that for each $i \in \omega$, for some $k \le \ell < \omega$*

$$t_i' \in \Sigma(t_{N_0+k}^p, \dots, t_{N_0+\ell}^p) \quad \& \quad \mathbf{nor}[t_i'] \ge \frac{1}{2} \min\{\mathbf{nor}[t_{N_0+k}^p], \dots, \mathbf{nor}[t_{N_0+\ell}^p]\}$$

and $(\exists w \in \mathrm{pos}(v, t))(z(v, \langle t_0', t_1', \dots \rangle) \trianglelefteq w)$.

PROOF. This is essentially [**Sh 207**, 2.14].
First note that if $i_0 < \dots < i_k$, $j \le j_0 < \dots < j_\ell < \omega$, $\{i_0, \dots, i_k\} \subseteq \{j_0, \dots, j_\ell\}$ (and $k \le \ell < \omega$), $w \in \mathrm{pos}(w^p, t_0^p, \dots, t_{j-1}^p)$ and $s_{j_n} \in \Sigma(t_{j_n}^p)$ (for $n \le \ell$) then

$$\langle w, v \rangle \in \mathbf{val}[\Sigma^{\mathrm{sum}}(s_{i_0} \restriction [m_{\mathrm{dn}}^{t_j^p}, m_{\mathrm{up}}^{s_{i_0}}), s_{i_1}, \dots, s_{i_k})]$$

implies

$$\langle w, v ^\frown \mathbf{0}_{[m_{\mathrm{up}}^{s_{i_k}}, m_{\mathrm{up}}^{s_{i_\ell}})} \rangle \in \mathbf{val}[\Sigma^{\mathrm{sum}}(s_{j_0} \restriction [m_{\mathrm{dn}}^{t_j^p}, m_{\mathrm{up}}^{s_{j_0}}), s_{j_1}, \dots, s_{j_\ell})].$$

Why? Suppose that $\langle w,v\rangle\in\mathbf{val}[\Sigma^{\rm sum}(s_{i_0}\restriction[m^{t^p_j}_{\rm dn},m^{s_{i_0}}_{\rm up}),s_{i_1},\ldots,s_{i_k})]$. If $j_0=i_0$ this immediately implies $\langle w,v\restriction m^{s_{j_0}}_{\rm up}\rangle\in\mathbf{val}[s_{j_0}\restriction[m^{t^p_j}_{\rm dn},m^{s_{j_0}}_{\rm up})]$. Otherwise necessarily $j_0<i_0$ and $v\restriction[m^{t^p_j}_{\rm dn},m^{s_{j_0}}_{\rm up})$ is constantly zero. Now, $w\in{\rm pos}(w^p,t^p_0,\ldots,t^p_{j-1})$, so using the assumptions that (K,Σ) is omittory and $(\otimes_1)$ we get $\langle w,v\restriction m^{s_{j_0}}_{\rm up}\rangle\in\mathbf{val}[s_{j_0}\restriction[m^{t^p_j}_{\rm dn},m^{s_{j_0}}_{\rm up})]$. Proceeding in this fashion further we get the desired conclusion.

Note that above we use "$\mathbf{val}[\Sigma^{\rm sum}(\ldots)]$" and not "${\rm pos}(w,\Sigma^{\rm sum}(\ldots))$" as we do not claim that $\Sigma^{\rm sum}(\ldots)$ is in K.

Let us define the function $H:[\omega]^{<\omega}\longrightarrow\omega$ by:

$H(u)\geq 0$: always,

$H(u)\geq 1$: if $|u\setminus N_0|>1$, $u\setminus N_0=\{i_0,\ldots,i_{k-1}\}$ (the increasing enumeration), and
if $s_\ell\in\Sigma({\rm half}(t^p_{i_\ell}))$, $\mathbf{nor}[s_\ell]>0$ (for $\ell<k$) and $v\in{\rm pos}(w^p,t^p_0,\ldots,t^p_{N_0-1})$
then there exists $\langle t'_0,t'_1,\ldots\rangle\in{\rm PC}(K,\Sigma)$ such that for each $i\in\omega$, for some $k\leq\ell<\omega$

$$t'_i\in\Sigma(t^p_{N_0+k},\ldots,t^p_{N_0+\ell})\quad\&\quad\mathbf{nor}[t'_i]\geq\frac{1}{2}\min\{\mathbf{nor}[t^p_{N_0+k}],\ldots,\mathbf{nor}[t^p_{N_0+\ell}]\}$$

and for some w

$$\langle v,w\rangle\in\mathbf{val}[\Sigma^{\rm sum}(s_0\restriction[m^{t^p_{N_0}}_{\rm dn},m^{t^p_{i_0}}_{\rm up}),s_1,\ldots,s_{k-1})]\quad\&\quad z(v,\langle t'_0,t'_1,t'_2,\ldots\rangle)\trianglelefteq w,$$

$H(u)\geq n+1$: if for every $u'\subseteq u$ either $H(u')\geq n$ or $H(u\setminus u')\geq n$ (for $n>0$).

Note that this defines correctly a nice pre–norm on $[\omega]^{<\omega}$; for monotonicity use the remark we started with. Thus (K,Σ) is saturated with respect to H.

Now it is enough to find $N_1>N_0$ such that $H([N_0,N_1))\geq m$ and then take $d\in D_H$ with $u_d=[N_0,N_1)$. Why? Suppose that we have such an N_1 (and the respective d) and let $t\in\Sigma(\Sigma^{\rm sum}_{d,u_d}({\rm half}(t^p_{N_0}),\ldots,{\rm half}(t^p_{N_1-1})))$, $\mathbf{nor}[t]>0$. By 4.2.4(2) (the second demand) we have that there are $d^*\in D_H$ and $s_i\in\Sigma({\rm half}(t^p_i))$ (for $i\in[N_0,N_1)$) such that $u_{d^*}\subseteq u_d$ and $\mathbf{nor}[s_i]>0$ for $i\in u_{d^*}$, and $\mathbf{val}[\Sigma^{\rm sum}_{d^*,u_{d^*}}(s_i:N_0\leq i<N_1)]\subseteq\mathbf{val}[t]$. But now look at the definition of H (and remember the definitions of $\Sigma^{\rm sum}$, $\Sigma^{\rm sum}_{d^*,u_{d^*}}$; note $H(u_{d^*})>0$).

As H is monotonic, it is enough to find a set $u\in[\omega\setminus N_0]^{<\omega}$ with $H(u)\geq m$. We will do this by induction on m for all p,N_0,z.

CASE 1: $m=1$
For $t\in\Sigma({\rm half}(t^p_{N_0+i}))$ with $\mathbf{nor}[t]>0$, $i\in\omega$ fix $s(t)\in\Sigma(t^p_{N_0+i})$ such that

$$\mathbf{nor}[s(t)]\geq\frac{1}{2}\mathbf{nor}[t]\quad\text{and}\quad(\forall w\in\mathbf{basis}(t))({\rm pos}(w,s(t))\subseteq{\rm pos}(w,t))$$

(possible by 2.2.7(2)(b$^-$)). We may additionally require that if $t\sim_\Sigma t'$ then $s(t)=s(t')$. Let

$$\mathcal{X}\stackrel{\rm def}{=}\{\langle t'_0,t'_1,\ldots\rangle\in{\rm PC}(K,\Sigma):(\forall i\in\omega)(t'_i\in\Sigma({\rm half}(t^p_{N_0+i}))\ \&\ \mathbf{nor}[t'_i]>0)\}.$$

As (K,Σ) is strongly finitary each $\Sigma({\rm half}(t^p_{N_0+i}))$ is finite (up to $\sim_\Sigma$–equivalence, but we may consider representatives only) and thus the space $\mathcal{X}$ equipped with the product topology (of discrete $\Sigma({\rm half}(t^p_{N_0+i}))$'s) is compact.

For $w \in \mathrm{pos}(w^p, t_0^p, \ldots, t_{N_0-1}^p)$ and $\langle t_0', t_1', \ldots \rangle \in \mathcal{X}$ let $M(w, \langle t_0', t_1', \ldots \rangle)$ be the unique $M < \omega$ such that

$$z(w, \langle s(t_0'), s(t_1'), \ldots \rangle) \in \mathrm{pos}(w, s(t_0'), \ldots, s(t_{M-1}')).$$

Note that the function $M : \mathrm{pos}(w^p, t_0^p, \ldots, t_{N_0-1}^p) \times \mathcal{X} \longrightarrow \omega$ is continuous. Why? Look at the definition of decision functions: if $\langle t_0'', t_1'', \ldots \rangle \in \mathcal{X}$ is such that $t_i'' \sim_\Sigma t_i'$ for all $i < M(w, \langle t_0', t_1', \ldots \rangle)$ then

$$z(w, \langle s(t_0'), s(t_1'), \ldots \rangle) = z(w, \langle s(t_0''), s(t_1''), \ldots \rangle).$$

Hence, by compactness of $\mathcal{X}$, the function M is bounded. Let $N_1' > 1$ be such that $\mathrm{rng}(M) \subseteq N_1'$ and let $N_1 = N_0 + N_1'$. We want to show $H([N_0, N_1)) \geq 1$. So suppose that $s_\ell \in \Sigma(\mathrm{half}(t_{N_0+\ell}^p))$, $\mathbf{nor}[s_\ell] > 0$ (for $\ell < N_1'$) and $v \in \mathrm{pos}(w^p, t_0^p, \ldots, t_{N_0-1}^p)$. Look at $\langle s_0, \ldots, s_{N_1'-1}, \mathrm{half}(t_{N_1}^p), \mathrm{half}(t_{N_1+1}^p), \ldots \rangle \in \mathcal{X}$, and let

$$\bar{\mathbf{s}} = \langle s(s_0), \ldots, s(s_{N_1'-1}), s(\mathrm{half}(t_{N_1}^p)), s(\mathrm{half}(t_{N_1+1}^p)), \ldots \rangle \in \mathrm{PC}(K, \Sigma).$$

By the choice of N_1' we know that for some $k < N_1'$:

$$z(v, \bar{\mathbf{s}}) \in \mathrm{pos}(v, s(s_0), \ldots, s(s_k)) \subseteq \mathrm{pos}(v, s_0, \ldots, s_k)$$

(by the choice of $s(t)$'s). Take $w^* \in \mathrm{pos}(v, s_0, \ldots, s_{N_1'-1})$ such that $z(v, \bar{\mathbf{s}}) \trianglelefteq w^*$. Applying $(\otimes_1)$ we get $\langle v, w^* \rangle \in \mathbf{val}[\Sigma^{\mathrm{sum}}(s_0, \ldots, s_{N_1'-1})]$. To finish this case note that $\mathbf{nor}[s(s_\ell)] \geq \frac{1}{2}\mathbf{nor}[t_{N_0+\ell}^p]$ for $\ell < N_1'$ and $\mathbf{nor}[s(\mathrm{half}(t_k^p))] \geq \frac{1}{2}\mathbf{nor}[t_k^p]$ for $k \geq N_1$.

CASE 2: $m' = m + 1 \geq 2$

Now suppose that we always can find a finite subset of ω of the pre–norm H at least m. Thus we find an increasing sequence $N_0 = \ell_0 < \ell_1 < \ldots < \omega$ such that $H([\ell_i, \ell_{i+1})) \geq m$ for each i. Consider the space of all increasing functions $\psi \in \omega^\omega$ such that $\psi \restriction N_0$ is the identity and $(\forall i \in \omega)(\psi(N_0 + i) \in [\ell_i, \ell_{i+1}))$ - it is a compact space. For each ψ from the space we may consider a condition

$$(w^p, t_0^p, \ldots, t_{N_0-1}^p, t_{\psi(N_0)}^p \restriction [m_{\mathrm{dn}}^{t_{\ell_0}^p}, m_{\mathrm{up}}^{t_{\psi(N_0)}^p}), t_{\psi(N_0+1)}^p \restriction [m_{\mathrm{up}}^{t_{\psi(N_0)}^p}, m_{\mathrm{up}}^{t_{\psi(N_0+1)}^p}), \ldots)$$

(and call it p_ψ) and the respective pre–norm H_ψ defined like H but for p_ψ, N_0, z (note that $p \leq p_\psi \in \mathbb{Q}_\emptyset^*(K, \Sigma)$, so z may be interpreted as a decision function for p_ψ).

CLAIM 4.3.4.1. *For each finite set $u \subseteq \omega$:*

$$H_\psi(u) \leq H(\{\psi(k) : k \in u\}).$$

Proof of the claim: Suppose $H_\psi(u) \geq 1$. We may assume that $u \subseteq \omega \setminus N_0$. Let $s_k \in \Sigma(\mathrm{half}(t_{\psi(k)}^p))$, $\mathbf{nor}[s_k] > 0$ for $k \in u$ and $v \in \mathrm{pos}(w^p, t_0^p, \ldots, t_{N_0-1}^p)$. By the assumption $(\otimes_2)$ we know that

$$s_k^* \stackrel{\mathrm{def}}{=} s_k \restriction [m_{\mathrm{dn}}^{t_k^{p_\psi}}, m_{\mathrm{up}}^{t_k^{p_\psi}}) \in \Sigma\big(\mathrm{half}(t_{\psi(k)}^p \restriction [m_{\mathrm{dn}}^{t_k^{p_\psi}}, m_{\mathrm{up}}^{t_k^{p_\psi}}))\big) = \Sigma(\mathrm{half}(t_k^{p_\psi})).$$

So, as $H_\psi(u) \geq 1$, we find $\langle t_0', t_1', \ldots \rangle \in \mathrm{PC}(K, \Sigma)$ such that

1. $t_i' \in \Sigma(t_{k_i}^{p_\psi}, \ldots, t_{n_i}^{p_\psi})$ (for some $N_0 \leq k_i \leq n_i < \omega$),
2. $\mathbf{nor}[t_i'] \geq \frac{1}{2}\min\{\mathbf{nor}[t_{k_i}^{p_\psi}], \ldots, \mathbf{nor}[t_{n_i}^{p_\psi}]\}$, and
3. for some w

$$\langle v, w \rangle \in \mathbf{val}\big[\Sigma^{\mathrm{sum}}(s_k^* : k \in u) \restriction [m_{\mathrm{dn}}^{t_{N_0}^{p_\psi}}, m_{\mathrm{up}}^{t_{\max(u)}^{p_\psi}})\big] \quad \& \quad z(v, \langle t_0', t_1', \ldots \rangle) \trianglelefteq w.$$

But then also $\langle v,w\rangle\in\mathbf{val}\big[\Sigma^{\rm sum}(s_k:k\in u)\restriction[m^{t^p_{N_0}}_{\rm dn},m^{t^p_{\psi(\max(u))}}_{\rm up})\big]$. As

$$\min\{\mathbf{nor}[t^{p_\psi}_{k_i}],\ldots,\mathbf{nor}[t^{p_\psi}_{n_i}]\}\geq\min\{\mathbf{nor}[t^p_{\psi(k_i-1)+1}],\ldots,\mathbf{nor}[t^p_{\psi(n_i)}]\}$$

and $t'_i\in\Sigma(t^p_{\psi(k_i-1)+1},\ldots,t^p_{\psi(n_i)})$, we may conclude that $1\leq H(\{\psi(k):k\in u\})$. Next we easily proceed by induction, finishing the proof of the claim.

By the induction hypothesis for each suitable function ψ we find $N_\psi>N_0$ such that $H_\psi([N_0,N_\psi))\geq m$. By the compactness of the space of all these functions we find one n such that $H_\psi([N_0,N_0+n))\geq m$ (for each ψ). Look at the interval $[N_0,\ell_n)$ – we claim that its H-norm is greater or equal than $m+1$. Why? By the choice of ℓ_i's we have that $H([N_0,\ell_n))\geq m$. Suppose that $u\subseteq[N_0,\ell_n)$. If $u\cap[\ell_k,\ell_{k+1})\neq\emptyset$ for each $k<n$ then we may take a function ψ from our space such that $\psi[[N_0,N_0+n)]\subseteq u$. But $H_\psi([N_0,N_0+n))\geq m$ and by 4.3.4.1

$$m\leq H(\{\psi(k):N_0\leq k<N_0+n\})\leq H(u).$$

If $u\cap[\ell_k,\ell_{k+1})=\emptyset$ for some $k<n$ then necessarily

$$m\leq H([\ell_k,\ell_{k+1}))\leq H([N_0,\ell_n)\setminus u).$$

This finishes the induction and the proof of the lemma. □

THEOREM 4.3.5. *Assume that (K,Σ) is a strongly finitary and omittory creating pair with the weak Halving Property. Further suppose that (K,Σ) is saturated with respect to nice pre–norms with values in ω and for each $t\in K$:*

($\otimes_1$) $(\forall s\in\Sigma(t))(\mathbf{val}[s]\subseteq\mathbf{val}[t])$,

($\otimes_2$) *if* $m_0\leq m^t_{\rm dn}<m^t_{\rm up}\leq m_1$, $s\in\Sigma(\mathrm{half}(t))$ *then* $s\restriction[m_0,m_1)\in\Sigma(\mathrm{half}(t\restriction[m_0,m_1)))$,

($\oplus_3$) $\Sigma(t\restriction[m_0,m_1))=\{s\restriction[m_0,m_1):s\in\Sigma(t)\}$.

Then the creating pair (K,Σ) is of the **AB**–*type. Consequently if (K,Σ) is additionally condensed then the forcing notion $\mathbb{Q}^*_{s\infty}(K,\Sigma)$ is almost ω^ω-bounding.*

PROOF. By 4.3.2, 4.3.4 and 4.3.3. □

REMARK 4.3.6. The assumptions of 4.3.5 may look very complicated, but in the real examples they are relatively easy to check and appear naturally. Sometimes it is easy to check directly that a creating pair is of the **AB**–type, but then it may happen that it is not condensed (this happens e.g. for $(K_{2.4.5},\Sigma_{2.4.5})$; see 4.4.1). To get that $\mathbb{Q}^*_{s\infty}(K,\Sigma)$ is almost ω^ω–bounding we do not have to require that (K,Σ) is condensed, but then we should strengthen the demands of 4.3.1(2) a little bit.

DEFINITION 4.3.7. We say that a creating pair (K,Σ) is *of the* $\mathbf{AB}^+$*–type* if it satisfies the demand $(\circledast)^0_{\mathbf{AB}}$ of 4.3.1 and the following strengthening of $(\circledast)^1_{\mathbf{AB}}$:

$(\circledast^+)^1_{\mathbf{AB}}$ *if* $p\in\mathbb{Q}^*_{s\infty}(K,\Sigma)$, $N_0\in\omega$, and

$$z:\mathrm{pos}(w^p,t^p_0,\ldots,t^p_{N_0-1})\times\mathrm{PC}(K,\Sigma)\longrightarrow\bigcup_{k\geq N_0}\mathrm{pos}(w^p,t^p_0,\ldots,t^p_{k-1})$$

is a decision function for $p,N_0,(K,\Sigma)$ *then* there are $N_1>N_0$ and $t^*\in\Sigma(t^p_{N_0},\ldots,t^p_{N_1-1})$ such that

$$\mathbf{nor}[t^*]\geq\frac{1}{2}\min\{\mathbf{nor}[t^p_{N_0}],\ldots,\mathbf{nor}[t^p_{N_1-1}]\}$$

and for each $v \in \mathrm{pos}(w^p, t_0^p, \ldots, t_{N_0-1}^p, t^*)$ such that

$$(\exists k \in [m_{\mathrm{dn}}^{t^*}, m_{\mathrm{up}}^{t^*}))(v(k) \neq 0)$$

there is $\langle t_0', t_1', \ldots \rangle \in \mathrm{PC}_{\mathrm{s}\infty}(K, \Sigma)$ such that $\langle t_{N_0}^p, t_{N_0+1}^p, \ldots \rangle \leq \langle t_0', t_1', \ldots \rangle$ and $z(v \restriction m_{\mathrm{dn}}^{t^*}, \langle t_0', t_1', t_2', \ldots \rangle) \trianglelefteq v$.

THEOREM 4.3.8. *Suppose that (K, Σ) is a finitary and omittory creating pair of the $\mathbf{AB}^+$-type such that for each $t \in K$*

($\oplus_0$) *if $\mathbf{nor}[t] > 1$ and $u \in \mathbf{basis}(t)$ then $|\mathrm{pos}(u, t)| > 2$ and*
($\oplus_3$) *$\Sigma(t \upharpoonright [m_0, m_1)) = \{s \upharpoonright [m_0, m_1) : s \in \Sigma(t)\}$.*

Then the forcing notion $\mathbb{Q}_{\mathrm{s}\infty}^*(K, \Sigma)$ is almost ω^ω-bounding.

PROOF. It is fully parallel to 4.3.3. First one proves that

CLAIM 4.3.8.1. *If $\dot{\tau}$ is a $\mathbb{Q}_{\mathrm{s}\infty}^*(K, \Sigma)$-name for an element of ω^ω, $q \in \mathbb{Q}_{\mathrm{s}\infty}^*(K, \Sigma)$ and $n \in \omega$, then there are a condition $p \in \mathbb{Q}_{\mathrm{s}\infty}^*(K, \Sigma)$ and an increasing $g \in \omega^\omega$ such that $q \leq_n^{\mathrm{s}\infty} p$ and for every $\ell \in \omega$*

($\boxplus_\ell^+$) *if $w \in \mathrm{pos}(w^p, t_0^p, \ldots, t_{n-1}^p, \ldots, t_{n+\ell}^p)$ is such that $w \restriction [m_{\mathrm{dn}}^{t_{n+\ell}^p}, m_{\mathrm{up}}^{t_{n+\ell}^p}) \neq \mathbf{0}$ then the condition $(w, t_{n+\ell+1}^p, t_{n+\ell+2}^p, \ldots)$ decides the value of $\dot{\tau}(g(\ell))$ and the decision is smaller than $g(\ell + 1)$.*

Proof of the claim: Repeat the proof of 4.3.3.1.

Next, assuming that $\dot{\tau}$ is a name for a strictly increasing function in ω^ω, $n = 0$, and $q \in \mathbb{Q}_{\mathrm{s}\infty}^*(K, \Sigma)$, we take the condition $p \geq_0 q$ and the function $g \in \omega^\omega$ given by 4.3.8.1. They have the property that for each $\ell \in \omega$

if $v \in \mathrm{pos}(w^p, t_0^p, \ldots, t_\ell^p)$ is such that $(\exists k \in [m_{\mathrm{dn}}^{t_\ell^p}, m_{\mathrm{up}}^{t_\ell^p}))(v(k) \neq 0)$
then $(v, t_{\ell+1}^p, t_{\ell+2}^p, \ldots) \Vdash$ "$\dot{\tau}(\ell) < g(\ell + 1)$".

To show that for every $A \in [\omega]^\omega$ there is $p' \geq p$ such that

$$p' \Vdash_{\mathbb{Q}_{\mathrm{s}\infty}^*(K,\Sigma)} (\exists^\infty k \in A)(\dot{\tau}(k) < g(k + 1))$$

we slightly modify the proof of 4.3.3.2. So suppose $A \in [\omega]^\omega$. Choose $0 = n_0 < n_1 < \ldots < \omega$ and $k_i \in A$ such that $n_i \leq k_i < n_{i+1}$ and let $t_i^{p'} = t_{k_i}^p \upharpoonright [m_{\mathrm{dn}}^{t_{n_i}^p}, m_{\mathrm{dn}}^{t_{n_{i+1}}^p})$. Look at the condition $p' = (w^p, t_0^{p'}, t_1^{p'}, \ldots)$. Assume that $p^+ \geq p'$, $k^* \in \omega$, $\ell g(w^{p^+}) > 1$ and $k_i > k^*$, where i is such that $w^{p^+} \in \mathrm{pos}(w^p, t_0^{p'}, \ldots, t_{i-1}^{p'})$. Take $j > i$ such that $t_0^{p^+} \in \Sigma(t_i^{p'}, \ldots, t_{j-1}^{p'})$. Choose $v \in \mathrm{pos}(w^{p^+}, t_0^{p^+})$ such that $v(\ell) \neq 0$ for some $\ell \in [m_{\mathrm{dn}}^{t_0^{p^+}}, m_{\mathrm{up}}^{t_0^{p^+}})$ (exists by ($\oplus_0$)). Let $i^* \in [i, j)$ be such that $\ell \in [m_{\mathrm{dn}}^{t_{i^*}^{p'}}, m_{\mathrm{up}}^{t_{i^*}^{p'}})$. By smoothness we know that $v \restriction m_{\mathrm{up}}^{t_{i^*}^{p'}} \in \mathrm{pos}(v \restriction m_{\mathrm{dn}}^{t_{i^*}^{p'}}, t_{i^*}^{p'})$, and therefore, by ($\oplus_3$) and the choice of $t_{i^*}^{p'}$ we get

$$v \restriction m_{\mathrm{dn}}^{t_{k_{i^*}}^p} \in \mathrm{pos}(w^p, t_0^p, \ldots, t_{k_{i^*}}^p) \quad \text{and} \quad m_{\mathrm{dn}}^{t_{k_{i^*}}^p} \leq \ell < m_{\mathrm{up}}^{t_{k_{i^*}}^p}.$$

Hence $(v \restriction m_{\mathrm{up}}^{t_{k_{i^*}}^p}, t_{k_{i^*}+1}^p, t_{k_{i^*}+2}^p, \ldots) \Vdash \dot{\tau}(k_{i^*}) < g(k_{i^*} + 1)$, so we are done. □

Let us finish this section by proving a parallel of 4.3.5 for the tree–like forcing notions.

THEOREM 4.3.9. *Suppose that* $\bigcup_{i\in\omega}\mathbf{H}(i)$ *is countable and* (K,Σ) *is a t-omittory tree creating pair for* $\mathbf{H}$. *Then the forcing notion* $\mathbb{Q}^{\rm tree}_0(K,\Sigma)$ *is almost* ω^ω*-bounding.*

PROOF. Let $p\in\mathbb{Q}^{\rm tree}_0(K,\Sigma)$ and let $\dot{f}$ be a $\mathbb{Q}^{\rm tree}_0(K,\Sigma)$–name for a function in ω^ω. For simplicity we may assume that for every $t\in T^p$ we have $|{\rm pos}(t)|=\omega$ or at least that above each $\nu\in T^p$ we find may an infinite front of $T^{p^{[\nu]}}$ (compare 3.1.1(2)).

Like in 2.3.6(2) we may construct a condition $q\in\mathbb{Q}^{\rm tree}_0(K,\Sigma)$ stronger than p and fronts F_n of T^q such that for all $n\in\omega$:

1. $F_n=\{\nu^s: s\in\omega^{n+1}\}$ (just a fixed enumeration), and for each $s\in\omega^{n+1}$:
2. if $m\in\omega$ then $\nu^s\vartriangleleft\nu^{s^\frown\langle m\rangle}$,
3. $\{\nu^{s^\frown\langle m\rangle}: m\in\omega\}$ is a front of $T^{q^{[\nu^s]}}$ and $\mathbf{nor}[t^q_{\nu^s}]\geq n+1$,
4. the condition $q^{[\nu^s]}$ decides the value of $\dot{f}{\restriction}\big(n+1+\sum\limits_{k\leq n}s(k)\big)$.

For $m\in\omega$ let $g(m)$ be

$$1+\max\{\ell<\omega: (\exists s\in\omega^{\leq m+1})(\sum_{k<\ell g(s)}s(k)\leq 4m\ \ \&\ \ q^{[\nu^s]}\Vdash\text{“ }\dot{f}(m)=\ell\text{ ”})\}.$$

Let $A\in[\omega]^\omega$. For $s\in\omega^{<\omega}$ choose $m(s)\in A$ such that $\ell g(s)+\sum\limits_{k<\ell g(s)}s(k)<m(s)$ and let $c(s)=s^\frown\langle m(s)\rangle$. Note that $q^{[\nu^{c(s)}]}\Vdash\dot{f}(m(s))<g(m(s))$. Now build inductively a condition $p'\in\mathbb{Q}^{\rm tree}_0(K,\Sigma)$ such that $q\leq^0_0 p'$, $F_0\subseteq T^{p'}$ and for each $n\in\omega$:

if $s\in\omega^{2n+1}$, $\nu^s\in F_{2n}\cap T^{p'}$ then

$t^{p'}_{\nu^s}\in\Sigma\big(t^p_\rho: (\exists\eta\in F_{2n+1})(\nu^s\trianglelefteq\rho\trianglelefteq\eta)\big)$, ${\rm pos}(t^{p'}_{\nu^s})\subseteq{\rm pos}(t^q_{\nu^{c(s)}})$ and

$\mathbf{nor}[t^{p'}_{\nu^s}]\geq 2n+1$

(possible as (K,Σ) is t-omittory and by the third requirement on F_n's). We claim that

$$p'\Vdash_{\mathbb{Q}^{\rm tree}_0(K,\Sigma)}(\exists^\infty m\in A)(\dot{f}(m)<g(m)).$$

To see this suppose that $p''\geq p'$, $N\in\omega$. Choose $s\in\omega^{2N+1}$ such that $\nu^s\in F_{2N}\cap{\rm dcl}(T^{p''})$. Then necessarily ${\rm pos}(t^{p'}_{\nu^s})\cap{\rm dcl}(T^{p''})\neq\emptyset$ so we may choose $\eta\in T^{p''}$ such that some initial segment of η is in ${\rm pos}(t^{p'}_{\nu^s})\subseteq{\rm pos}(t^q_{\nu^{c(s)}})$ (see the construction of p'). But now we conclude

$$(p'')^{[\eta]}\Vdash_{\mathbb{Q}^{\rm tree}_0(K,\Sigma)}\dot{f}(m(s))<g(m(s))$$

what finishes the proof as $N<m(s)\in A$. □

4.4. Examples

Let us start with noting that the Blass–Shelah forcing notion $\mathbb{Q}^*_{{\rm s}\infty}(K^*_{2.4.5},\Sigma^*_{2.4.5})$ is a good application of the notions introduced in this section.

PROPOSITION 4.4.1. *The creating pair* $(K^*_{2.4.5},\Sigma^*_{2.4.5})$ *(see the end of the construction for 2.4.5) is meagering, of the* $\mathbf{AB}^+$*–type and satisfies the demands* $(\oplus_0)$, $(\oplus_3)$ *of 4.3.8. Consequently, the forcing notion* $\mathbb{Q}^*_{{\rm s}\infty}(K^*_{2.4.5},\Sigma^*_{2.4.5})$:

(α) *makes ground model reals meager,*

(β) *adds an unbounded real,*
(γ) *is almost ω^ω-bounding,*
(δ) *does not add Cohen reals.*

PROOF. To show that $(K^*_{2.4.5},\Sigma^*_{2.4.5})$ is meagering assume that $(t_0,\ldots,t_{n-1})\in \mathrm{PFC}(K^*_{2.4.5},\Sigma^*_{2.4.5})$, $t\in\Sigma^*_{2.4.5}(t_0,\ldots,t_{n-1})$ and $\langle k_i: i<n\rangle$ are such that $\mathbf{nor}[t_i]\geq 2$, $m^{t_i}_{\rm dn}\leq k_i<m^{t_i}_{\rm up}$ (for $i<n$) and $\mathbf{nor}[t]\geq 2$. Fix $u\in\prod\limits_{i<m^{t_0}_{\rm dn}}\mathbf{H}(i)$. By the definition of $K^*_{2.4.5}$ (see clause (γ) there) we know that $2\leq\mathbf{nor}[t]\leq \mathrm{dp}^1_0(A^t_u)$. Moreover, by the definition of $\Sigma^*_{2.4.5}$ we know that no element of A^t_u is included in $\{k_i: i<n\}$ (by clause (α) of 2.4.5; remember $a\cap[m^{t_i}_{\rm dn},m^{t_i}_{\rm up})\in A^{t_i}_{u\cup(a\cap m^{t_i}_{\rm dn})}\cup\{\emptyset\}$ for all $a\in A^t_u$). Consequently, if $A^s_u=\{a\in A^t_u: a\cap\{k_i:i<n\}=\emptyset\}$, then $\mathrm{dp}^1(A^s_u)\geq\mathrm{dp}^1(A^t_u)-1$ and $\mathrm{dp}^1_0(A^s_u)\geq\mathbf{nor}[t]-1$. This determines a condition $s\in\Sigma^*_{2.4.5}(t)$ which is as required in 4.1.2(1).

It should be clear that $(K^*_{2.4.5},\Sigma^*_{2.4.5})$ satisfies the conditions $(\oplus_0)$, $(\oplus_3)$ of 4.3.8 (actually, $(\oplus_3)$ is satisfied if interpreted "modulo $\sim_{\Sigma^*_{2.4.5}}$", but this makes no problems). The proof that $(K^*_{2.4.5},\Sigma^*_{2.4.5})$ is of the $\mathbf{AB}^+$–type follows exactly the lines of [**BsSh 242**, 2.6] (see [**BaJu95**, 7.4.20] too) and is left to the reader.

Consequently, the assertion (α) follows from 4.1.3(1), clause (β) is a consequence of 4.1.1 and (γ) follows from 4.3.8. To show (δ) one uses 2.2.6, or see 6.3.8.

Note, that if $\dot W$ is interpreted as a name for an infinite subset of ω, then

$$\Vdash_{\mathbb{Q}^*_{\rm s\infty}(K^*_{2.4.5},\Sigma^*_{2.4.5})}(\forall X\in[\omega]^\omega\cap\mathbf{V})(|\dot W\cap X|<\omega\ \text{ or }\ |\dot W\setminus X|<\omega).$$

Thus forcing with $\mathbb{Q}^*_{\rm s\infty}(K^*_{2.4.5},\Sigma^*_{2.4.5})$ makes ground model reals null too. □

Now we will present an application of forcing notions determined by omittory creating pairs with the weak Halving Properties to questions coming from localizing subsets of ω. These problems were studied in [**RoSh 501**] and our example is built in a manner similar to that of the forcing notion constructed in [**RoSh 501**, 2.4]. Moreover, all these examples are relatives of the forcing notion presented in [**Sh 207**]. The creating pair constructed there can be build like $(K^\psi_{4.4.2},\Sigma^\psi_{4.4.2})$ for $\psi\equiv 1$.

EXAMPLE 4.4.2. Let $\psi\in\omega^\omega$ be a non-decreasing function, $\psi(0)>0$. We construct a creating pair $(K^\psi_{4.4.2},\Sigma^\psi_{4.4.2})$ which:
(α) is strongly finitary, forgetful and omittory,
(β) has the weak Halving Property,
(γ) is saturated with respect to nice pre-norms with values in ω,
(δ) is condensed and satisfies the demands $(\otimes_1)$, $(\otimes_2)$ and $(\oplus_3)$ of 4.3.5 [thus, by 4.3.5, $(K^\psi_{4.4.2},\Sigma^\psi_{4.4.2})$ is of the $\mathbf{AB}$–type],
(ε) is anti-big and meagering.

CONSTRUCTION. Let $\mathbf{H}(m)=2$ for $m\in\omega$. First we describe which creatures $t\in\mathrm{CR}[\mathbf{H}]$ are taken to be in $K^\psi_{4.4.2}$. So, $t=(\mathbf{nor}[t],\mathbf{val}[t],\mathbf{dis}[t])\in K^\psi_{4.4.2}$ if:

- $\mathbf{dis}[t]=(T[t],L[t],R[t],D[t],\mathrm{NOR}[t],m^t_{\rm dn},m^t_{\rm up})$, where, letting $T=T[t]$, $L=L[t]$, $R=R[t]$, $D=D[t]$ and $\mathrm{NOR}=\mathrm{NOR}[t]$:
  (a) T is a finite tree, $D\subseteq\{\nu\in T:\mathrm{succ}_T(\nu)\neq\emptyset\}$ and
    (i) $(\forall\nu\in T\setminus D)(\mathrm{succ}_T(\nu)=\emptyset$ or $|\mathrm{succ}_T(\nu)|=\psi(L(\nu)))$,
    (ii) if $\nu\in T\setminus D$ and $\eta\in\mathrm{succ}_T(\nu)$ then either $\eta\in D$ or $\mathrm{succ}_T(\eta)=\emptyset$,

(b) $L, R : T \longrightarrow [m^t_{\rm dn}, m^t_{\rm up})$ are such that for each $\nu \in T$
   (i) $L(\nu) \leq R(\nu)$,
   (ii) if $\eta \in \mathrm{succ}_T(\nu)$ then $L(\nu) \leq L(\eta) \leq R(\eta) \leq R(\nu)$,
   (iii) if $\mathrm{succ}_T(\nu) = \emptyset$ then $L(\nu) = R(\nu)$,
   (iv) if $\eta_0, \eta_1 \in \mathrm{succ}_T(\nu)$, $\eta_0 \neq \eta_1$ then

$$[L(\eta_0), R(\eta_0)] \cap [L(\eta_1), R(\eta_1)] = \emptyset,$$

(c) NOR is a function on D such that for each $\nu \in D$, $\mathrm{NOR}(\nu)$ is a nice pre-norm on $\mathrm{succ}_T(\nu)$ with values in ω,

- if $\langle\rangle \notin D$ and $(\exists \nu \in \mathrm{succ}_T(\langle\rangle))(\mathrm{succ}_T(\nu) = \emptyset)$, or $D = \emptyset$ then $\mathbf{nor}[t] = 0$, otherwise $\mathbf{nor}[t] = \min\{\mathrm{NOR}(\nu)(\mathrm{succ}_T(\nu)) : \nu \in D\}$,
- $\mathbf{val}[t] = \{\langle u, v\rangle \in 2^{m^t_{\rm dn}} \times 2^{m^t_{\rm up}} : u \lhd v \ \&\ \{i \in [m^t_{\rm dn}, m^t_{\rm up}) : v(i) = 1\} \subseteq \{L(\nu) : \nu \in T \ \&\ \mathrm{succ}_T(\nu) = \emptyset\}\}$.

For $t \in K^{\psi}_{4.4.2}$ we define

$$\mathbf{nor}^0[t] = \begin{cases} \min\{\mathrm{NOR}[t](\nu)(\mathrm{succ}_{T[t]}(\nu)) : \nu \in D[t]\} & \text{if } D[t] \neq \emptyset, \\ 0 & \text{otherwise.} \end{cases}$$

Note that $\mathbf{nor}[t] \leq \mathbf{nor}^0[t]$ and in most cases they agree. One could use $\mathbf{nor}^0[t]$ as the norm of t and get the same forcing notion. We take $\mathbf{nor}[t]$ for technical reasons only. Now we are going to describe a composition operation $\Sigma^{\psi}_{4.4.2}$ on $K^{\psi}_{4.4.2}$ by giving basic operations which may be applied to creatures from $K^{\psi}_{4.4.2}$.

(1) For a creature $t \in K_{4.4.2}$ let $\mathrm{half}(t) \in K^{\psi}_{4.4.2}$ be such that

- if $\mathbf{nor}[t] < 2$ then $\mathrm{half}(t) = t$,
- if $\mathbf{nor}[t] \geq 2$ then $\mathbf{val}[\mathrm{half}(t)] = \mathbf{val}[t]$, $T[\mathrm{half}(t)] = T[t]$, $L[\mathrm{half}(t)] = L[t]$, $R[\mathrm{half}(t)] = R[t]$, $D[\mathrm{half}(t)] = D[t]$ and
  if $\nu \in D[\mathrm{half}(t)]$, $A \subseteq \mathrm{succ}_{T[\mathrm{half}(t)]}(\nu)$ then

$$\mathrm{NOR}[\mathrm{half}(t)](\nu)(A) = \max\{0, \mathrm{NOR}[t](\nu)(A) - \lfloor \frac{\mathbf{nor}[t]}{2} \rfloor\}.$$

[Thus $m^{\mathrm{half}(t)}_{\rm dn} = m^t_{\rm dn}$, $m^{\mathrm{half}(t)}_{\rm up} = m^t_{\rm up}$ and $\mathbf{nor}[\mathrm{half}(t)] = \mathbf{nor}[t] - \lfloor \frac{\mathbf{nor}[t]}{2} \rfloor$ when $\mathbf{nor}[t] \geq 2$.]

(2) For $t \in K^{\psi}_{4.4.2}$, $m_0 \leq m^t_{\rm dn}$, $m_1 \geq m^t_{\rm up}$ let $s = S_{m_0,m_1}(t) \in K^{\psi}_{4.4.2}$ be a creature such that $\mathbf{dis}[s] = (T[t], L[t], R[t], D[t], \mathrm{NOR}[t], m_0, m_1)$, $\mathbf{nor}[s] = \mathbf{nor}[t]$ and

$$\mathbf{val}[s] = \big\{\langle u, v\rangle \in 2^{m_0} \times 2^{m_1} : u \lhd v \ \&\ v{\restriction}[m_0, m^t_{\rm dn}) = \mathbf{0} \ \&\ v{\restriction}[m^t_{\rm up}, m_1) = \mathbf{0} \ \&\ \&\ \langle v{\restriction}m^t_{\rm dn}, v{\restriction}m^t_{\rm up}\rangle \in \mathbf{val}[t]\big\}.$$

Thus, essentially, $S_{m_0,m_1}(t) = t \restriction^{*} [m_0, m_1)$, the small difference in the definition of **dis** is immaterial.

(3) For $t \in K_{4.4.2}$ let $\Sigma^*_{4.4.2}(t)$ consist of all $s \in K^{\psi}_{4.4.2}$ such that
   (i) $m^s_{\rm dn} = m^t_{\rm dn}$, $m^s_{\rm up} = m^t_{\rm up}$, $T[s] \subseteq T[t]$, $D[s] = D[t] \cap T[s]$,
   (ii) $(\forall \nu \in T[s])(\mathrm{succ}_{T[s]}(\nu) = \emptyset \ \Leftrightarrow\ \mathrm{succ}_{T[t]}(\nu) = \emptyset)$ (thus if $\nu \in T[s] \setminus D[s]$ then $\mathrm{succ}_{T[s]}(\nu) = \mathrm{succ}_{T[t]}(\nu)$),
   (iii) $L[s] = L[t]{\restriction}T[s]$ and $R[s] = R[t]{\restriction}T[s]$,
   (iv) if $\nu \in D[s]$, $A \subseteq \mathrm{succ}_{T[s]}(\nu)$ then $\mathrm{NOR}[s](\nu)(A) = \mathrm{NOR}[t](\nu)(A)$.

(4) Suppose that $t_0, \ldots, t_{k-1} \in K^{\psi}_{4.4.2}$ are such that $k = \psi(L[t_0](\langle\rangle))$, $m^{t_i}_{\rm up} \leq m^{t_{i+1}}_{\rm dn}$ (for $i < k-1$) and

$$(\forall i < k)(\langle\rangle \in D[t_i] \ \text{ or } \ \mathrm{succ}_{T[t_i]}(\langle\rangle) = \emptyset).$$

Let $s = S^*(t_0,\dots,t_{k-1}) \in K^\psi_{4.4.2}$ be a creature such that

(i) $m^s_{\rm dn} = m^{t_0}_{\rm dn}$, $m^s_{\rm up} = m^{t_{k-1}}_{\rm up}$ and $T[s]$ is a tree such that $|{\rm succ}_{T[s]}(\langle\rangle)| = k$ and for every $\nu \in {\rm succ}_{T[s]}(\langle\rangle)$ there is a unique $i = i(\nu) < k$ such that

$$\{\eta \in T[s]\colon \nu \trianglelefteq \eta\} = \{\nu^\frown\eta^*\colon \eta^* \in T[t_i]\}\ \&\ L[s](\nu) = L[t_i](\langle\rangle)\ \&\ R[s](\nu) = R[t_i](\langle\rangle),$$

(ii) $D[s] = \{\nu^\frown\eta^* : \nu \in {\rm succ}_{T[s]}(\langle\rangle)\ \&\ \eta^* \in D[t_{i(\nu)}]\}$,

(iii) $L[s](\langle\rangle) = L[t_0](\langle\rangle)$, $R[s](\langle\rangle) = R[t_{k-1}](\langle\rangle)$ and for all $\nu \in {\rm succ}_{T[s]}(\langle\rangle)$

$$(\forall\eta^* \in T[t_{i(\nu)}])(L[s](\nu^\frown\eta^*) = L[t_{i(\nu)}](\eta^*)\ \&\ R[s](\nu^\frown\eta^*) = R[t_{i(\nu)}](\eta^*)),$$

(iv) if $\nu \in {\rm succ}_{T[s]}(\langle\rangle)$, $\eta^* \in D[t_{i(\nu)}]$ and $A \subseteq {\rm succ}_{T[t_{i(\nu)}]}(\eta^*)$ then

$${\rm NOR}[s](\nu^\frown\eta^*)(\{\nu^\frown\eta' : \eta' \in A\}) = {\rm NOR}[t_{i(\nu)}](\eta^*)(A).$$

(5) Suppose that $H : \mathcal{P}(m) \longrightarrow \omega$ is a nice pre-norm and $t_0,\dots,t_{m-1} \in K^\psi_{4.4.2}$ are such that $m^{t_i}_{\rm up} \le m^{t_{i+1}}_{\rm dn}$ for $i < m-1$. Let $s = S^{**}_H(t_0,\dots,t_{m-1}) \in K^\psi_{4.4.2}$ be a creature such that

(i) $m^s_{\rm dn} = m^{t_0}_{\rm dn}$, $m^s_{\rm up} = m^{t_{m-1}}_{\rm up}$ and $T[s]$ is a tree such that $|{\rm succ}_{T[s]}(\langle\rangle)| = m$ and for every $\nu \in {\rm succ}_{T[s]}(\langle\rangle)$ there is a unique $j = j(\nu) < m$ such that

$$\{\eta \in T[s]\colon \nu \trianglelefteq \eta\} = \{\nu^\frown\eta^*\colon \eta^* \in T[t_j]\}\ \&\ L[s](\nu) = L[t_j](\langle\rangle)\ \&\ R[s](\nu) = R[t_j](\langle\rangle),$$

(ii) $D[s] = \{\langle\rangle\} \cup \{\nu^\frown\eta^* : \nu \in {\rm succ}_{T[s]}(\langle\rangle)\ \&\ \eta^* \in D[t_{j(\nu)}]\}$,

(iii) $L[s](\langle\rangle) = L[t_0](\langle\rangle)$, $R[s](\langle\rangle) = R[t_{m-1}](\langle\rangle)$ and for every $\nu \in {\rm succ}_{T[s]}(\langle\rangle)$

$$(\forall\eta^* \in T[t_{j(\nu)}])(L[s](\nu^\frown\eta^*) = L[t_{j(\nu)}](\eta^*)\ \&\ R[s](\nu^\frown\eta^*) = R[t_{j(\nu)}](\eta^*)),$$

(iv) if $A \subseteq {\rm succ}_{T[s]}(\langle\rangle)$ then ${\rm NOR}[s](\langle\rangle)(A) = H(\{j(\nu) : \nu \in A\})$,

(v) if $\nu \in {\rm succ}_{T[s]}(\langle\rangle)$, $\eta^* \in D[t_{j(\nu)}]$ and $A \subseteq {\rm succ}_{T[t_{j(\nu)}]}(\eta^*)$ then

$${\rm NOR}[s](\nu^\frown\eta^*)(\{\nu^\frown\eta' : \eta' \in A\}) = {\rm NOR}[t_{j(\nu)}](\eta^*)(A).$$

Note that, under the respective assumptions, the procedures described in (1)–(5) above determine creatures in $K^\psi_{4.4.2}$, though (in cases (4) and (5)) not uniquely: there is some freedom in defining ${\rm succ}_{T[s]}(\langle\rangle)$. However, this freedom becomes irrelevant when we identify creatures that look the same. The last operation ($\Sigma^{**}_{4.4.2}$ below) is a way to describe which creatures are identified.

(6) For $t \in K^\psi_{4.4.2}$, let $\Sigma^{**}_{4.4.2}(t)$ consist of all creatures $s \in K^\psi_{4.4.2}$ such that $m^s_{\rm dn} = m^t_{\rm dn}$, $m^s_{\rm up} = m^t_{\rm up}$ and there is an (order) isomorphism $\pi : T[s] \longrightarrow T[t]$ which preserves L, R, D and NOR.

Finally, if $t_0,\dots,t_{m-1} \in K^\psi_{4.4.2}$ are such that $m^{t_i}_{\rm up} = m^{t_{i+1}}_{\rm dn}$ (for $i < m-1$) then $\Sigma^\psi_{4.4.2}(t_0,\dots,t_{m-1})$ consists of all creatures $s \in K^\psi_{4.4.2}$ such that $m^s_{\rm dn} = m^{t_0}_{\rm dn}$, $m^s_{\rm up} = m^{t_{m-1}}_{\rm up}$ and s may be obtained from $t_0,\dots,t_{m-1}$ by use of the operations half, S_{m_0,m_1}, S^*, S^{**}_H, $\Sigma^*_{4.4.2}$ and $\Sigma^{**}_{4.4.2}$ (with suitable parameters).

Let us check that $(K^\psi_{4.4.2}, \Sigma^\psi_{4.4.2})$ has the required properties. It should be clear that $(K^\psi_{4.4.2}, \Sigma^\psi_{4.4.2})$ is a finitary, forgetful and omittory creating pair. The relation $\sim_{\Sigma^\psi_{4.4.2}}$ (see 1.1.4(3)) is an equivalence relation on $K^\psi_{4.4.2}$ and $\Sigma^\psi_{4.4.2}$ depends on $\sim_{\Sigma^\psi_{4.4.2}}$–equivalence classes only (remember the definition of $\Sigma^{**}_{4.4.2}$; note that $\Sigma^\psi_{4.4.2}(t_0) = \Sigma^\psi_{4.4.2}(t_1)$ implies that t_0, t_1 are the same up to the isomorphism of the trees $T[t_0]$, $T[t_1]$). Thus the value of $\Sigma^\psi_{4.4.2}(t_0,\dots,t_{m-1})$ does not depend on the particular representation of the trees $T[t_i]$ (for $i < m$). Hence, if $t \in \Sigma^\psi_{4.4.2}(t_0,\dots,t_{m-1})$ then we may think that $T[t]$ is a tree built of $T[t_0],\dots,T[t_{m-1}]$

in the following sense. There are $s_i \in \Sigma^*_{4.4.2}(t_i)$ (for $i < m$), a front F of $T[t]$ and a one-to-one mapping $\varphi : F \longrightarrow m$ such that for $\nu \in F$

$$\{\eta \in T[t] : \nu \trianglelefteq \eta\} = \{\nu^\frown \eta' : \eta' \in T[s_{\varphi(\nu)}]\}$$

and the $L[t], R[t], D[t]$ above ν in $T[t]$, $\nu \in F$, look like $L[s_{\varphi(\nu)}], R[s_{\varphi(\nu)}], D[s_{\varphi(\nu)}]$ (but the norms given by NOR may be substantially different, still their values may be only smaller). Now it should be clear that $(K^\psi_{4.4.2}, \Sigma^\psi_{4.4.2})$ is strongly finitary. The pair $(K^\psi_{4.4.2}, \Sigma^\psi_{4.4.2})$ has the weak Halving Property as witnessed by the function half defined in (1) above. [Why? Note that if $t_0 \in K^\psi_{4.4.2}$, $\mathbf{nor}[t_0] \geq 2$, $t \in \Sigma^\psi_{4.4.2}(\mathrm{half}(t_0))$, $\mathbf{nor}[t] > 0$ (so $\mathbf{nor}[t] \geq 1$) then t is obtained from t_0 by alternate applications of half and shrinking (i.e. $\Sigma^*_{4.4.2}$). Look at the tree $T[t]$ with $L[t], R[t], D[t]$: necessarily the last three objects are the restrictions of $L[t_0], R[t_0], D[t_0]$ to $T[t]$. Let $s \in \Sigma^*_{4.4.2}(t_0)$ be such that $T[s] = T[t]$ (it should be clear clear that there is one; actually the s is uniquely determined by the tree $T[s]$). Since in the process of building t the norms were decreased only, and we started with $\mathrm{half}(t_0)$, we may conclude that $\mathbf{nor}[s] \geq \mathbf{nor}[t] + \lfloor \frac{1}{2}\mathbf{nor}[t_0] \rfloor \geq \frac{1}{2}\mathbf{nor}[t_0]$. Clearly $\mathbf{val}[s] = \mathbf{val}[t]$.]

Suppose that $H_0 : \mathcal{P}(\omega) \longrightarrow \omega$ is a nice pre-norm, $d \in D_{H_0}$ (see 4.2.4). Note that $\Sigma^{\mathrm{sum}}_{d,u_d}(t_i : n_0 \leq i < n_1)$ (for $\langle t_i : n_0 \leq i < n_1 \rangle \in \mathrm{PFC}(K^\psi_{4.4.2}, \Sigma^\psi_{4.4.2})$, $u_d \subseteq [n_0, n_1)$) corresponds to a creature t obtained from $t_{n_0}, \ldots, t_{n_1 - 1}$ by suitable applications of S_{m_0,m_1} and S^{**}_H (the last for the pre-norm $H = H_0 \restriction \mathcal{P}(u_d)$, the first is to omit creatures t_i for $i \in [n_0, n_1) \setminus u_d$). The difference is in **dis**, but this causes no real problem as we may read $H_0 \restriction \mathcal{P}(u_d)$ from the resulting creature (and it is essentially the $\mathrm{NOR}[t](\langle\rangle)$). We could have changed the definition of $(K^\psi_{4.4.2}, \Sigma^\psi_{4.4.2})$ to make this correspondence more literal, but that would result in unnecessary complications in the definition. Now note that

if $t \in \Sigma^\psi_{4.4.2}(\Sigma^{\mathrm{sum}}_{d,u_d}(t_0, \ldots, t_{m-1}))$, $\mathbf{nor}[t] > 0$ and $d^* \in D_{H_0}$ is such that $u_{d^*} \approx \mathrm{succ}_{T[t]}(\langle\rangle)$ and $s_i \in \Sigma^*_{4.4.2}(t_i)$ for $i < m$ are such that

$$\nu \in \mathrm{succ}_{T[t]}(\langle\rangle) \quad \Rightarrow \quad \{\nu^\frown \eta' : \eta' \in T[s_{j(\nu)}]\} = \{\eta \in T[t] : \nu \trianglelefteq \eta\},$$

then $\mathbf{val}[t] = \mathbf{val}[\Sigma^{\mathrm{sum}}_{d^*,u_{d^*}}(s_0, \ldots, s_{m-1})]$. Moreover, if $\mathbf{nor}[t_i] > 0$ for $i < m$ then $\mathbf{nor}[s_i] > 0$ for $i \in u_{d^*}$.

This shows that $(K^\psi_{4.4.2}, \Sigma^\psi_{4.4.2})$ is saturated with respect to nice pre-norms with values in ω.

Suppose that $t \in \Sigma^\psi_{4.4.2}(t_0, \ldots, t_{n-1})$, $\mathbf{nor}[t] > 0$, $\mathbf{nor}[t_i] > 0$ (for $i < n$). Then for some $i < n$ and $\nu \in T[t]$ we find $s \in \Sigma^*_{4.4.2}(t_i)$ such that

$$\{\eta \in T[t] : \nu \trianglelefteq \eta\} = \{\nu^\frown \eta' : \eta' \in T[s]\}$$

and $L[t], R[t]$ and $D[t]$ above ν are like $L[s], R[s], D[s]$, but remember that $\mathrm{NOR}[t]$ may have nothing in common with $\mathrm{NOR}[s]$: the operation half may be involved. However, by the definition of $\mathbf{nor}[t]$, we know that if $\eta' \in D[s]$ then

$$\mathrm{NOR}[t](\nu^\frown \eta')(\mathrm{succ}_{T[t]}(\nu^\frown \eta')) > 0,$$

and this is enough to conclude that $\mathrm{NOR}[s](\eta')(\mathrm{succ}_{T[s]}(\eta')) > 0$. Moreover $D[s] \neq \emptyset$ as $\mathbf{nor}[t_i] > 0$. Consequently $\mathbf{nor}[s] > 0$. Since

$$v \restriction [m^{t_0}_{\mathrm{dn}}, m^s_{\mathrm{dn}}) = \mathbf{0} \ \& \ \langle v, u \rangle \in \mathbf{val}[s] \quad \Rightarrow \quad (\exists u^*)(\langle v, u^* \rangle \in \mathbf{val}[t] \ \& \ u \trianglelefteq u^*)$$

we conclude that $(K^\psi_{4.4.2}, \Sigma^\psi_{4.4.2})$ is condensed.

One easily checks that the demands ($\otimes_1$), ($\otimes_2$) and ($\oplus_3$) of 4.3.5 are satisfied (remember that $t \restriction [m_0, m_1)$ is, basically, $S_{m_0,m_1}(t)$).

Finally, let us check that $(K^{\psi}_{4.4.2}, \Sigma^{\psi}_{4.4.2})$ is meagering and anti-big. For the first note that if $t \in K^{\psi}_{4.4.2}$, $\mathbf{nor}[t] > 2$ and $\nu \in T[t]$ is a maximal node of $T[t]$ then there is $s \in \Sigma^*_{4.4.2}(t)$ such that $\nu \notin T[s]$ and $\mathbf{nor}[s] \geq \mathbf{nor}[t] - 1$. [Why? Take the shortest $\eta \lhd \nu$ such that $\eta \in D[t]$ – there must be one as $\mathbf{nor}[t] > 0$ – and choose $s \in \Sigma^*_{4.4.2}$ so that $T[s] = \{\rho \in T[t] : \eta \lhd \rho \ \Rightarrow\ \neg(\nu^* \unlhd \rho)\}$, where $\nu^* \in \mathrm{succ}_{T[t]}(\eta)$ is such that $\nu^* \unlhd \nu$. Since $\mathrm{NOR}[t](\eta)$ is a nice pre-norm we have

$$\mathrm{NOR}[t](\eta)(\mathrm{succ}_{T[s]}(\eta)) \geq \mathrm{NOR}[t](\eta)(\mathrm{succ}_{T[t]}(\eta)) - 1$$

and hence s is as required.] Now suppose that $(t_0, \dots, t_{n-1}) \in \mathrm{PFC}(K^{\psi}_{4.4.2}, \Sigma^{\psi}_{4.4.2})$, $\mathbf{nor}[t_i] > 2$, $k_i \in [m^{t_i}_{\mathrm{dn}}, m^{t_i}_{\mathrm{up}})$ and $t \in \Sigma^{\psi}_{4.4.2}(t_0, \dots, t_{n-1})$, $\mathbf{nor}[t] > 2$. Then there is a front F of $T[t]$ such that for every $\nu \in F$, for a unique $i = i(\nu) < n$ and some $s \in \Sigma^*_{4.4.2}(t_i)$:

$$\{\eta \in T[t] : \nu \unlhd \eta\} = \{\nu^\frown\eta^* : \eta^* \in T[s_i]\}$$

and $D[t], L[t], R[t]$ above ν look like $D[s_i], L[s_i], R[s_i]$ (but $\mathrm{NOR}[t]$ might be different than that of s_i: the values may be smaller). Now apply the previous remark to choose $s^*_i \in \Sigma^*_{4.4.2}(s_i)$ such that $\mathbf{nor}[s^*_i] \geq \mathbf{nor}[s_i] - 1$ and $k_i \notin \{L[s^*_i](\eta) : \eta \in T[s^*_i]\ \&\ \mathrm{succ}_{T[s^*_i]}(\eta) = \emptyset\}$. Finally let $s \in \Sigma^*_{4.4.2}(t)$ be such that $F \subseteq T[s]$ and for each $\nu \in F$

$$\{\eta \in T[s] : \nu \unlhd \eta\} = \{\nu^\frown\eta^* : \eta^* \in T[s^*_{i(\nu)}]\}.$$

It is easy to check that this determines a creature s as required in 4.1.2(1) for $\langle k_i, t_i : i < n\rangle, t$.

To verify that $(K^{\psi}_{4.4.2}, \Sigma^{\psi}_{4.4.2})$ is anti-big define colourings

$$c_t : \bigcup_{u \in \mathbf{basis}(t)} \mathrm{pos}(u, t) \longrightarrow 3$$

for $t \in K^{\psi}_{4.4.2}$ by:

$$c_t(v) = \begin{cases} 0 & \text{if } |\{k \in [m^t_{\mathrm{dn}}, m^t_{\mathrm{up}}) : v(k) = 1\}| \text{ is even } > 0, \\ 1 & \text{if } |\{k \in [m^t_{\mathrm{dn}}, m^t_{\mathrm{up}}) : v(k) = 1\}| \text{ is odd }, \\ 2 & \text{if } v \restriction [m^t_{\mathrm{dn}}, m^t_{\mathrm{up}}) = \mathbf{0}_{[m^t_{\mathrm{dn}}, m^t_{\mathrm{up}})}. \end{cases}$$

Suppose $(t_0, \dots, t_{n-1}) \in \mathrm{PFC}(K^{\psi}_{4.4.2}, \Sigma^{\psi}_{4.4.2})$, $t \in \Sigma^{\psi}_{4.4.2}(t_0, \dots, t_{n-1})$, $\mathbf{nor}[t_i] > 1$ and $\mathbf{nor}[t] > 1$. Clearly for some $i < n$

$$|\{L[t](\eta) : L[t_i](\langle\rangle) \leq L[t](\eta) \leq R[t_i](\langle\rangle) \text{ and } \mathrm{succ}_{T[t]}(\eta) = \emptyset\}| \geq 2.$$

Let $u \in \mathbf{basis}(t_0) = \prod\limits_{i < m^t_{\mathrm{dn}}} 2$. Take $v_0, v_1 \in \prod\limits_{i < m^t_{\mathrm{up}}} 2$ such that for $\ell = 0, 1$

$$\begin{array}{l} u \lhd v, \qquad |\{k \in [m^t_{\mathrm{dn}}, m^t_{\mathrm{up}}) : v_\ell(k) = 1\}| = \ell + 1, \qquad \text{and} \\ \{k \in [m^t_{\mathrm{dn}}, m^t_{\mathrm{up}}) : v_\ell(k) = 1\} \quad \text{is contained in} \\ \{L[t](\eta) : L[t_i](\langle\rangle) \leq L[t](\eta) \leq R[t_i](\langle\rangle)\ \&\ \eta \in T[t]\ \&\ \mathrm{succ}_{T[t]}(\eta) = \emptyset\}. \end{array}$$

Now check that the v_0, v_1 are as required in 4.1.2(2) for $\langle t_i : i < n\rangle, t$. □

COROLLARY 4.4.3. *$\mathbb{Q}^*_{\mathrm{s}\infty}(K^{\psi}_{4.4.2}, \Sigma^{\psi}_{4.4.2})$ is a proper almost ω^ω-bounding forcing notion which makes ground reals meager and adds a Cohen real.*

PROOF. By 4.4.2, 4.3.5 and 4.1.3. □

DEFINITION 4.4.4. 1. For an infinite set $X \in [\omega]^\omega$ let $\mu_X : \omega \longrightarrow X$ be the increasing enumeration of X.

2. Let $\psi \in \omega^\omega$ be non-decreasing. We define relations $S^\psi_{\rm dn}, S^\psi_{\rm up} \subseteq [\omega]^\omega \times [\omega]^\omega$ by

$(X,Y) \in S^\psi_{\rm dn}$ if and only if
$(\exists^\infty n \in \omega)(\forall i < \psi(\mu_Y(n)))(|[\mu_Y(n+i), \mu_Y(n+i+1)) \cap X| \geq 2)$,
$(X,Y) \in S^\psi_{\rm up}$ if and only if
$(\exists^\infty n \in \omega)(\forall i < \psi(\mu_Y(n+1)))(|[\mu_Y(n+i), \mu_Y(n+i+1)) \cap X| \geq 2)$.

If ψ is a constant function, say $\psi \equiv k$, then S^ψ_* may be called S_k.

REMARK 4.4.5. We will consider the notions of $S^\psi_{\rm dn}$– and $S^\psi_{\rm up}$–localizations as given by 0.2.2 for these relations as well as the corresponding dominating numbers $\mathfrak{d}(S^\psi_{\rm dn})$ and $\mathfrak{d}(S^\psi_{\rm up})$. Note that for a non-decreasing $\psi \in \omega^\omega$, $S^\psi_{\rm up}$–localization implies $S^\psi_{\rm dn}$–localization (and $\mathfrak{d}(S^\psi_{\rm dn}) \leq \mathfrak{d}(S^\psi_{\rm up})$). If $(\forall^\infty n \in \omega)(\psi(n) \leq \varphi(n))$ then S^φ_*–localization implies S^ψ_*–localization and for eventually constant ψ, $S^\psi_{\rm up}$–localization is the same as $S^\psi_{\rm dn}$–localization.

PROPOSITION 4.4.6. *Suppose $\varphi, \psi \in \omega^\omega$ are non-decreasing functions such that $(\forall^\infty n \in \omega)(1 \leq \psi(n) < \varphi(n))$. Then the forcing notion $\mathbb{Q}^*_{\rm s\infty}(K^\psi_{4.4.2}, \Sigma^\psi_{4.4.2})$ does not have the $S^\varphi_{\rm up}$–localization property.*

PROOF. This is parallel to [**RoSh 501**, 2.4.3]. The main step is done by the following claim, which is essentially a repetition of [**RoSh 501**, 2.4.2].

CLAIM 4.4.6.1. *Suppose that $t \in K^\psi_{4.4.2}$ is such that $\mathbf{nor}^0[t] \geq 11$. Let $Y \in [\omega]^\omega$. Then there is $s \in \Sigma^*_{4.4.2}(t)$ such that $\mathbf{nor}^0[s] \geq \mathbf{nor}^0[t] - 10$ and for every $n \in \omega$ there is $i \leq \psi(\mu_Y(n+1))$ such that*

$$\{L(s)(\nu) : \nu \in T[s] \ \&\ \mathrm{succ}_{T[s]}(\nu) = \emptyset\} \cap [\mu_Y(n+i), \mu_Y(n+i+1)) = \emptyset.$$

Proof of the claim: The proof is by induction on the height of the tree $T[t]$. One could try just to apply the inductive hypothesis to creatures determined by $\{\eta \in T[t] : \nu \trianglelefteq \eta\}$ for each $\nu \in \mathrm{succ}_{T[t]}(\langle\rangle)$. However, this would not be enough. What we need to do is to shrink t to separate the sets

$$\{L[t](\eta) : \nu \trianglelefteq \eta \ \&\ \eta \in T[t] \ \&\ \mathrm{succ}_{T[t]}(\eta) = \emptyset\}$$

for distinct $\nu \in \mathrm{succ}_{T[t]}(\langle\rangle)$ by intervals $[\mu_Y(m), \mu_Y(m+1))$. This will prevent "bad events" occurring above distinct $\nu \in \mathrm{succ}_{T[t]}(\langle\rangle)$ from accumulating. The shrinking procedure depends on the character of $\langle\rangle$ in $T[t]$, so the arguments brake into two cases.

CASE 1: $\langle\rangle \in D[t]$.
Let

$$A_0 = \{\nu \in \mathrm{succ}_{T[t]}(\langle\rangle) : (\exists m \in \omega)(\mu_Y(2m) \leq L[t](\nu) \leq R[t](\nu) < \mu_Y(2m+1))\},$$
$$A_1 = \{\nu \in \mathrm{succ}_{T[t]}(\langle\rangle) : (\exists m \in \omega)(\mu_Y(2m-1) \leq L[t](\nu) \leq R[t](\nu) < \mu_Y(2m))\},$$
$$A_2 = \{\nu \in \mathrm{succ}_{T[t]}(\langle\rangle) : (L[t](\nu), R[t](\nu)] \cap Y \neq \emptyset\}.$$

Note that $A_0 \cup A_1 \cup A_2 = \mathrm{succ}_{T[t]}(\langle\rangle)$ and $\mathrm{NOR}[t](\langle\rangle)(\mathrm{succ}_{T[t]}(\langle\rangle)) \geq \mathbf{nor}^0[t]$. As $\mathrm{NOR}[t](\langle\rangle)$ is a nice pre-norm, at least one of the following holds:

(0) $\mathrm{NOR}[t](\langle\rangle)(A_0) \geq \mathbf{nor}^0[t] - 2$,
(1) $\mathrm{NOR}[t](\langle\rangle)(A_1) \geq \mathbf{nor}^0[t] - 2$,

(2) $\mathrm{NOR}[t](\langle\rangle)(A_2) \geq \mathbf{nor}^0[t] - 2$.

Suppose that (0) holds. Let $s \in \Sigma^*_{4.4.2}(t)$ be a creature such that

$$T[s] = \{\eta \in T[t] : (\exists \nu \in A_0)(\eta \trianglelefteq \nu \text{ or } \nu \triangleleft \eta)\}.$$

[It should be clear that this uniquely defines the creature s and $\mathbf{nor}^0[s] \geq \mathbf{nor}^0[t] - 2$.] Note that by the definition of the set A_0, if $\nu_1, \nu_2 \in A_0$ are distinct and $R[t](\nu_1) \leq L[t](\nu_2)$ then there are $m_1 < m_2 < \omega$ such that

$$\mu_Y(2m_1) \leq L[s](\nu_1) \leq R[s](\nu_1) < \mu_Y(2m_1+1) < \mu_Y(2m_2) \leq L[s](\nu_2) \leq R[s](\nu_2) < \mu_Y(2m_2+1).$$

Hence we may conclude that for every $n \in \omega$ there is $i < 2$ such that

$$\{L[s](\eta) : \eta \in T[s] \ \&\ \mathrm{succ}_{T[s]}(\eta) = \emptyset\} \cap [\mu_Y(n+i), \mu_Y(n+i+1)) = \emptyset,$$

and thus s is as required in the assertion of the claim.

If (1) holds then we may proceed in the same manner (considering the set A_1).

So suppose that (2) holds true. Divide the set A_2 into three sets A_2^0, A_2^1, A_2^2 such that for each $i < 3$ and $\nu_1, \nu_2 \in A_2^i$ with $R[t](\nu_1) < L[t](\nu_2)$, there is $m \in \omega$ such that

$$R[t](\nu_1) < \mu_Y(m) < \mu_Y(m+1) < L[t](\nu_2)$$

[e.g. each A_2^i contains every third element of A_2 counting according to the values of $L[t](\nu)$]. For some $i < 3$ we have

$$\mathrm{NOR}[t](\langle\rangle)(A_2^i) \geq \mathrm{NOR}[t](\langle\rangle)(A_2) - 2 \geq \mathbf{nor}^0[t] - 4.$$

Fix $\nu \in A_2^i$ for a moment and let $T_\nu = \{\eta \in T[t] : \nu \triangleleft \eta \text{ or } \eta \trianglelefteq \nu\}$. If $T_\nu \cap D[t] = \{\langle\rangle\}$ then necessarily T_ν does not contain sequences of length > 2 (remember clause **(a)(ii)** of the definition of $K^\psi_{4.4.2}$) and

$$|\{\eta \in T_\nu : \mathrm{succ}_{T[t]}(\eta) = \emptyset\}| \in \{1, \psi(L[t](\nu))\}.$$

Let $n^* \in \omega$ be maximal such that $\mu_Y(n^*) \leq L[t](\nu)$ (if there is no such n^* the we let $n^* = -1$). Then, for every $n \geq n^*$, $\psi(\mu_Y(n+1)) \geq \psi(\mu_Y(n^*+1)) \geq \psi(L[t](\nu))$ (as ψ is non-decreasing). Hence, letting $T^*_\nu = T_\nu$ we have that for each $n \in \omega$ there is $i \leq \psi(\mu_Y(n+1))$ such that

$$\{L[t](\eta) : \eta \in T^*_\nu \ \&\ \mathrm{succ}_{T[t]}(\eta) = \emptyset\} \cap [\mu_Y(n+i), \mu_Y(n+i+1)) = \emptyset.$$

If $T_\nu \cap D[t] \neq \{\langle\rangle\}$ then we may look at a creature $t^* \in K^\psi_{4.4.2}$ such that $T[t^*] = \{\eta' : \nu^\frown\eta' \in T[t]\}$ and $L[t^*], R[t^*], D[t^*], \mathrm{NOR}[t^*]$ are copied in a suitable manner from t (so they are restrictions of the corresponding objects to T_ν) and $m^{t^*}_{\mathrm{dn}} = m^t_{\mathrm{dn}}$, $m^{t^*}_{\mathrm{up}} = m^t_{\mathrm{up}}$. Clearly $\mathbf{nor}^0[t^*] \geq \mathbf{nor}^0[t]$ and the height of $T[t^*]$ is smaller than that of $T[t]$. Thus we may apply the inductive hypothesis and we find $s^* \in \Sigma^*_{4.4.2}(t^*)$ such that $\mathbf{nor}^0[s^*] \geq \mathbf{nor}^0[t^*] - 10 \geq \mathbf{nor}^0[t] - 10$ and for all $n \in \omega$ there is $i \leq \psi(\mu_Y(n+1))$ such that

$$\{L[s^*](\eta) : \eta \in T[s^*] \ \&\ \mathrm{succ}_{T[s^*]}(\eta) = \emptyset\} \cap [\mu_Y(n+i), \mu_Y(n+i+1)) = \emptyset.$$

Let $T^*_\nu = \{\nu^\frown\eta' : \eta' \in T[s^*]\}$.

Look at the tree $T^* = \bigcup\limits_{\nu \in A_2^i} T^*_\nu$. It determines a creature $s \in \Sigma^*_{4.4.2}(t)$ (i.e. s is such that $T[s] = T^*$, $D[s] = D[t] \cap T^*$ etc). Clearly $\mathbf{nor}^0[s] \geq \mathbf{nor}^0[t] - 10$ and it is easy to check that s is as required (remember the choice of the A_2^i and the T^*_ν's).

CASE 2: $\langle\rangle \notin D[t]$.
Since $T[t] \neq \{\langle\rangle\}$ (as $\mathbf{nor}^0[t] > 0$) we have $|\mathrm{succ}_{T[t]}(\langle\rangle)| = \psi(L[t](\langle\rangle))$. Moreover, for each $\nu \in \mathrm{succ}_{T[t]}(\langle\rangle)$

$$\text{either} \quad \nu \in D[t] \quad \text{or} \quad \mathrm{succ}_{T[t]}(\nu) = \emptyset$$

(remember the demand **(a)(ii)** of the definition of $K^{\psi}_{4.4.2}$). Note that necessarily $D[t] \cap \mathrm{succ}_{T[t]}(\langle\rangle) \neq \emptyset$ (as $\mathbf{nor}^0[t] > 0$). Fix $\nu \in D[t] \cap \mathrm{succ}_{T[t]}(\langle\rangle)$.
Choose a set $A_\nu \subseteq \mathrm{succ}_{T[t]}(\nu)$ such that $\mathrm{NOR}[t](\nu)(A_\nu) \geq \mathbf{nor}^0[t] - 5$ and one of the following holds:

- $(\exists m \in \omega)(\forall \eta \in A_\nu)(\mu_Y(m) \leq L[t](\eta) \leq R[t](\eta) < \mu_Y(m+1))$,
- there are $m_0 < m_1 < m_2 < m_3 < \omega$ such that

$$\begin{array}{ll} (\forall \eta \in A_\nu)(\mu_Y(m_1) \leq L[t](\eta) \leq R[t](\eta) < \mu_Y(m_2)) & \text{and} \\ (\exists \eta \in \mathrm{succ}_{T[t]}(\nu))(L[t](\eta) \leq \mu_Y(m_0)) & \text{and} \\ (\exists \eta \in \mathrm{succ}_{T[t]}(\nu))(R[t](\eta) \geq \mu_Y(m_3)). & \end{array}$$

Why is the choice of A_ν possible? For $m \in \omega$ let

$$B_0^m = \{\eta \in \mathrm{succ}_{T[t]}(\nu) : \mu_Y(m) \leq L[t](\eta) \leq R[t](\eta) < \mu_Y(m+1)\}.$$

If there is m such that $\mathrm{NOR}[t](\langle\rangle)(B_0^m) \geq \mathbf{nor}^0[t] - 5$ then we may take the respective B_0^m as A_ν. So suppose that

$$(\forall m \in \omega)(\mathrm{NOR}[t](\langle\rangle)(B_0^m) < \mathbf{nor}^0[t] - 5).$$

Let $B_0 = \bigcup\limits_{m \in \omega} B_0^m$ and suppose that $\mathrm{NOR}[t](\nu)(B_0) \geq \mathbf{nor}^0[t] - 1$. Let k_0, k_1 be the two smallest elements of $\{m : B_0^m \neq \emptyset\}$ and let k_2, k_3 be the two largest elements of this set (note that $|\{m : B_0^m \neq \emptyset\}| \geq 6$; remember that $\mathrm{NOR}[t](\nu)$ is a nice pre-norm). We let

$$A_\nu = \{\eta \in \mathrm{succ}_{T[t]}(\nu) : \mu_Y(k_1 + 1) \leq L[t](\eta) \leq R[t](\eta) < \mu_Y(k_2)\},$$

and $m_0 = k_0 + 1$, $m_1 = k_1 + 1$, $m_2 = k_2$, $m_3 = k_3$. Easily $\mathrm{NOR}[t](\nu)(A_\nu) \geq \mathrm{NOR}[t](\nu)(B_0) - 4 \geq \mathbf{nor}^0[t] - 5$ and since $B_0^{k_0} \neq \emptyset$, $B_0^{k_3} \neq \emptyset$ we see that A_ν is as required. So we are left with the possibility that $\mathrm{NOR}[t](\nu)(B_0) < \mathbf{nor}^0[t] - 1$. In this case we have

$$\mathrm{NOR}[t](\nu)(\{\eta \in \mathrm{succ}_{T[t]}(\nu) : (L[t](\eta), R[t](\eta)] \cap Y \neq \emptyset\}) \geq \mathbf{nor}^0[t] - 1.$$

Let $\eta_0, \eta_1, \eta_2, \eta_3 \in \mathrm{succ}_{T[t]}(\nu) \setminus B_0$ be such that $L[t](\eta_0) < L[t](\eta_1)$ are the first two members of $\{L[t](\eta) : \eta \in \mathrm{succ}_{T[t]}(\nu) \setminus B_0\}$ and $L[t](\eta_2) < L[t](\eta_3)$ are the last two members of this set. Let m_i be such that $\mu_Y(m_i) \in (L[t](\eta_i), R[t](\eta_i)]$ (for $i < 4$) and let

$$A_\nu = \{\eta \in \mathrm{succ}_{T[t]}(\nu) \setminus B_0 : \mu_Y(m_1) \leq L[t](\eta) \leq R[t](\eta) < \mu_Y(m_2)\}.$$

Since

$$\mathrm{NOR}[t](\nu)(\mathrm{succ}_{T[t]}(\nu) \setminus B_0 \setminus \{\eta_0, \eta_1, \eta_2, \eta_3\}) \geq \mathbf{nor}^0[t] - 5,$$

one easily checks that A_ν is as required.

Let $T^\nu = \{\eta^* : \nu^\frown\eta^* \in T[t] \ \&\ (\exists \eta \in A_\nu)(\eta \vartriangleleft \nu^\frown\eta^* \text{ or } \nu^\frown\eta^* \trianglelefteq \eta)\}$. We would like to apply the inductive hypothesis to the creature determined by T^ν (with L, R, D and NOR copied in a suitable way from t). However, this creature may have too small norm: it may happen that $\mathrm{NOR}[t](\nu)(A_\nu) < 11$. But we may repeat the procedure of CASE 1, noticing that the inductive hypothesis was applied there above some elements of $\mathrm{succ}(\langle\rangle)$. Here, this corresponds to applying the inductive hypothesis to creatures determined by $\{\eta^* : \eta^\frown\eta^* \in T[t]\}$ for some $\eta \in A_\nu$, and these creatures have norms not smaller than $\mathbf{nor}^0[t]$. Consequently we will get a tree

$$T_*^\nu \subseteq \{\eta' : \nu \lhd \eta' \ \& \ (\exists \eta \in A_\nu)(\eta \unlhd \eta')\}$$

corresponding to a creature s_ν with $\mathbf{nor}^0[s_\nu] \geq \mathbf{nor}^0[t] - 10$ and such that for each $n \in \omega$, for some $i \leq \psi(\mu_Y(n+1))$ we have

$$\{L[t](\eta') : \eta' \in T_*^\nu \ \& \ \mathrm{succ}_{T[t]}(\eta') = \emptyset\} \cap [\mu_Y(n+i), \mu_Y(n+i+1)) = \emptyset.$$

[Note that the procedure of CASE 1 may involve further shrinking of A_ν and dropping the norm by 4. Still, $5+4 < 10$ so the norm of s_ν is above $\mathbf{nor}^0[t] - 10$.] Next let

$$T^* = \bigcup \{T_*^\nu : \nu \in \mathrm{succ}_{T[t]}(\langle\rangle) \cap D[t]\} \cup \mathrm{succ}_{T[t]}(\langle\rangle) \cup \{\langle\rangle\},$$

and let s be the restriction of the creature t to T^*. Check that $s \in \Sigma^*_{4.4.2}(t)$ is as required. This finishes the proof of the claim.

Now we may prove the proposition. We are going to show that, in the model $\mathbf{V}^{\mathbb{Q}^*_{s\infty}(K^\psi_{4.4.2}, \Sigma^\psi_{4.4.2})}$, the set $\{m \in \omega : \dot{W}(m) = 1\}$ witnesses that the S^φ_{up}–localization fails. So suppose that $p = (w, t_0, t_1, t_2 \dots) \in \mathbb{Q}^*_{s\infty}(K^\psi_{4.4.2}, \Sigma^\psi_{4.4.2})$ and $Y \in [\omega]^\omega$. Since $(K^\psi_{4.4.2}, \Sigma^\psi_{4.4.2})$ is omittory we may assume that $\psi(k) < \varphi(k)$ for $k \geq \ell g(w)$ and for each $i \in \omega$

(α) $\mathbf{nor}[t_i] \geq 11 + i + m^{t_i}_{\mathrm{dn}}$, $\langle\rangle \in D[t_i]$ and

(β) $|(R[t_i](\langle\rangle), L[t_{i+1}](\langle\rangle)) \cap Y| > 2$, $|(\ell g(w), L[t_0]) \cap Y| > 2$.

Apply 4.4.6.1 to get $s_i \in \Sigma^*_{4.4.2}(t_i)$ such that $\mathbf{nor}^0[s_i] \geq \mathbf{nor}^0[t_i] - 10$ and for every $n \in \omega$ there is $j \leq \psi(\mu_Y(n+1))$ such that

$$\{L[s_i](\nu) : \nu \in T[s_i] \quad \& \quad \mathrm{succ}_{T[s_i]}(\nu) = \emptyset\} \cap [\mu_Y(n+j), \mu_Y(n+j+1)) = \emptyset.$$

Note that by (α) and the definition of $\Sigma^*_{4.4.2}$ and $\mathbf{nor}^0$ we have

$$\mathbf{nor}^0[t_i] = \mathbf{nor}[t_i], \quad \mathbf{nor}^0[s_i] = \mathbf{nor}[s_i].$$

Hence, letting $q = (w, s_0, s_1, s_2, \dots)$ we will have $p \leq q \in \mathbb{Q}^*_{s\infty}(K^\psi_{4.4.2}, \Sigma^\psi_{4.4.2})$ and for every $n > \ell g(w)$

$$q \Vdash (\exists j < \varphi(\mu_Y(n+1)))([\mu_Y(n+j), \mu_Y(n+j+1)) \cap \{m \in \omega : \dot{W}(m) = 1\} = \emptyset);$$

remember that q forces that $\{m \in \omega : \dot{W}(m) = 1\}$ is a subset of

$$\{m < m^{s_0}_{\mathrm{dn}} : w(m) = 1\} \cup \{L[s_i](\nu) : i < \omega \ \& \ \nu \in T[s_i] \ \& \ \mathrm{succ}_{T[s_i]}(\nu) = \emptyset\}.$$

This finishes the proof. □

PROPOSITION 4.4.7. *Let $\varphi, \psi \in \omega^\omega$ be non-decreasing, $\varphi(0), \psi(0) > 0$ and $\lim\limits_{n\to\infty} \varphi(n) \leq \lim\limits_{n\to\infty} \psi(n)$. Suppose that $N \prec (\mathcal{H}(\chi), \in, <^*_\chi)$ is countable, $p \in N \cap \mathbb{Q}^*_{s\infty}(K^\psi_{4.4.2}, \Sigma^\psi_{4.4.2})$, $\varphi, \psi \in N$ and $Y \in [\omega]^\omega$ is such that*

$$(\forall X \in [\omega]^\omega \cap N)(\exists^\infty n \in \omega)(\forall i < \varphi(\mu_Y(n)))(|[\mu_Y(n+i), \mu_Y(n+i+1)) \cap X| \geq 2).$$

Then there is an $(N, \mathbb{Q}^*_{s\infty}(K^\psi_{4.4.2}, \Sigma^\psi_{4.4.2}))$–generic condition q stronger than p and such that

$$q \Vdash \text{“for every } X \in [\omega]^\omega \cap N[\Gamma_{\mathbb{Q}^*_{s\infty}(K^\psi_{4.4.2}, \Sigma^\psi_{4.4.2})}], \\ (\exists^\infty n \in \omega)(\forall i < \varphi(\mu_Y(n)))(|[\mu_Y(n+i), \mu_Y(n+i+1)) \cap X| \geq 2)\text{”}.$$

Consequently, if ψ is unbounded then the forcing notion $\mathbb{Q}^*_{s\infty}(K^\psi_{4.4.2},\Sigma^\psi_{4.4.2})$ has the S^φ_{dn}-localization property for every non-decreasing φ. If ψ is bounded, say $\lim\limits_{n\to\infty}\psi(n)=k$, then $\mathbb{Q}^*_{s\infty}(K^\psi_{4.4.2},\Sigma^\psi_{4.4.2})$ has the S_k-localization property.

PROOF. This is like [**RoSh 501**, 2.4.5]. We will deal with the case $\lim\limits_{n\to\infty}\psi(n)=\infty$ (if ψ is bounded then the arguments are similar).

Suppose that φ,ψ,N,p,Y are as in the assumptions. Let $\langle\dot\sigma_n : n<\omega\rangle$ enumerate all $\mathbb{Q}^*_{s\infty}(K^\psi_{4.4.2},\Sigma^\psi_{4.4.2})$–names from N for ordinals and let $\langle\dot X_n : n<\omega\rangle$ list all names for infinite subsets of ω. Further, for $n\in\omega$, let $\dot\tau_n\in N$ be a name for a function in ω^ω such that

$$\Vdash_{\mathbb{Q}^*_{s\infty}(K^\psi_{4.4.2},\Sigma^\psi_{4.4.2})} (\forall m\in\omega)(\forall i\leq n)(|[m,\dot\tau_n(m))\cap \dot X_i|>2).$$

We inductively construct sequences $\langle p_n,p^*_n : n<\omega\rangle$, $\langle g_n : n<\omega\rangle$ and $\langle i_n : n<\omega\rangle$ such that for each $n\in\omega$:

(a) $p_n,p^*_n\in\mathbb{Q}^*_{s\infty}(K^\psi_{4.4.2},\Sigma^\psi_{4.4.2})\cap N$, $g_n\in\omega^\omega\cap N$ is strictly increasing, $i_n\in\omega$,
(b) $p\leq^{s\infty}_0 p_n\leq^{s\infty}_n p^*_n\leq^{s\infty}_{n+1} p_{n+1}\leq^{s\infty}_{n+1} p^*_{n+1}$, $\quad i_n<i_{n+1}$,
(c) p_n approximates $\dot\sigma_n$ at each $t^p_{n+\ell}$ for $\ell\in\omega$,
(d) for each $\ell\in\omega$ and $v\in\mathrm{pos}(w^p,t^{p_n}_0,\ldots,t^{p_n}_{n+\ell-1})$,
 if $t\in\Sigma^\psi_{4.4.2}(t^{p_n}_{n+\ell})$, $\mathbf{nor}[t]>0$ then there is $w\in\mathrm{pos}(v,t)$ such that $(w,t^{p_n}_{n+\ell+1},t^{p_n}_{n+\ell+2},\ldots)\Vdash$ "$\dot\tau_n(g_n(\ell))<g_n(\ell+1)$",
(e) $\varphi(g_n(\ell))<\psi(L[t^{p_n}_{n+\ell}](\langle\rangle))$ for every $\ell\in\omega$,
(f) $\langle\rangle\in D[t^{p_n}_m]$ for all $m\in\omega$,
(g) i_n is such that $(\forall i<\varphi(\mu_Y(i_n)))(|[\mu_Y(i_n+i),\mu_Y(i_n+i+1))\cap\mathrm{rng}(g_n)|\geq 2)$,
(h) $t^{p^*_n}_m=t^{p_n}_m$ for $m<n$,
(i) for $i<\varphi(\mu_Y(i_n))$ let $j^n(i)=\min\{j\in\omega : g_n(j)\in[\mu_Y(i_n+i),\mu_Y(i_n+i+1))\}$ and for $i\in[\varphi(\mu_Y(i_n)),\psi(L[t^{p_n}_{n+j^n(0)}](\langle\rangle)))$ let $j^n(i)=j^n(\varphi(\mu_Y(i_n))-1)+i$;
 then $t^{p^*_n}_n = S_{m_0,m_1}(S^*(t^{p_n}_{n+j^n(i)} : i<\psi(L[t^{p_n}_{n+j^n(0)}](\langle\rangle))))$ (see clauses (2), (4) of the definition of $(K^\psi_{4.4.2},\Sigma^\psi_{4.4.2})$ in 4.4.2, remember **(f)** above), where $m_0=m_{\mathrm{up}}^{t^{p_n}_{n-1}}$, $m_1=m_{\mathrm{up}}^{t^{p_n}_k}$, $k=n+j^n(\psi(L[t^{p_n}_{n+j^n(0)}](\langle\rangle))-1)$,
(j) $t^{p^*_n}_{n+1+m}=t^{p_n}_{k+1+m}$ for every $m\in\omega$ (where k is as in **(i)** above).

The construction is quite straightforward and essentially described by the requirements **(a)–(j)** above. Having defined p^*_n we first choose a condition $p'_{n+1}\in N$ such that it approximates $\dot\sigma_{n+1}$ at each $m\geq n+1$ and $p^*_n\leq^{s\infty}_{n+1}p'_{n+1}$ (by 2.1.4). Then we use 4.3.3.1 *inside* N to find a condition $p_{n+1}\geq_{n+1}p'_{n+1}$, $p_{n+1}\in N$ and a function $g_{n+1}\in N$ such that the demands **(a)**, **(d)–(f)** are satisfied. Note here, that the creatures t^p_ℓ constructed in the proof of 4.3.3.1 came from the application of $\circledast^1_{\mathbf{AB}}$ of 4.3.1. This condition, in turn, is exemplified in our case by the use of the operation S^{**}_H (see item (5) of the definition of $(K^\psi_{4.4.2},\Sigma^\psi_{4.4.2})$, see 4.3.4, 4.3.9). Consequently we may ensure that **(f)** holds. As far as **(e)** is concerned, note that in the inductive construction of the condition p in 4.3.3.1, when choosing p^*_ℓ, we may require additionally that $\varphi(g(\ell))<\psi(L[t^{p^*_\ell}_\ell](\langle\rangle))$ (remember $(K^\psi_{4.4.2},\Sigma^\psi_{4.4.2})$ is omittory, ψ is unbounded). Thus we have defined p_{n+1},g_{n+1} satisfying **(a)–(f)**. By the assumptions on Y we know

$$(\exists^\infty m\in\omega)(\forall i<\varphi(\mu_Y(m)))(|[\mu_Y(m+i),\mu_Y(m+i+1))\cap\mathrm{rng}(g_{n+1})|\geq 2)$$

(remember $\mathrm{rng}(g_{n+1}) \in [\omega]^\omega \cap N$). Hence we may choose $i_{n+1} > i_n$ as required in **(g)**. Clauses **(h)**–**(j)**, **(b)** fully describe the condition

$$p^*_{n+1} = (w^p, t_0^{p_{n+1}}, \ldots, t_n^{p_{n+1}}, t_{n+1}^{p^*_{n+1}}, t_{n+2}^{p^*_{n+1}}, \ldots) \geq_{n+1} p_{n+1}.$$

Note here that

$$\varphi(\mu_Y(i_{n+1})) \leq \varphi(g_{n+1}(j^{n+1}(0))) < \psi(L[t^{p_{n+1}}_{n+1+j^{n+1}(0)}](\langle\rangle))$$

(by **(e)**; remember ϕ is non-decreasing), so there are no problems with the definition of $t_{n+1}^{p^*_{n+1}}$ in clause **(i)**. Moreover, $t_{n+1}^{p^*_{n+1}} \in \Sigma^\psi_{4.4.2}(t_{n+1}^{p_{n+1}}, \ldots, t_k^{p_{n+1}})$, where $k = n+1+j^{n+1}(\psi(L[t^{p_{n+1}}_{n+1+j^{n+1}(0)}](\langle\rangle)) - 1$, and $\mathbf{nor}[t_{n+1}^{p^*_{n+1}}] > m_{\mathrm{dn}}^{t_{n+1}^{p^*_{n+1}}}$. Clearly $p^*_{n+1} \in N$.

Now let $q = \lim\limits_n p_n = \lim\limits_n p^*_n \in \mathbb{Q}^*_{\mathrm{s}\infty}(K^\psi_{4.4.2}, \Sigma^\psi_{4.4.2})$ (see 1.2.13). We claim that q is as required in the assertion of the proposition. Clearly it is stronger than p (by **(b)**) and is $(N, \mathbb{Q}^*_{\mathrm{s}\infty}(K^\psi_{4.4.2}, \Sigma^\psi_{4.4.2}))$–generic (by **(c)** and 1.2.10). The proof will be finished if we show that for each $k \in \omega$

$$q \Vdash (\exists^\infty m \in \omega)(\forall i < \varphi(\mu_Y(m)))(|[\mu_Y(m+i), \mu_Y(m+i+1)) \cap \dot{X}_k| \geq 2).$$

To this end suppose that $k, \ell \in \omega$, $q' \in \mathbb{Q}^*_{\mathrm{s}\infty}(K^\psi_{4.4.2}, \Sigma^\psi_{4.4.2})$, $q \leq q'$. Passing to an extension of q' we may assume that for some $n > k$ we have $w^{q'} \in \mathrm{pos}(w^q, t_0^q, \ldots, t_{n-1}^q)$ and $\ell < i_n$ (remember **(b)**). Let $m \geq n$ be such that $t_0^{q'} \in \Sigma^\psi_{4.4.2}(t_n^q, \ldots, t_m^q)$. Since $\mathbf{nor}[t_0^{q'}] \geq m_{\mathrm{dn}}^{t_0^{q'}} > 0$ we find $n_0 \in [n, m]$ and $s \in \Sigma^\psi_{4.4.2}(t^p_{n_0})$ such that $\mathbf{nor}[s] > 0$ and

$$\{L[s](\eta) \colon \eta \in T[s] \ \&\ \mathrm{succ}_{T[s]}(\eta) = \emptyset\} \subseteq \{L[t_0^{q'}](\eta) \colon \eta \in T[t_0^{q'}] \ \&\ \mathrm{succ}_{T[t_0^{q'}]}(\eta) = \emptyset\}$$

(compare the proof that $(K^\psi_{4.4.2}, \Sigma^\psi_{4.4.2})$ is condensed). Look at the creature $t^q_{n_0}$: it is $t_{n_0}^{p^*_{n_0}}$. So look at the clause **(i)**: the creature $t_{n_0}^{p^*_{n_0}}$ was obtained from $t^{p_{n_0}}_{n_0+j^{n_0}(i)}$ (for $i < \psi(L[t^{p_{n_0}}_{n_0+j^{n_0}(0)}](\langle\rangle))$) by applying the operation S^* and S_{m_0,m_1}. Since $s \in \Sigma^\psi_{4.4.2}(t^q_{n_0})$ we have

$$|\mathrm{succ}_{T[s]}(\langle\rangle)| = \psi(L[t^{p_{n_0}}_{n_0+j^{n_0}(0)}](\langle\rangle)) = \psi(L[s](\langle\rangle))$$

and for each $\nu \in \mathrm{succ}_{T[s]}(\langle\rangle)$ for unique $i = i(\nu) < \psi(L[t^{p_{n_0}}_{n_0+j^{n_0}(0)}](\langle\rangle))$, the tree $T[s]$ above ν and $L[s], R[s]$ and $D[s]$ look exactly like $T[s_i], L[s_i], R[s_i]$ and $D[s_i]$ for some $s_i \in \Sigma^*_{4.4.2}(t^{p_{n_0}}_{n_0+j^{n_0}(i)})$ with $\mathbf{nor}[s_i] > 0$. Now, applying successively clause **(d)** to each $t^{p_{n_0}}_{n_0+j^{n_0}(i)}$, s_i (for $i < \psi(L[s](\langle\rangle))$) we find $w \in \mathrm{pos}(w^{q'}, t_0^{q'})$ such that for every $\nu \in \mathrm{succ}_{T[s]}(\langle\rangle)$

$$(w, t_j^{p_{n_0}}, t_{j+1}^{p_{n_0}}, \ldots) \Vdash_{\mathbb{Q}^*_{\mathrm{s}\infty}(K^\psi_{4.4.2}, \Sigma^\psi_{4.4.2})} \text{“ } \dot{\tau}_{n_0}(g_{n_0}(j^{n_0}(i(\nu)))) < g_{n_0}(j^{n_0}(i(\nu)) + 1) \text{ ”},$$

where j is such that $m_{\mathrm{dn}}^{t_j^{p_{n_0}}} = m_{\mathrm{up}}^{t_0^{q'}}$. By the definition of the name $\dot{\tau}_{n_0}$ and the fact that $k < n \leq n_0$ we get that for each $\nu \in \mathrm{succ}_{T[s]}(\langle\rangle)$

$$(w, t_j^{p_{n_0}}, t_{j+1}^{p_{n_0}}, \ldots) \Vdash \text{“ } |[g_{n_0}(j^{n_0}(i(\nu))), g_{n_0}(j^{n_0}(i(\nu)) + 1) \cap \dot{X}_k| \geq 2 \text{ ”}.$$

By the choice of i_{n_0}, $j^{n_0}(i)$ we know that

$$\mu_Y(i_{n_0} + i) \leq g_{n_0}(j^{n_0}(i)) < g_{n_0}(j^{n_0}(i) + 1) < \mu_Y(i_{n_0} + i + 1)$$

for each $i < \varphi(\mu_Y(i_{n_0})) \le \varphi(g_{n_0}(j^{n_0}(0))) \le \psi(L[s])(\langle\rangle)$. Therefore

$(w, t_j^{p_{n_0}}, t_{j+1}^{p_{n_0}}, \dots) \Vdash$ "$(\forall i < \varphi(\mu_Y(i_{n_0})))(|[\mu_Y(i_{n_0}+i), \mu_Y(i_{n_0}+i+1)) \cap \dot{X}_k| \ge 2)$".

We finish noticing that $(w, t_j^{p_{n_0}}, t_{j+1}^{p_{n_0}}, \dots) \le (w, t_1^{q'}, t_2^{q'}, \dots)$ and $\ell < i_n \le i_{n_0}$. □

COROLLARY 4.4.8. *The following is consistent with ZFC:*

$\mathfrak{d} = \mathbf{cov}(\mathcal{M}) = \mathbf{non}(\mathcal{M}) = \aleph_2$ +
for every non-decreasing unbounded $\psi \in \omega^\omega$, $\mathfrak{d}(S^\psi_{\rm up}) = \aleph_2$ +
for every non-decreasing $\varphi \in \omega^\omega$, $\mathfrak{d}(S^\varphi_{\rm dn}) = \aleph_1$.

PROOF. Start with a model for CH and build a countable support iteration $\langle \mathbb{P}_\alpha, \dot{\mathbb{Q}}_\alpha : \alpha < \omega_2\rangle$ and a sequence $\langle \dot\psi_\alpha : \alpha < \omega_2\rangle$ such that

(a) $\dot\psi_\alpha$ is a $\mathbb{P}_\alpha$–name for a non-decreasing unbounded function in ω^ω,
(b) $\Vdash_{\mathbb{P}_\alpha}$ " $\dot{\mathbb{Q}}_\alpha = \mathbb{Q}^*_{\rm s\infty}(K^{\dot\psi_\alpha}_{4.4.2}, \Sigma^{\dot\psi_\alpha}_{4.4.2})$ ",
(c) for each $\mathbb{P}_{\omega_2}$–name $\dot\psi$ for a non-decreasing unbounded function in ω^ω, for ω_2 many $\alpha < \omega_2$, $\Vdash_{\mathbb{P}_\alpha}$ " $\dot\psi = \dot\psi_\alpha$ ".

By 4.4.3 we have

$$\Vdash_{\mathbb{P}_{\omega_2}} \text{`` } \mathbf{cov}(\mathcal{M}) = \mathbf{non}(\mathcal{M}) = \aleph_2 = \mathfrak{c} \ \ \& \ \ \mathfrak{b} = \aleph_1 \text{ ''}$$

and by 4.4.6 we get

$$\Vdash_{\mathbb{P}_{\omega_2}} \text{`` } \mathfrak{d} = \aleph_2 + \text{for each non-decreasing unbounded } \psi \in \omega^\omega,\ \mathfrak{d}(S^\psi_{\rm up}) = \aleph_2 \text{ ''}.$$

To show that

$$\Vdash_{\mathbb{P}_{\omega_2}} \text{`` } \mathfrak{d}(S^\varphi_{\rm dn}) = \aleph_1 \text{ for every non-decreasing } \varphi \in \omega^\omega \text{ ''}$$

we use 4.4.7 and [**Sh:f**, Ch XVIII, 3.6] and we show that the property described in 4.4.7 is preserved in countable support iterations.

So suppose that $\varphi \in \omega^\omega$ is non-decreasing and define a context $(R^\varphi, S^\varphi, \mathbf{g}^\varphi)$ (see [**Sh:f**, Ch XVIII, 3.1]) as follows. First, for $\eta \in \omega^\omega$ let $X_\eta \in [\omega]^\omega$ be such that $\mu_{X_\eta}(n) = \sum\limits_{i \le n} (\eta(i)+1)$. Next we let (in the ground model $\mathbf{V}$):

- S^φ is the collection of all $N \cap \mathcal{H}(\aleph_1)$ for N a countable elementary submodel of $(\mathcal{H}(\chi), \in, <^*_\chi)$,
- for each $a \in S^\varphi$, $d[a] = c[a] = \omega = d'[a] = c'[a]$,
- $\alpha^* = 1$,
- $R^\varphi = R^\varphi_0$ is the relation determined by the $S^\varphi_{\rm dn}$–localization:
  $\eta\ R^\varphi\ g$ if and only if ($\eta, g \in \omega^\omega$ and)
  $(\exists^\infty n \in \omega)(\forall i < \varphi(\mu_{X_g}(n)))(|[\mu_{X_g}(n+i), \mu_{X_g}(n+i+1)) \cap X_\eta| \ge 2)$
  (remember that $\alpha^* = 1$ so we have R^φ_0 only),
- $\mathbf{g}^\varphi = \langle \mathbf{g}_a : a \in S^\varphi\rangle \subseteq \omega^\omega$ is such that for every $a \in S^\varphi$, for each $\eta \in a \cap \omega^\omega$ we have $\eta\ R^\varphi\ \mathbf{g}_a$ (exists as each a is countable, e.g. one may take as $\mathbf{g}_a$ any real dominating a).

By the choice of $(R^\varphi, S^\varphi, \mathbf{g}^\varphi)$ we know that it covers in $\mathbf{V}$ (see [**Sh:f**, Ch XVIII, 3.2]).

CLAIM 4.4.8.1. *Let* $\mathbb{P}$ *be a proper forcing notion such that*

$$\Vdash_{\mathbb{P}} \text{`` } (R^\varphi, S^\varphi, \mathbf{g}^\varphi) \text{ covers''}.$$

Then:

1. $\Vdash_{\mathbb{P}}$ *" $(R^{\varphi}, S^{\varphi}, \mathbf{g}^{\varphi})$ strongly covers by Possibility B" (see* [**Sh:f**, Ch XVIII, 3.3]*),*
2. *for every $\mathbb{P}$-name $\dot{\psi}$ for a non-decreasing unbounded function in ω^{ω},*

$$\Vdash_{\mathbb{P}} \text{“ the forcing notion } \mathbb{Q}^*_{\mathrm{s}\infty}(K^{\dot{\psi}}_{4.4.2}, \Sigma^{\dot{\psi}}_{4.4.2}) \text{ is } (R^{\varphi}, S^{\varphi}, \mathbf{g}^{\varphi})\text{-preserving”}.$$

Proof of the claim: 1) As $\alpha^* = 1$ it is enough to show that the second player has an absolute (for extensions by proper forcing notions) winning strategy in the following game G_a (for each $a \in S^{\varphi}$).

The play lasts ω moves.

Player I, in his n^{th} move chooses functions $f^n_0, \dots, f^n_n \in \omega^{\omega}$ such that

$$f^n_\ell{\restriction}b_{n-1} = f^{n-1}_\ell{\restriction}b_{n-1} \text{ for } \ell < n \text{ and } \quad f^n_\ell \; R^{\varphi} \; \mathbf{g}_a \text{ for each } \ell \le n.$$

Player II answers choosing a finite set $b_n \subseteq \omega$, $b_{n-1} \subseteq b_n$.

At the end the second player wins if and only if for each $\ell \in \omega$

$$\bigcup_{n \ge \ell} f^n_\ell{\restriction}b_n \; R^{\varphi} \; \mathbf{g}_a.$$

But it should be clear that Player II has a (nice) winning strategy in this game. In his n^{th} move he chooses as b_n a sufficiently long initial segment of ω to provide new "witnesses" for the quantifier $(\exists^{\infty} n \in \omega)$ in the definition of R^{φ} (for all $f^n_0, \dots, f^n_n$).
2) Since $\alpha^* = 1$ what we have to prove is exactly the statement of 4.4.7 (see [**Sh:f**, Ch XVIII, 3.4A]), so we are done with the claim.

Now we may use [**Sh:f**, Ch XVIII, 3.5, 3.6] (and 4.4.8.1) to conclude that $\Vdash_{\mathbb{P}_{\omega_2}}$ "$(R^{\varphi}, S^{\varphi}, \mathbf{g}^{\varphi})$ covers" and hence immediately

$$\Vdash_{\mathbb{P}_{\omega_2}} \mathfrak{d}(S^{\varphi}_{\mathrm{dn}}) = \aleph_1.$$

As every function from $\omega^{\omega} \cap \mathbf{V}^{\mathbb{P}_{\omega_2}}$ appears in $\omega^{\omega} \cap \mathbf{V}^{\mathbb{P}_{\alpha}}$ for some $\alpha < \omega_2$ we finish the proof. □

CHAPTER 5

Around not adding Cohen reals

The starting point for this chapter was the following request of Bartoszyński (see [**Ba94**, Problem 4]): construct a proper forcing notion $\mathbb{P}$ such that:

1. $\mathbb{P}$ is ω^ω–bounding,
2. $\mathbb{P}$ preserves non–meager sets,
3. $\mathbb{P}$ makes ground reals to have measure zero,
4. $\mathbb{P}$ has the Laver property,
5. countable support iterations of $\mathbb{P}$ with Laver forcing, random real forcing and Miller's rational perfect set forcing do not add Cohen reals.

A forcing notion with these properties would correspond to the invariant $\mathbf{non}(\mathcal{N})$ (the minimal size of a non-null set; see [**BaJu95**, 7.3C]). Forcing notions with properties (1)–(4) were known. The fourth property is a kind of technical assumption and might be replaced by

4^-. $\mathbb{P}$ is (f,g)–bounding for some $f,g \in \omega^\omega$ (with $g(n) \ll f(n)$, of course).

At least we believe that that was the intension (see an example presented in [**BaJu95**, 7.3C], see 5.4.3 here too). However it was not clear how one should take care of the last required property. The problem comes from the fact that we do not have any good (meaning: iterable and sufficiently weak) condition for "not adding Cohen reals". The difficulty starts already at the level of compositions of forcing notions: adding first a dominating real and then "infinitely often equal real below it" one produces a Cohen real. Various iterable properties implying "no Cohen reals" are in use, but the point is to find one capturing as many of them as possible. The first section deals with (f,g)–bounding property. We generalize this property in the following section (a special case of the methods developed there is presented in [**BaJu95**, 7.2E], however not fully). The "$(\bar{t},\bar{\mathcal{F}})$–bounding" property seems to be still not weak enough to capture the measure algebra. So we weaken this further and we present a good candidate for a property "responsible" for not adding Cohen reals in the third part of this chapter (see 5.4.2 too). The tools developed in this section are very general and will be used later too.

5.1. (f,g)-bounding

Let us recall that a proper forcing notion $\mathbb{P}$ is (f,g)–bounding (for some increasing $f,g \in \omega^\omega$) if

$$\Vdash_{\mathbb{P}} (\forall x \in \prod_{i\in\omega} f(i))(\exists S \in \mathbf{V} \cap \prod_{i\in\omega} [\omega]^{<\omega})(\forall i \in \omega)(|S(i)| \leq g(i) \ \ \& \ \ x(i) \in S(i)).$$

It is almost obvious that (f,g)–bounding forcing notions add neither Cohen reals nor random reals (see e.g. [**BaJu95**, 7.2.15]). For the treatment of this property in countable support iterations see [**Sh 326**, A2.5] or [**Sh:f**, Ch VI, 2.11A-C].

DEFINITION 5.1.1. Let $f \in \omega^\omega$. We say that a weak creating pair (K,Σ) for $\mathbf{H}$ is *essentially f-big* if

$(\otimes^f_{5.1.1})$ for every weak creature $t \in K$ and $u \in \mathbf{basis}(t)$ such that $0 < \ell g(u)$ and $\mathbf{nor}[t] > f(0)$ and each function $h : \mathrm{pos}(u,t) \longrightarrow f(\ell g(u))$ *there is* $s \in \Sigma(t)$ such that $u \in \mathbf{basis}(s)$, $h{\restriction}\mathrm{pos}(u,s)$ is constant and $\mathbf{nor}[s] \geq \mathbf{nor}[t] - \frac{f(0)}{\ell g(u)}$.

REMARK 5.1.2. Definition 5.1.1 may be thought of as a kind of strengthening of 2.2.1 and 2.3.2.

LEMMA 5.1.3. *Let $f \in \omega^\omega$ be increasing. Suppose that (K,Σ) is a finitary essentially f-big tree-creating pair, $p \in \mathbb{Q}^{\mathrm{tree}}_1(K,\Sigma)$ and $\dot\tau$ is a $\mathbb{Q}^{\mathrm{tree}}_1(K,\Sigma)$-name such that*

$$p \Vdash_{\mathbb{Q}^{\mathrm{tree}}_1(K,\Sigma)} \text{“}\dot\tau \in \omega^\omega \text{ is such that } (\forall n \in \omega)(\dot\tau(n) < f(n))\text{”}.$$

Then there is a condition $q \in \mathbb{Q}^{\mathrm{tree}}_1(K,\Sigma)$ stronger than p and such that for every $\rho \in T^q$ the condition $q^{[\rho]}$ forces a value to $\dot\tau{\restriction}(\ell g(\rho)+1)$.

PROOF. First note that the essential f–bigness of (K,Σ) implies that if $t \in K$, $\mathbf{nor}[t] > f(0)$ and $\ell g(\mathrm{root}(t)) > f(0)+1$ then the tree creature t is 2-big. This is more than enough to carry out the proofs of 2.3.6(2) and 2.3.7(2) and thus we find a condition $q_0 \geq p$, $\ell g(\mathrm{root}(q_0)) > f(0)+1$, and fronts $F_0, F_1, F_2, \ldots$ of T^{q_0} such that

$$(\forall n \in \omega)(\forall \eta \in F_n)(\ell g(\eta) > n \ \text{ and } \ q_0^{[\eta]} \text{ decides } \dot\tau(n))$$

and $(\forall \eta \in T^{q_0})(\mathbf{nor}[t^{q_0}_\eta] > 2\cdot f(0)+1)$. For each $n \in \omega$ we have a function $h_n : F_n \longrightarrow f(n)$ such that

$$(\forall \eta \in F_n)(q_0^{[\eta]} \Vdash_{\mathbb{Q}^{\mathrm{tree}}_1(K,\Sigma)} \dot\tau(n) = h_n(\eta)).$$

Suppose that $\nu \in T^{q_0}$ is such that $\mathrm{pos}(t^{q_0}_\nu) \subseteq F_n$ and $\ell g(\nu) \geq n$ (note that there are $\nu \in T^{q_0}$ such that $\mathrm{pos}(t^{q_0}_\nu) \subseteq F_n$ as F_n is finite). Then we may apply $(\otimes^f_{5.1.1})$ and we find $s \in \Sigma(t^{q_0}_\nu)$ such that $h_n{\restriction}\mathrm{pos}(s)$ is constant and $\mathbf{nor}[s] \geq \mathbf{nor}[t^{q_0}_\nu] - \frac{f(0)}{\ell g(\nu)}$. Repeating this process downward and for all $n \in \omega$ we find a quasi tree $T^* \subseteq T^{q_0}$ and $s_\nu \in \Sigma(t^{q_0}_\nu)$ for $\nu \in T^*$ such that

(α) $\mathrm{root}(T^*) = \mathrm{root}(q_0)$,

(β) $\mathbf{nor}[s_\nu] \geq \mathbf{nor}[t^{q_0}_\nu] - (\ell g(\nu)+1)\cdot\frac{f(0)}{\ell g(\nu)} \geq \mathbf{nor}[t^{q_0}_\nu] - (f(0)+1)$,

(γ) $\mathrm{pos}(s_\nu) = \mathrm{succ}_{T^*}(\nu)$,

(δ) if $\nu, \eta_0, \eta_1 \in T^*$, $n \leq \ell g(\nu)$, $\nu \vartriangleleft \eta_0$, $\nu \vartriangleleft \eta_1$, $\eta_0, \eta_1 \in F_n$ then $h_n(\eta_0) = h_n(\eta_1)$.

This defines a condition $q^* \in \mathbb{Q}^{\mathrm{tree}}_1(K,\Sigma)$. Clearly it is stronger than q_0 and, by (δ) above, it has the required property. □

REMARK 5.1.4. If (K,Σ) is a local tree-creating pair (see 1.4.3), $p \in \mathbb{Q}^{\mathrm{tree}}_e(K,\Sigma)$ then

$$(\forall \nu \in \mathrm{dcl}(T^p))(\mathrm{root}(p) \trianglelefteq \nu \ \Rightarrow\ \nu \in T^p).$$

CONCLUSION 5.1.5. Suppose that $f \in \omega^\omega$ is increasing, (K,Σ) is essentially f-big finitary and local tree creating pair. Then the condition $q \in \mathbb{Q}^{\mathrm{tree}}_1(K,\Sigma)$ provided by the assertion of 5.1.3, gives at most $|T^q \cap \prod\limits_{m<n} \mathbf{H}(m)|$ possible values to $\dot\tau{\restriction}(n+1)$ (for each n). Hence, if $g(n) = \prod\limits_{m<n} |\mathbf{H}(m)|$ then $\mathbb{Q}^{\mathrm{tree}}_1(K,\Sigma)$ is (f,g)-bounding.

DEFINITION 5.1.6. A weak creating pair is *reducible* if for each $t \in K$ with $\mathbf{nor}[t] > 3$ there is $s \in \Sigma(t)$ such that $\frac{\mathbf{nor}[t]}{2} \leq \mathbf{nor}[s] \leq \mathbf{nor}[t] - 1$.

DEFINITION 5.1.7. Let $h : \omega \times \omega \longrightarrow \omega$. We say that a weak creating pair (K, Σ) is *h–limited* whenever

if $t \in K$, $u \in \mathbf{basis}(t)$, $\ell g(u) \leq m_0$ and $\mathbf{nor}[t] \leq m_1$ then $|\mathrm{pos}(u,t)| \leq h(m_0, m_1)$.

If the function h does not depend on the first coordinate (i.e. $h(m_0, m_1) = h_0(m_1)$) then we say that (K, Σ) is *h–norm-limited.* We may say then that (K, Σ) is h_0–norm-limited or just h_0–limited.

THEOREM 5.1.8. *Suppose that $f, g \in \omega^\omega$ are increasing, $h : \omega \times \omega \longrightarrow \omega$ and*

$$(\forall^\infty n \in \omega)(\prod_{m<n} h(m,m) < g(n) < f(n)).$$

Assume that (K, Σ) is a reducible finitary tree creating pair which is h-limited and essentially f-big. Then the forcing notion $\mathbb{Q}_1^{\mathrm{tree}}(K, \Sigma)$ is (f, g)-bounding.

PROOF. Let N be such that $(\forall n \geq N)(\prod\limits_{m<n} h(m,m) < g(n) < f(n))$. Suppose that $p \Vdash_{\mathbb{Q}_1^{\mathrm{tree}}(K,\Sigma)} \dot{\tau} \in \prod\limits_{n<\omega} f(n)$. By 5.1.3 we find $q \geq p$ such that for every $\eta \in T^q$ the condition $q^{[\eta]}$ decides $\dot{\tau}{\restriction}(\ell g(\eta)+1)$. As (K, Σ) is reducible we may assume that $(\forall \eta \in T^q)(\mathbf{nor}[t^q_\eta] \leq \ell g(\eta))$ and $\ell g(\mathrm{root}(q)) > N$. For $n \in \omega$ let

$$F^*_n \stackrel{\mathrm{def}}{=} \{\eta \in T^q : \ell g(\eta) \geq n \ \text{ and } \ (\forall \nu \in T^q)(\nu \lhd \eta \quad \Rightarrow \quad \ell g(\nu) < n)\}.$$

Clearly each F^*_n is a front of T^q and if $\eta \in F^*_n$ then $q^{[\eta]}$ decides the value of $\dot{\tau}(n)$. Now note that $|F^*_n| = 1$ for $n \leq \ell g(\mathrm{root}(q))$ and $|F^*_n| \leq \prod\limits_{m<n} h(m,m) < g(n)$ for all other n. This allows us to finish the proof. □

THEOREM 5.1.9. *Assume that (K, Σ) is a finitary and reducible creating pair which is h-limited for some function h. Further suppose that (K, Σ) is* either *growing and big* or *omittory and omittory–big. Then the forcing notion $\mathbb{Q}^*_{\mathrm{s}\infty}(K, \Sigma)$ is (f, g)–bounding for any strictly increasing functions $f, g \in \omega^\omega$.*

PROOF. Suppose that $\dot{\tau}$ is a $\mathbb{Q}^*_{\mathrm{s}\infty}(K, \Sigma)$–name for a function in $\prod\limits_{n\in\omega} f(n)$ and $p \in \mathbb{Q}^*_{\mathrm{s}\infty}(K, \Sigma)$. Applying repeatedly 2.2.3 (or 2.2.6 in the second case) we may construct inductively an increasing sequence $n_0 < n_1 < \ldots < \omega$ and a condition $q = (w^q, t^q_0, t^q_1, \ldots) \in \mathbb{Q}^*_{\mathrm{s}\infty}(K, \Sigma)$ such that $p \leq^{\mathrm{s}\infty}_0 q$ (so $w^q = w^p$) and for all $k \in \omega$:

($\oplus_0$) $g(n_k) > \prod\limits_{i \leq k} h(m^{t^q_i}_{\mathrm{dn}}, 2 \cdot m^{t^q_i}_{\mathrm{dn}} + 1)$,

($\oplus_1$) $\mathbf{nor}[t^q_k] \leq 2 \cdot m^{t^q_k}_{\mathrm{dn}} + 1$, and

($\oplus_2$) if $w \in \mathrm{pos}(w^q, t^q_0, \ldots, t^q_{k-1})$ then the condition $(w, t^q_k, t^q_{k+1}, \ldots)$ decides the value of $\dot{\tau}{\restriction}(n_k + 1)$.

This is straightforward; to get ($\oplus_1$) we use the assumption that (K, Σ) is reducible. Now we note that for each k:

$$|\mathrm{pos}(w^q, t^q_0, \ldots, t^q_k)| \leq \prod_{i \leq k} h(m^{t^q_i}_{\mathrm{dn}}, 2 \cdot m^{t^q_i}_{\mathrm{dn}} + 1)$$

and so the condition q allows less than $g(n_k)$ candidates for values of $\dot{\tau}$ on the interval $(n_k, n_{k+1}]$. □

THEOREM 5.1.10. *Let (K,Σ) be a reducible and finitary creating pair. Suppose that increasing functions $f,g\in\omega^\omega$ and a function $h:\omega\times\omega\longrightarrow\omega$ are such that*

1. *(K,Σ) is h–limited,*
2. *(K,Σ) is essentially f–big,*
3. $(\forall^\infty n)(\prod\limits_{m<n} h(m,m)<g(n)<f(n))$.

Lastly assume that (K,Σ) captures singletons. Then the forcing notion $\mathbb{Q}^*_{\mathrm{w}\infty}(K,\Sigma)$ is (f,g)–bounding.

PROOF. Take $N\in\omega$ such that $\prod\limits_{m<n} h(m,m)<g(n)<f(n)$ for all $n\geq N$. Let $\dot{\tau}$ be a $\mathbb{Q}^*_{\mathrm{w}\infty}(K,\Sigma)$–name for a function in $\prod\limits_{n\in\omega} f(n)$, and let $p\in\mathbb{Q}^*_{\mathrm{w}\infty}(K,\Sigma)$. First note that, as (K,Σ) captures singletons, for each $t\in K$ we may find $s\in\Sigma(t)$ such that for some $u\in\mathbf{basis}(s)$ (equivalently: for each u, remember (K,Σ) is forgetful) we have $|\mathrm{pos}(u,s)|=1$. Using this remark and 2.1.12 we find a condition $(w^p,s_0,s_1,s_2,\dots)\in\mathbb{Q}^*_{\mathrm{w}\infty}(K,\Sigma)$ and a sequence $0\leq\ell_0<\ell_1<\ell_2<\ldots<\omega$ such that:

(α) $\mathbf{nor}[s_{\ell_0}]\geq 2\cdot f(0)+2$, $\mathbf{nor}[s_{\ell_{i+1}}]\geq 2\cdot f(0)\cdot|\mathrm{pos}(w^p,s_0,\dots,s_{\ell_i})|+2(i+1)$,

(β) if $n\in\omega\setminus\{\ell_0,\ell_1,\ell_2,\dots\}$ then for some $u\in\mathbf{basis}(s_n)$ we have $|\mathrm{pos}(u,s_n)|=1$,

(γ) $N+4<m^{s_{\ell_0}}_{\mathrm{dn}}$, $\quad p\leq(w^p,s_0,s_1,s_2,\dots)$,

(δ) for each $i\in\omega$, $u\in\mathrm{pos}(w^p,s_0,\dots,s_{\ell_i})$ the condition $(u,s_{\ell_i+1},s_{\ell_i+2},\dots)$ decides $\dot{\tau}\restriction(m^{s_{\ell_i}}_{\mathrm{up}}+1)$.

Next we slightly correct creatures s_{ℓ_i} to ensure that the value of $\dot{\tau}\restriction(m^{s_{\ell_i}}_{\mathrm{dn}}+1)$ is decided by any $u\in\mathrm{pos}(w^p,s_0,\dots,s_{\ell_i-1})$. For this we use the procedure similar to that in the proof of 5.1.1 (and based on the assumption that (K,Σ) is essentially f–big). Thus we get creatures $t_{\ell_i}\in\Sigma(s_{\ell_i})$ such that (for $i\in\omega$):

$$(\forall u\in\mathrm{pos}(w^p,s_0,\dots,s_{\ell_i-1}))\big((u,t_{\ell_i},s_{\ell_i+1},s_{\ell_i+2},\dots)\text{ decides }\dot{\tau}\restriction(m^{s_{\ell_i}}_{\mathrm{dn}}+1)\big)\quad\text{and}$$

$$\mathbf{nor}[t_{\ell_i}]\geq\mathbf{nor}[s_{\ell_i}]-(m^{s_{\ell_i}}_{\mathrm{dn}}+1)\cdot\frac{f(0)}{m^{s_{\ell_i}}_{\mathrm{dn}}}\cdot|\mathrm{pos}(w^p,s_0,\dots,s_{\ell_i-1})|.$$

Note that $|\mathrm{pos}(w^p,s_0,\dots,s_{\ell_0-1})|=1$ and hence $\mathbf{nor}[t_{\ell_0}]\geq 1$. Moreover, for each $i\in\omega$, $|\mathrm{pos}(w^p,s_0,\dots,s_{\ell_i})|=|\mathrm{pos}(w^p,s_0,\dots,s_{\ell_i},\dots,s_{\ell_{i+1}-1})|$ and therefore $\mathbf{nor}[t_{\ell_{i+1}}]\geq 2(i+1)$. Finally, as (K,Σ) is reducible, we may choose $t^*_{\ell_i}\in\Sigma(t_{\ell_i})$ (for $i\in\omega$) such that $\frac{i+2}{2}\leq\mathbf{nor}[t^*_{\ell_i}]\leq m^{t^*_{\ell_i}}_{\mathrm{dn}}$. Now we let

$$w^q=w^p,\qquad t^p_m=\begin{cases} s_m & \text{if } m\in\omega\setminus\{\ell_0,\ell_1,\dots\},\\ t^*_{\ell_i} & \text{if } m=\ell_i,\ i\in\omega.\end{cases}$$

This defines a condition $q\in\mathbb{Q}^*_{\mathrm{w}\infty}(K,\Sigma)$ stronger than p and such that for each $i\in\omega$:

(a) $(\forall u\in\mathrm{pos}(w^q,t^q_0,\dots,t^q_{\ell_i-1}))\big((u,t^q_{\ell_i},t^q_{\ell_i+1},\dots)$ decides $\dot{\tau}\restriction(m^{t^q_{\ell_i}}_{\mathrm{dn}}+1)\big)$,

(b) $|\mathrm{pos}(w^q,t^q_0,\dots,t^q_{\ell_{i+1}-1})|\leq\prod\limits_{j\leq i}h(m^{t^q_{\ell_j}}_{\mathrm{dn}},m^{t^q_{\ell_j}}_{\mathrm{dn}})<g(m^{t^q_{\ell_i}}_{\mathrm{dn}}+1)$,

(c) $|\mathrm{pos}(w^q,t^q_0,\dots,t^q_{\ell_0-1})|=1$.

Now we easily finish. □

DEFINITION 5.1.11. Let $\mathbf{H}$ be finitary, $F \in \omega^\omega$ be increasing. A function $f: \omega \times \omega \longrightarrow \omega$ is called *$(\mathbf{H}, F)$–fast* if it is $\mathbf{H}$–fast (see 1.1.12) and additionally

$$(\forall n, \ell \in \omega)(f(n+1,\ell) > f(n,\ell) + F(\ell) \cdot \varphi_{\mathbf{H}}(\ell) \cdot \ell).$$

THEOREM 5.1.12. *Suppose that (K,Σ) is a finitary local and $\bar{2}$–big creating pair for $\mathbf{H}$ which has the (weak) Halving Property. Let $F \in \omega^\omega$ be increasing and $f: \omega \times \omega \longrightarrow \omega$ be $(\mathbf{H}, F)$–fast. Then the forcing notion $\mathbb{Q}^*_f(K,\Sigma)$ is $(F, \varphi_{\mathbf{H}})$–bounding.*

PROOF. Suppose that $\dot{\tau}$ is a $\mathbb{Q}^*_f(K,\Sigma)$–name for an element of $\prod\limits_{m<\omega} F(m)$ and $p \in \mathbb{Q}^*_f(K,\Sigma)$. By 2.2.11 we find a condition $q \geq p$ which essentially decides all the values $\dot{\tau}(m)$ (for $m \in \omega$). We may assume that $(\forall i \in \omega)(\mathbf{nor}[t^q_i] > f(2, m^{t^q_i}_{\rm dn}))$. Applying, in a standard by now way, the bigness (like in 2.2.3 or 5.1.1) we build a condition $r \geq_0 q$ such that $t^r_i \in \Sigma(t^q_i)$, $\mathbf{nor}[t^r_i] \geq \mathbf{nor}[t^q_i] - F(m^{t^r_i}_{\rm dn}) \cdot \varphi_{\mathbf{H}}(m^{t^r_i}_{\rm dn}) \cdot m^{t^r_i}_{\rm dn}$ and for each $i \in \omega$ and $u \in \mathrm{pos}(w^p, t^r_0, \ldots, t^r_{i-1})$ the condition $(u, t^r_i, t^r_{i+1}, \ldots)$ decides the value of $\dot{\tau}{\restriction}(\ell g(u)+1)$. Now we easily finish (remembering that (K,Σ) is local). $\square$

5.2. $(\bar{t}, \bar{\mathcal{F}})$–bounding

Here we introduce and deal with a property which, in our context, is a natural generalization of the notion of (f,g)–bounding forcing notions. This is a first step toward handling "not adding Cohen reals" and, in some sense, it will be developed in the next parts of this chapter. After we formulate and prove some basic results we show how one may treat this property in countable support iterations.

A particular case of this machinery was presented in [**BaJu95**, 7.2E].

DEFINITION 5.2.1. Let (K,Σ) be a creating pair, $\bar{t} = \langle t_n : n \in \omega \rangle \in \mathrm{PC}(K,\Sigma)$.

1. For a function $h \in \omega^\omega$ we define $U_h(\bar{t})$ as the set

   $$\{\bar{s} = \langle s_n : n \in \omega \rangle \in \mathrm{PC}(K,\Sigma) : \bar{t} \leq \bar{s} \ \&\ (\forall n \in \omega)(\mathbf{nor}[s_n] \leq h(m^{s_n}_{\rm dn}))\}.$$

   For $n \in \omega$ and $h \in \omega^\omega$ we let

   $$V^n_h(\bar{t}) = \{s \in \Sigma(t_n) : \mathbf{nor}[s] \leq h(m^s_{\rm dn})\}.$$

2. Let $h_1, h_2 \in \omega^\omega$. We say that a forcing notion $\mathbb{P}$ is $(\bar{t}, h_1, h_2)$–bounding if

   $$\Vdash_{\mathbb{P}} (\forall \bar{s} \in U_{h_1}(\bar{t}))(\exists \bar{s}^* \in U_{h_2}(\bar{t}) \cap \mathbf{V})(\bar{s}^* \leq \bar{s}).$$

REMARK 5.2.2. 1. We will be interested in the notions introduced in 5.2.1 only for $\bar{t} \in \mathrm{PC}_\infty(K,\Sigma)$ (i.e. $\lim\limits_{n\to\infty} \mathbf{nor}[t_n] = \infty$) and $(\forall^\infty n)(h_1(n) < h_2(n))$, $\lim\limits_{n\to\infty} h_1(n) = \lim\limits_{n\to\infty} h_2(n) = \infty$.

2. Note that if (K,Σ) is nice and simple (see 2.1.7) then

   $$U_h(\bar{t}) = \prod_{n\in\omega} V^n_h(\bar{t}).$$

DEFINITION 5.2.3. For a creating pair (K,Σ) on $\mathbf{H}$ we say that:

1. (K,Σ) is *monotonic* if for each $t \in K$, $s \in \Sigma(t)$ we have $\mathbf{val}[s] \subseteq \mathbf{val}[t]$.
2. (K,Σ) is *strictly monotonic* if it is monotonic and for all $n \in \omega$, $t_0, \ldots, t_n \in K$ and $s \in \Sigma(t_0, \ldots, t_n)$ such that $\mathbf{nor}[s] \leq \max\{\mathbf{nor}[t_\ell] - 1 : \ell \leq n\}$ we have:

   $$(\forall u \in \mathbf{basis}(t_0))(\mathrm{pos}(u,s) \subsetneq \mathrm{pos}(u, t_0, \ldots, t_n)).$$

3. (K,Σ) is *spread* if for each $t\in K$, $u\in \mathbf{basis}(t)$ and $v\in \mathrm{pos}(u,t)$ there is $s\in\Sigma(t)$ such that
$$\mathbf{nor}[s]\le \frac{1}{2}\mathbf{nor}[t]\quad\text{ and }\quad v\in \mathrm{pos}(u,s).$$

PROPOSITION 5.2.4. *Let $(K,\Sigma)\in\mathcal{H}(\aleph_1)$ be a strictly monotonic and spread creating pair for $\mathbf{H}$. Suppose that $\bar{t}=\langle t_n : n\in\omega\rangle\in \mathrm{PC}_\infty(K,\Sigma)$ and $h_1,h_2\in\omega^\omega$ are such that*
$$(\forall n)(0<h_1(m^{t_n}_{\rm dn})\le h_2(m^{t_n}_{\rm dn})\le \mathbf{nor}[t_n])\quad\textit{and}\quad (\forall^\infty n)(h_2(m^{t_n}_{\rm dn})\le \mathbf{nor}[t_n]-1).$$
Then every $(\bar{t},h_1,h_2)$–bounding forcing notion does not add Cohen reals.

PROOF. Let $w\in\mathbf{basis}(t_0)$ be such that
$$(\forall n\in\omega)\big(\mathrm{pos}(w,t_0,\ldots,t_{n-1})\subseteq \mathbf{basis}(t_n)\big).$$
Look at the space
$$\mathcal{X}=\{x\in\prod_{m\in\omega}\mathbf{H}(m):(\forall n\in\omega)(x{\restriction}m^{t_n}_{\rm up}\in \mathrm{pos}(w,t_0,\ldots,t_n))\}$$
equipped with the natural (product) topology. It is a perfect Polish space (note that as (K,Σ) is strictly monotonic and spread, by $\lim\limits_{n\to\infty}\mathbf{nor}[t_n]=\infty$, for sufficiently large n, for each $u\in\mathbf{basis}(t_n)$ we find two distinct $v_0,v_1\in\mathrm{pos}(u,t_n)$). Thus, if a forcing notion $\mathbb{P}$ adds a Cohen real then it adds a Cohen real $c\in\mathcal{X}$. In $\mathbf{V}[c]$, choose (e.g. inductively) a sequence $\bar{s}=\langle s_n:n\in\omega\rangle\in U_{h_1}(\bar{t})$ such that
$$(\forall n\in\omega)\big(s_n\in\Sigma(t_n)\ \ \&\ \ c{\restriction}m^{s_n}_{\rm dn}\in \mathrm{pos}(w,s_0,\ldots,s_{n-1})\subseteq\mathbf{basis}(s_n)\big)$$
(possible by 5.2.3(3), remember that (K,Σ) is nice). We claim that there is no $\bar{s}^*\in U_{h_2}(\bar{t})\cap\mathbf{V}$ with $\bar{s}^*\le\bar{s}$. Why? Suppose that $\bar{s}^*\in U_{h_2}(\bar{t})\cap\mathbf{V}$. Working in $\mathbf{V}$, consider the set
$$\mathcal{O}\stackrel{\rm def}{=}\{x\in\mathcal{X}:(\exists n\in\omega)(x{\restriction}m^{s^*_n}_{\rm dn}\notin\mathrm{pos}(w,s^*_0,\ldots,s^*_{n-1}))\}.$$
This set is open dense in $\mathcal{X}$ (for the density use strict monotonicity of (K,Σ); remember that for sufficiently large $n\in\omega$, $\mathbf{nor}[s^*_n]\le h_2(m^{s^*_n}_{\rm dn})\le\mathbf{nor}[t_m]-1$, where m is such that $m^{t_m}_{\rm dn}=m^{s^*_n}_{\rm dn}$). Consequently, in $\mathbf{V}[c]$, $c\in\mathcal{O}$ and $\bar{s}^*\not\le\bar{s}$. □

DEFINITION 5.2.5. Let (K,Σ) be a weak creating pair.

1. We say that a weak creature $t\in K$ is (n,m)*–additive* if for all $t_0,\ldots,t_{n-1}\in\Sigma(t)$ such that $\mathbf{nor}[t_i]\le m$ (for $i<n$) there is $s\in\Sigma(t)$ such that
$$t_0,\ldots,t_{n-1}\in\Sigma(s)\quad\text{ and }\quad \mathbf{nor}[s]\le\max\{\mathbf{nor}[t_\ell]:\ell<n\}+1.$$
2. *m–additivity of a weak creature $t\in K$* is defined as
$$\mathbf{add}_m(t)=\sup\{k<\omega: t\text{ is }(k,m)\text{–additive}\}.$$
[Note that each t is at least $(1,m)$–additive.]
3. We say that (K,Σ) is (g,h)–additive (for $g,h\in\omega^\omega$) if $\mathbf{add}_{h(m_{\rm dn}(t))}(t)\ge g(m_{\rm dn}(t))$ for all $t\in K$.
Similarly, if (K,Σ) is a creating pair and $\bar{t}\in\mathrm{PC}(K,\Sigma)$ then we say that $\bar{t}$ *is (g,h)–additive* if $(\forall n\in\omega)(\mathbf{add}_{h(m^{t_n}_{\rm dn})}(t_n)\ge g(m^{t_n}_{\rm dn}))$.
4. If the function g is constant, say $g\equiv n$, then instead of "(g,h)–additive" we may say "(n,h)–additive" etc.

[Note that for creatures we have $m_{\rm dn}(t)=m^t_{\rm dn}$.]

REMARK 5.2.6. The notion of additivity of a weak creature is very close to that of bigness: in most applications they coincide. One can easily formulate conditions under which (k, m)–additivity is equivalent to k–bigness.

Let us recall that a forcing notion $\mathbb{P}$ has *the Laver property* if it is (f, g^*)–bounding for every increasing function $f \in \omega^\omega$, where $g^*(n) = 2^n$ (g^* may be replaced by any other fixed increasing function in ω^ω).

PROPOSITION 5.2.7. *Assume that (K, Σ) is a strongly finitary (see 3.3.4) and simple (see 2.1.7) creating pair, $\bar{t} = \langle t_n : n < \omega \rangle \in \mathrm{PC}(K, \Sigma)$ and $h_1, h_2 \in \omega^\omega$ are such that*

$$(\forall n \in \omega)(h_1(m^{t_n}_{\mathrm{dn}}) \leq h_2(m^{t_n}_{\mathrm{dn}})) \quad \textit{and} \quad (\forall^\infty n)(h_1(m^{t_n}_{\mathrm{dn}}) + 1 \leq h_2(m^{t_n}_{\mathrm{dn}})).$$

1. *If $f \in \omega^\omega$ is such that $(\forall n \in \omega)(|V^n_{h_1}(\bar{t})| \leq f(m^{t_n}_{\mathrm{dn}}))$, $g \in \omega^\omega$ is strictly increasing and $\bar{t}$ is (g, h_1)–additive then every (f, g)–bounding forcing notion is $(\bar{t}, h_1, h_2)$–bounding.*
2. *If $\bar{t}$ is (g^*, h_1)–additive (where $g^*(n) = 2^n$) then every forcing notion with Laver property is $(\bar{t}, h_1, h_2)$–bounding.*

PROOF. 1) Suppose that $\langle \dot{s}_n : n \in \omega \rangle$ is a $\mathbb{P}$-name for an element of $U_{h_1}(\bar{t})$, $p \in \mathbb{P}$. Since (K, Σ) is simple we know that $p \Vdash \dot{s}_n \in \Sigma(t_n)$. Consequently, we may apply the assumption that $\mathbb{P}$ is (f, g)–bounding (remember the property of f) and we get a condition $p_0 \geq p$ and a sequence $\langle s^+_{n,\ell} : \ell < g^+(m^{t_n}_{\mathrm{dn}}),\ n < \omega \rangle$ such that

$$\begin{array}{ll} (\forall n < \omega)(\forall \ell < g(m^{t_n}_{\mathrm{dn}}))(s^+_{n,\ell} \in V^n_{h_1}(\bar{t})) & \text{and} \\ p_0 \Vdash_{\mathbb{P}} (\forall n \in \omega)(\dot{s}_n \in \{s^+_{n,\ell} : \ell < g^+(m^{t_n}_{\mathrm{dn}})\}), & \end{array}$$

where $g^+ \in \omega^\omega$ is such that for each $n \in \omega$:

$$g^+(m^{t_n}_{\mathrm{dn}}) = \begin{cases} g(m^{t_n}_{\mathrm{dn}}) & \text{if } h_1(m^{t_n}_{\mathrm{dn}}) + 1 \leq h_2(m^{t_n}_{\mathrm{dn}}), \\ 1 & \text{otherwise.} \end{cases}$$

Since $\bar{t}$ is (g, h_1)-additive we find $\langle s^*_n : n \in \omega \rangle \in U_{h_2}(\bar{t})$ such that for each $n \in \omega$

$$(\forall \ell < g^+(m^{t_n}_{\mathrm{dn}}))(s^+_{n,\ell} \in \Sigma(s^*_n)).$$

Clearly, $p_0 \Vdash \langle s^*_n : n < \omega \rangle \leq \langle \dot{s}_n : n < \omega \rangle$.

2) Similarly. □

DEFINITION 5.2.8. Let (K, Σ) be a creating pair and $\bar{t} = \langle t_n : n \in \omega \rangle \in \mathrm{PC}_\infty(K, \Sigma)$.

1. We say that a partial ordering $\bar{\mathcal{F}} = (\mathcal{F}, <^*_{\mathcal{F}})$ on $\mathcal{F} \subseteq \omega^\omega$ is $\bar{t}$*–good* if:
   (a) $\bar{\mathcal{F}}$ is a dense partial order with no maximal and minimal elements,
   (b) for each $h \in \mathcal{F}$

$$(\forall n \in \omega)(1 < h(m^{t_n}_{\mathrm{dn}}) \leq \mathbf{nor}[t_n]) \quad \text{and} \quad \lim_{n \to \infty} [\mathbf{nor}[t_n] - h(m^{t_n}_{\mathrm{dn}})] = \infty,$$

   (c) if $h_1, h_2 \in \mathcal{F}$, $h_1 <^*_{\mathcal{F}} h_2$ then

$$(\forall n \in \omega)(h_1(m^{t_n}_{\mathrm{dn}}) \leq h_2(m^{t_n}_{\mathrm{dn}})) \quad \text{and} \quad \lim_{n \to \infty} [h_2(m^{t_n}_{\mathrm{dn}}) - h_1(m^{t_n}_{\mathrm{dn}})] = \infty.$$

2. Let $\bar{\mathcal{F}}$ be a $\bar{t}$–good partial order (on $\mathcal{F} \subseteq \omega^\omega$). We say that a forcing notion $\mathbb{P}$ is $(\bar{t}, \bar{\mathcal{F}})$–bounding if $\mathbb{P}$ is $(\bar{t}, h_1, h_2)$–bounding for all $h_1, h_2 \in \mathcal{F}$ such that $h_1 <^*_{\mathcal{F}} h_2$.

It should be clear that if $\bar{\mathcal{F}}$ is $\bar{t}$–good, $\bar{t}\in \mathrm{PC}_\infty(K,\Sigma)$ then the composition of $(\bar{t},\bar{\mathcal{F}})$–bounding forcing notions is $(\bar{t},\bar{\mathcal{F}})$–bounding. To deal with the limit stages (in countable support iterations) we have to apply the technique of [**Sh:f**, Ch VI, §1]. Theorem 5.2.9 below fulfills the promise of [**BaJu95**, 7.2.29].

THEOREM 5.2.9. *Let $(K,\Sigma)\in\mathcal{H}(\aleph_1)$ be a simple and reducible creating pair and let $\bar{t}=\langle t_n : n<\omega\rangle\in \mathrm{PC}_\infty(K,\Sigma)$, $\mathbf{nor}[t_n]\geq 2$. Suppose that $\bar{\mathcal{F}}=(\mathcal{F},<^*_{\mathcal{F}})$ is a $\bar{t}$–good partial order such that $\bar{t}$ is $(2,h)$–additive for all $h\in\mathcal{F}$. Assume that $\langle \mathbb{P}_\alpha,\dot{\mathbb{Q}}_\alpha:\alpha<\beta\rangle$ is a countable support iteration of proper forcing notions such that for each $\alpha<\beta$:*

$$\Vdash_{\mathbb{P}_\alpha} \text{“}\dot{\mathbb{Q}}_\alpha \textit{ is } (\bar{t},\bar{\mathcal{F}})\textit{–bounding”}.$$

Then $\mathbb{P}_\beta$ is $(\bar{t},\bar{\mathcal{F}})$–bounding.

PROOF. We are going to use [**Sh:f**, Ch VI, 1.13A] and therefore we will closely follow the notation and terminology of [**Sh:f**, Ch VI, §1], checking all necessary assumptions. First we define a fine covering model $(D^{\bar{\mathcal{F}},\bar{t}},R^{\bar{\mathcal{F}},\bar{t}},<^{\bar{\mathcal{F}},\bar{t}})$ (see [**Sh:f**, Ch VI, 1.2]).

For $h\in\mathcal{F}$ and $n\in\omega$ we fix a mapping $\psi^h_n:\omega\stackrel{\rm onto}{\longrightarrow}V^n_h(\bar{t})$. Next, for $h\in\mathcal{F}$ and $\eta\in\omega^\omega$ we let $\psi^h(\eta)=\langle\psi^h_n(\eta(n)):n<\omega\rangle$. Note that, as (K,Σ) is nice, $\psi^h(\eta)\in U_h(\bar{t})$ (for all $\eta\in\omega^\omega$). Further, for $h\in\mathcal{F}$ and for $\bar{s}=\langle s_n:n<\omega\rangle\geq\bar{t}$ we let

$$T_{\bar{s},h}=\{\nu\in\omega^{<\omega}:(\forall n<\ell g(\nu))(\psi^h_n(\nu(n))\in\Sigma(s_n))\}.$$

Clearly each $T_{\bar{s},h}$ is a subtree of $\omega^{<\omega}$ and any node in $T_{\bar{s},h}$ has a proper extension in $T_{\bar{s},h}$ (remember that (K,Σ) is reducible). We define:

- $D^{\bar{\mathcal{F}},\bar{t}}$ is $\mathcal{H}(\aleph_1)=\mathcal{H}(\aleph_1)^{\mathbf{V}}$ (we want to underline here that – in the iteration – $D^{\bar{\mathcal{F}},\bar{t}}$ is fixed and consists of elements of the ground universe),
- for $x,T\in D^{\bar{\mathcal{F}},\bar{t}}$ we say that $x\ R^{\bar{\mathcal{F}},\bar{t}}\ T$ if and only if $x=\langle h^*,h\rangle$ and $T=T_{\bar{s},h^*}$ for some $h^*,h\in\mathcal{F}$, $h^*<^*_{\mathcal{F}}h$, $\bar{s}\in U_h(\bar{t})\cap D^{\bar{\mathcal{F}},\bar{t}}$,
- for $\langle h^*,h\rangle,\langle h^{**},h'\rangle\in\mathrm{dom}(R^{\bar{\mathcal{F}},\bar{t}})$ we say that $\langle h^*,h\rangle<^{\bar{\mathcal{F}},\bar{t}}\langle h^{**},h'\rangle$ if and only if $h^*=h^{**}<^*_{\mathcal{F}}h<^*_{\mathcal{F}}h'$.

CLAIM 5.2.9.1. *$(D^{\bar{\mathcal{F}},\bar{t}},R^{\bar{\mathcal{F}},\bar{t}})$ is a weak covering model in $\mathbf{V}$ (see* [**Sh:f**, Ch VI, 1.1]*).*

Proof of the claim: The demand (a) of the definition [**Sh:f**, Ch VI, 1.1] of weak covering models holds by the way we defined $R^{\bar{\mathcal{F}},\bar{t}}$. The clause (b) there is satisfied as for each $\eta\in\omega^\omega$ and $x=\langle h^*,h\rangle$ such that $h^*,h\in\mathcal{F}$, $h^*<^*_{\mathcal{F}}h$, we have $\psi^{h^*}(\eta)\in U_{h^*}(\bar{t})\subseteq U_h(\bar{t})$, $x\ R^{\bar{\mathcal{F}},\bar{t}}\ T_{\psi^{h^*}(\eta),h^*}$ and $\eta\in\lim(T_{\psi^{h^*}(\eta),h^*})$.

CLAIM 5.2.9.2. *A forcing notion $\mathbb{P}$ is $(\bar{t},\bar{\mathcal{F}})$–bounding if and only if it is $(D^{\bar{\mathcal{F}},\bar{t}},R^{\bar{\mathcal{F}},\bar{t}})$–preserving (see* [**Sh:f**, Ch VI, 1.5]*).*

Proof of the claim: Suppose $\mathbb{P}$ is $(\bar{t},\bar{\mathcal{F}})$–bounding. Let $\dot{\eta}$ be a $\mathbb{P}$–name for an element of ω^ω and $x=\langle h^*,h\rangle\in\mathrm{dom}(R^{\bar{\mathcal{F}},\bar{t}})$ (so $h^*,h\in\mathcal{F}$ and $h^*<^*_{\mathcal{F}}h$). Then $\Vdash_{\mathbb{P}}\psi^{h^*}(\dot{\eta})\in U_{h^*}(\bar{t})$ and, as by 5.2.8(2) $\mathbb{P}$ is $(\bar{t},h^*,h)$–bounding,

$$\Vdash_{\mathbb{P}}(\exists\bar{s}\in U_h(\bar{t})\cap\mathbf{V})(\bar{s}\leq\psi^{h^*}(\dot{\eta}))$$

and hence

$$\Vdash_{\mathbb{P}}(\exists T\in D^{\bar{\mathcal{F}},\bar{t}})(x\ R^{\bar{\mathcal{F}},\bar{t}}\ T\quad\&\quad\dot{\eta}\in\lim(T))$$

(so $(D^{\bar{\mathcal{F}},\bar{t}}, R^{\bar{\mathcal{F}},\bar{t}})$ covers in $\mathbf{V}^{\mathbb{P}}$).

On the other hand suppose that $\mathbb{P}$ is $(D^{\bar{\mathcal{F}},\bar{t}}, R^{\bar{\mathcal{F}},\bar{t}})$–preserving. Take $h^*, h \in \mathcal{F}$ such that $h^* <^*_{\mathcal{F}} h$ and let $\dot{\bar{s}}$ be a $\mathbb{P}$–name for an element of $U_{h^*}(\bar{t})$. Let $\dot{\eta}$ be a $\mathbb{P}$–name for an element of ω^ω such that $\Vdash_{\mathbb{P}} \psi^{h^*}(\dot{\eta}) = \dot{\bar{s}}$. As, in $\mathbf{V}^{\mathbb{P}}$, $(D^{\bar{\mathcal{F}},\bar{t}}, R^{\bar{\mathcal{F}},\bar{t}})$ still covers (and $x = \langle h^*, h\rangle \in \mathrm{dom}(R^{\bar{\mathcal{F}},\bar{t}})$) we have

$$\Vdash_{\mathbb{P}} (\exists \bar{s} \in D^{\bar{\mathcal{F}},\bar{t}})(\bar{s} \in U_h(\bar{t})\ \&\ \dot{\eta} \in \lim(T_{\bar{s},h^*})).$$

Hence we may conclude that $\mathbb{P}$ is $(\bar{t}, h^*, h)$–bounding.

CLAIM 5.2.9.3. *$(D^{\bar{\mathcal{F}},\bar{t}}, R^{\bar{\mathcal{F}},\bar{t}}, <^{\bar{\mathcal{F}},\bar{t}})$ is a fine covering model (see* [**Sh:f**, Ch VI, 1.2]*).*

Proof of the claim: We have to check the requirements of [**Sh:f**, Ch VI, 1.2(1)]. We will comment on each of them, referring to the enumeration there.

(α) $(D^{\bar{\mathcal{F}},\bar{t}}, R^{\bar{\mathcal{F}},\bar{t}})$ is a weak covering model (by 5.2.9.1).

(β) $<^{\bar{\mathcal{F}},\bar{t}}$ is a partial order on $\mathrm{dom}(R^{\bar{\mathcal{F}},\bar{t}}) = \{\langle h^*, h\rangle \in \mathcal{F} \times \mathcal{F} : h^* <^*_{\mathcal{F}} h\}$ such that:

  (i) there is no minimal element in $<^{\bar{\mathcal{F}},\bar{t}}$,

  (ii) $<^{\bar{\mathcal{F}},\bar{t}}$ is dense as $<^*_{\mathcal{F}}$ is such,

  (iii) if $x_1 <^{\bar{\mathcal{F}},\bar{t}} x_2$ (so $x_1 = \langle h^*, h_1\rangle$, $x_2 = \langle h^*, h_2\rangle$ and $h^* <^*_{\mathcal{F}} h_1 <^*_{\mathcal{F}} h_2$) and $x_1\ R^{\bar{\mathcal{F}},\bar{t}}\ T$ then there is $T^* \in D^{\bar{\mathcal{F}},\bar{t}}$ such that $T \subseteq T^*$ and $x_2\ R^{\bar{\mathcal{F}},\bar{t}}\ T$ – namely T itself may serve as T^*,

  (iv) if $x_1, x_2 \in \mathrm{dom}(R^{\bar{\mathcal{F}},\bar{t}})$, $x_1 <^{\bar{\mathcal{F}},\bar{t}} x_2$ and $x_1\ R^{\bar{\mathcal{F}},\bar{t}}\ T_1$, $x_1\ R^{\bar{\mathcal{F}},\bar{t}}\ T_2$ then there is $T \in D^{\bar{\mathcal{F}},\bar{t}}$ such that

$$x_2\ R^{\bar{\mathcal{F}},\bar{t}}\ T, \quad T_1 \subseteq T, \quad \text{and} \quad (\exists n \in \omega)(\forall \nu \in T_2)(\nu{\restriction}n \in T_1 \Rightarrow \nu \in T).$$

For (β)(**iv**) we use the assumption that $\bar{t}$ is $(2,h)$–additive for $h \in \mathcal{F}$. Let $x_1 = \langle h^*, h_1\rangle$, $x_2 = \langle h^*, h_2\rangle$ (so $h^* <^*_{\mathcal{F}} h_1 <^*_{\mathcal{F}} h_2$) and let $\bar{s}_\ell = \langle s_{\ell,m} : m < \omega\rangle \in U_{h_1}(\bar{t})$ be such that $T_\ell = T_{\bar{s}_\ell,h^*}$ (for $\ell = 1,2$). By 5.2.8(1c) we find $n < \omega$ such that

$$(\forall m \geq n)(h_1(m^{t_n}_{\mathrm{dn}}) + 1 < h_2(m^{t_n}_{\mathrm{dn}})).$$

For each $m \geq n$ we choose $s_m \in \Sigma(t_m)$ such that

$$s_{1,m}, s_{2,m} \in \Sigma(s_m) \quad \text{and} \quad \mathbf{nor}[s_m] \leq \max\{\mathbf{nor}[s_{1,m}], \mathbf{nor}[s_{2,m}]\} + 1 < h_2(m^{t_n}_{\mathrm{dn}})$$

(remember that t_m is $(2, h_1(m^{t_m}_{\mathrm{dn}}))$–additive). For $m < n$ we let $s_m = s_{1,m} \in V^m_{h_1}(\bar{t}) \subseteq V^m_{h_2}(\bar{t})$. Finally let $\bar{s} = \langle s_m : m < \omega\rangle$. As (K, Σ) is nice, and by the choice of the s_m's we have $\bar{s} \in U_{h_2}(\bar{t})$. Look at $T \stackrel{\mathrm{def}}{=} T_{\bar{s},h^*} \in D^{\bar{\mathcal{F}},\bar{t}}$. By definitions, $x_2\ R^{\bar{\mathcal{F}},\bar{t}}\ T$, $T_1 \subseteq T$ (remember that $\Sigma(s_{1,m}) \subseteq \Sigma(s_m)$ for all $m \in \omega$) and if $\nu \in T_2$, $\nu{\restriction}n \in T_1$ then $\nu \in T$ (as $\Sigma(s_{2,m}) \subseteq \Sigma(s_m)$ for $m \geq n$ and $s_m = s_{1,m}$ for $m < n$).

Now comes the main part: conditions (γ) and (δ) of [**Sh:f**, Ch VI, 1.2(1)]. As the second one is stronger, we will verify it only. Let us state what we have to show.

(δ) If $\mathbf{V}^*$ is a generic extension of $\mathbf{V}$ and, in $\mathbf{V}^*$, $(D^{\bar{\mathcal{F}},\bar{t}}, R^{\bar{\mathcal{F}},\bar{t}})$ is a weak covering model (i.e. it still covers) then the following two requirements are satisfied (in $\mathbf{V}^*$).

  (a) If $x, x^+, x_n \in \mathrm{dom}(R^{\bar{\mathcal{F}},\bar{t}})$ and $T_n \in D^{\bar{\mathcal{F}},\bar{t}}$ are such that for each $n \in \omega$:

$$x_n <^{\bar{\mathcal{F}},\bar{t}} x_{n+1} <^{\bar{\mathcal{F}},\bar{t}} x^+ <^{\bar{\mathcal{F}},\bar{t}} x \quad \text{and} \quad x_n\ R^{\bar{\mathcal{F}},\bar{t}}\ T_n$$

then there are $T^*\in D^{\bar{\mathcal{F}},\bar{t}}$ and $W\in[\omega]^{\omega}$ such that $x\ R^{\bar{\mathcal{F}},\bar{t}}\ T^*$ and

$$\{\eta\in\omega^{\omega}\colon(\forall i\in W)(\eta\restriction\min(W\setminus(i+1))\in\bigcup_{\substack{j<i,\\ j\in W}}T_j\cup T_0)\}\subseteq\lim(T^*).$$

(b) If $\eta,\eta_n\in\omega^{\omega}$ are such that $\eta\restriction n=\eta_n\restriction n$ for every $n\in\omega$ and $x\in\operatorname{dom}(R^{\bar{\mathcal{F}},\bar{t}})$ then

$$(\exists T\in D^{\bar{\mathcal{F}},\bar{t}})(\exists^{\infty}n)\big(x\ R^{\bar{\mathcal{F}},\bar{t}}\ T\quad\&\quad\eta_n\in\lim(T)\big).$$

So suppose that $\mathbf{V}\subseteq\mathbf{V}^*$ is a generic extension and $\mathbf{V}^*\models$"$(D^{\bar{\mathcal{F}},\bar{t}},R^{\bar{\mathcal{F}},\bar{t}})$ covers". We work in $\mathbf{V}^*$.

(δ)**(a)** Let $h^*,h_n,h^+,h\in\mathcal{F}$ be such that $x_n=\langle h^*,h_n\rangle$, $x=\langle h^*,h\rangle$, $x^+=\langle h^*,h^+\rangle$ and $h^*<^*_{\mathcal{F}}h_n<^*_{\mathcal{F}}h_{n+1}<^*_{\mathcal{F}}h^+<^*_{\mathcal{F}}h$. Choose inductively an increasing sequence $0<n_0<n_1<\ldots<\omega$ such that

$$(\forall m\geq n_0)(h_0(m^{t_m}_{\rm dn})+1<h_1(m^{t_m}_{\rm dn})\leq h^+(m^{t_m}_{\rm dn}))\qquad\text{and}$$

$$(\forall i\in\omega)(\forall m\geq n_{i+1})(h_{n_i}(m^{t_m}_{\rm dn})+i+2<h_{n_i+1}(m^{t_m}_{\rm dn})\leq h^+(m^{t_m}_{\rm dn}))$$

(possible by 5.2.8(1c)). Let $\bar{s}_n=\langle s_{n,m}:m<\omega\rangle\in U_{h_n}(\bar{t})\cap D^{\bar{\mathcal{F}},\bar{t}}$ be such that $T_n=T_{\bar{s}_n,h^*}$. (Note: each $\bar{s}_n$, h_n is in $\mathbf{V}$ but the sequences $\langle\bar{s}_n:n<\omega\rangle$, $\langle h_n:n<\omega\rangle$ do not have to be there.) Using $(2,h^+)$–additivity of $\bar{t}$ we choose $s^+_m\in\Sigma(t_m)$ such that for all $m\in\omega$ we have $\mathbf{nor}[s^+_m]\leq h^+(m^{s_m}_{\rm dn})$ and

if $m<n_1$ then $s^+_m=s_{0,m}$ (so $s_{0,m}\in\Sigma(s^+_m)$),

if $n_{i+1}\leq m<n_{i+2}$, $i<\omega$ then $s_{0,m},s_{n_0,m},\ldots,s_{n_i,m}\in\Sigma(s^+_m)$

(remember the choice of the n_i's). As (K,Σ) is nice we have $\bar{s}^+=\langle s^+_m:m<\omega\rangle\in U_{h^+}(\bar{t})$. Since $h^+<^*_{\mathcal{F}}h$ and $\mathbf{V}^*\models$"$(D^{\bar{\mathcal{F}},\bar{t}},R^{\bar{\mathcal{F}},\bar{t}})$ covers" we find $\bar{s}\in U_h(\bar{t})\cap D^{\bar{\mathcal{F}},\bar{t}}$ such that $\bar{s}\leq\bar{s}^+$ (compare the proof of 5.2.9.2). Look at $T^*\stackrel{\rm def}{=}T_{\bar{s},h^*}\in D^{\bar{\mathcal{F}},\bar{t}}$. By the definitions, $x\ R^{\bar{\mathcal{F}},\bar{t}}\ T^*$. Let $W\stackrel{\rm def}{=}\{n_i:i\in\omega\}$. Suppose that $\eta\in\omega^{\omega}$ is such that

$$(\forall i\in\omega)(\eta\restriction n_{i+1}\in\bigcup_{j<i}T_{n_j}\cup T_0).$$

This means that for each $m\in\omega$:

if $m<n_1$ then $\eta\restriction(m+1)\in T_0=T_{\bar{s}_0,h^*}$ and so

$$\psi^{h^*}_m(\eta(m))\in\Sigma(s_{0,m})\subseteq\Sigma(s^+_m)\subseteq\Sigma(s_m),$$

if $n_{i+1}\leq m<n_{i+2}$, $i\in\omega$ then $\eta\restriction(m+1)\in T_0\cup T_{n_0}\cup\ldots\cup T_{n_i}$ and so

$$\psi^{h^*}_m(\eta(m))\in\Sigma(s_{0,m})\cup\Sigma(s_{n_0,m})\cup\ldots\cup\Sigma(s_{n_i,m})\subseteq\Sigma(s^+_m)\subseteq\Sigma(s_m).$$

Hence $(\forall m\in\omega)(\psi^{h^*}_m(\eta(m))\in\Sigma(s_m))$ what implies that $\eta\in\lim(T_{\bar{s},h^*})=\lim(T^*)$, finishing the proof of (δ)**(a)**.

(δ)**(b)** This is somewhat similar to (δ)**(a)**. Let $h^*,h\in\mathcal{F}$ be such that $x=\langle h^*,h\rangle$ (so $h^*<^*_{\mathcal{F}}h$). As $\bar{\mathcal{F}}$ is a dense partial order we may take $h^+\in\mathcal{F}$ such that $h^*<^*_{\mathcal{F}}h^+<^*_{\mathcal{F}}h$. Choose $0<n_0<n_1<\ldots<\omega$ such that

$$(\forall i\in\omega)(\forall m\geq n_i)(h^*(m^{t_m}_{\rm dn})+i+2<h^+(m^{t_m}_{\rm dn})).$$

Now take $s^+_m\in\Sigma(t_m)$ such that for $m\in\omega$ we have $\mathbf{nor}[s^+_m]\leq h^+(m^{s_m}_{\rm dn})$ and

if $m < n_0$ then $s^+_m = \psi^{h^*}_m(\eta_{n_0}(m))$ (so $\psi^{h^*}_m(\eta_{n_0}(m)) \in \Sigma(s^+_m)$), and
if $n_i \le m < n_{i+1}$, $i < \omega$ then $\psi^{h^*}_m(\eta_{n_0}(m)), \ldots, \psi^{h^*}_m(\eta_{n_{i+1}}(m)) \in \Sigma(s^+_m)$

(possible as $\bar{t}$ is $(2, h^+)$–additive; remember the choice of the n_i's). Note that, since $\eta{\restriction}n = \eta_n{\restriction}n$ for all $n \in \omega$, we have that

$$(\forall i \in \omega)(\forall m \in \omega)(\psi^{h^*}_m(\eta_{n_i}(m)) \in \Sigma(s^+_m)).$$

Let $\bar{s}^+ = \langle s^+_m : m < \omega\rangle$. Thus $\bar{s}^+ \in U_{h^+}(\bar{t})$ and, as $(D^{\bar{\mathcal{F}},\bar{t}}, R^{\bar{\mathcal{F}},\bar{t}})$ covers in $\mathbf{V}^*$ and $h^+ <^*_{\mathcal{F}} h$, we find $\bar{s} = \langle s_m : m < \omega\rangle \in U_h(\bar{t}) \cap D^{\bar{\mathcal{F}},\bar{t}}$ such that $\bar{s} \le \bar{s}^+$. Now we have

$$(\forall i \in \omega)(\forall m \in \omega)(\psi^{h^*}_m(\eta_{n_i}(m)) \in \Sigma(s_m)),$$

and therefore $(\forall i \in \omega)(\eta_{n_i} \in \lim(T_{\bar{s},h^*}))$. As $x\ R^{\bar{\mathcal{F}},\bar{t}}\ T_{\bar{s},h^*}$, we finish the proof of the claim.

Now, to finish the proof of the theorem we put together 5.2.9.2, 5.2.9.3 and [**Sh:f**, Ch VI, 1.13A]. □

5.3. Quasi-generic Γ and preserving them

Here we will develop the technique announced in [**Sh:f**, Ch XVIII, 3.14, 3.15], putting it in a slightly more general setting, more suitable for our context. We will get a reasonably weak, but still easily iterable, condition for not adding Cohen reals – this will be used in 5.4.3, 5.4.4. But the general schema presented here will be applied in the next chapter too (to preserve some ultrafilters on ω).

DEFINITION 5.3.1. Suppose that (K, Σ) is a creating pair, $\bar{t} = \langle t_k : k < \omega\rangle \in \mathrm{PC}(K, \Sigma)$.

1. A function $W : \omega \times \omega \times \omega^{\underline{\omega}} \longrightarrow \mathcal{P}(K)$ is called a *$\bar{t}$–system for* (K, Σ) if:
   (a) if $k \le \ell < \omega$ and $\sigma : [m^{t_k}_{\rm dn}, m^{t_\ell}_{\rm up}) \longrightarrow \omega$ then $W(m^{t_k}_{\rm dn}, m^{t_\ell}_{\rm up}, \sigma) \subseteq \Sigma(t_k, \ldots, t_\ell)$, in all other instances $W(m_0, m_1, \sigma)$ is empty,
   (b) if $s \in \Sigma(t_k, \ldots, t_\ell)$, $k \le \ell < \omega$ then there is $n = n_W(s) \in [k, \ell]$ such that for each $\sigma_0, \sigma_1 : [m^{t_k}_{\rm dn}, m^{t_\ell}_{\rm up}) \longrightarrow \omega$
   if $\quad \sigma_0{\restriction}[m^{t_n}_{\rm dn}, m^{t_n}_{\rm up}) = \sigma_1{\restriction}[m^{t_n}_{\rm dn}, m^{t_n}_{\rm up})$
   then $\quad s \in W(m^s_{\rm dn}, m^s_{\rm up}, \sigma_0) \quad \Leftrightarrow \quad s \in W(m^s_{\rm dn}, m^s_{\rm up}, \sigma_1)$,
   (c) if $k_0 < k_1 < \ldots < k_i$, $s_j \in \Sigma(t_{k_j}, \ldots, t_{k_{j+1}-1})$ for $j < i$, $s \in \Sigma(s_0, \ldots, s_{i-1})$ and $j_0 < i$ is such that $n_W(s) \in [k_{j_0}, k_{j_0+1})$ (see (b) above) and $\sigma : [m^s_{\rm dn}, m^s_{\rm up}) \longrightarrow \omega$ then

$$s_{j_0} \in W(m^{s_{j_0}}_{\rm dn}, m^{s_{j_0}}_{\rm up}, \sigma{\restriction}[m^{s_{j_0}}_{\rm dn}, m^{s_{j_0}}_{\rm up})) \quad \Rightarrow \quad s \in W(m^s_{\rm dn}, m^s_{\rm up}, \sigma),$$

   (d) for some unbounded non-decreasing function $G : (1, \infty) \longrightarrow \mathbb{R}^{\ge 0}$ (called sometimes *the weight of* W), for every $s \in \Sigma(t_k, \ldots, t_\ell)$, $k \le \ell < \omega$, $\mathbf{nor}[s] > 1$, and each $\sigma : [m^{t_k}_{\rm dn}, m^{t_\ell}_{\rm dn}) \longrightarrow \omega$ there is $t \in W(m^{t_k}_{\rm dn}, m^{t_\ell}_{\rm up}, \sigma)$ such that

$$t \in \Sigma(s) \quad \text{and} \quad \mathbf{nor}[t] \ge G(\mathbf{nor}[s]).$$

   If the function G might be $G(x) = x - 1$ then we call the $\bar{t}$–system W *regular*.

2. For a norm condition $\mathcal{C}(\mathbf{nor})$ we let

$$\mathbb{P}^*_{\mathcal{C}(\mathbf{nor})}(\bar{t},(K,\Sigma)) \stackrel{\rm def}{=} \{\bar{s}\in \mathrm{PC}_{\mathcal{C}(\mathbf{nor})}(K,\Sigma) : \bar{t}\leq \bar{s}\}.$$

It is equipped with the partial order $\leq$ inherited from $\mathrm{PC}_{\mathcal{C}(\mathbf{nor})}(K,\Sigma)$. We introduce another relation $\preceq_{\mathcal{C}(\mathbf{nor})}$ on $\mathbb{P}^*_{\mathcal{C}(\mathbf{nor})}(\bar{t},(K,\Sigma))$ letting

$\bar{s}_0 \preceq_{\mathcal{C}(\mathbf{nor})} \bar{s}_1$ if and only if
there is $\bar{s}_2\in\mathbb{P}^*_{\mathcal{C}(\mathbf{nor})}(\bar{t},(K,\Sigma))$ such that $\bar{s}_0\leq\bar{s}_2$ and the sequence $\bar{s}_2$ is eventually equal to $\bar{s}_1$.

If the norm condition $\mathcal{C}(\mathbf{nor})$ is clear we may omit the index to $\preceq$.

3. For a $\bar{t}$–system W (for (K,Σ)) and $\Gamma\subseteq\mathbb{P}^*_{\mathcal{C}(\mathbf{nor})}(\bar{t},(K,\Sigma))$ we say that Γ is *quasi-W-generic in $\mathbb{P}^*_{\mathcal{C}(\mathbf{nor})}(\bar{t},(K,\Sigma))$* if
   (a) $(\Gamma,\preceq)$ is directed (i.e. $(\forall\bar{s}_0,\bar{s}_1\in\Gamma)(\exists\bar{s}\in\Gamma)(\bar{s}_0\preceq\bar{s}\ \&\ \bar{s}_1\preceq\bar{s})$) and countably closed (i.e. if $\langle\bar{s}_n : n<\omega\rangle\subseteq\Gamma$ is $\preceq$–increasing then there is $\bar{s}\in\Gamma$ such that $(\forall n\in\omega)(\bar{s}_n\preceq\bar{s})$),
   (b) for every function $\eta\in\omega^\omega$ there is $\bar{s}=\langle s_m : m<\omega\rangle\in\Gamma$ such that

$$(\forall^\infty m)(s_m\in W(m^{s_m}_{\rm dn},m^{s_m}_{\rm up},\eta{\restriction}[m^{s_m}_{\rm dn},m^{s_m}_{\rm up}))).$$

REMARK 5.3.2. The demand 5.3.1(1d) is to ensure the existence of quasi-W-generic sets Γ (see 5.3.4(2) below). Conditions 5.3.1(1b) and 5.3.1(1c) are to preserve quasi-W-genericity in countable support iterations. As formulated, they will be crucial in the proof of 5.3.12.2(2).

Natural applications of the notions introduced in 5.3.1 will be when (K,Σ) is simple or "simple plus at most omitting". In both cases it will be easy to check demands 5.3.1(1b,c). In the first case they are trivial, see 5.3.3 below.

PROPOSITION 5.3.3. *Suppose that (K,Σ) is a creating pair and $\bar{t}\in\mathrm{PC}(K,\Sigma)$.*

1. *If (K,Σ) is simple then the condition 5.3.1(1b) is empty (so may be omitted) and the condition 5.3.1(1c) is equivalent to*
   $(c)^-$ *if $t\in W(m^{t_n}_{\rm dn},m^{t_n}_{\rm up},\sigma)$, $\sigma:[m^{t_n}_{\rm dn},m^{t_n}_{\rm up})\longrightarrow\omega$ and $s\in\Sigma(t)$ then $s\in W(m^{t_n}_{\rm dn},m^{t_n}_{\rm up},\sigma)$.*
2. *If W is a $\bar{t}$–system for (K,Σ), $\eta\in\omega^\omega$ and $\bar{s}_\ell=\langle s_{\ell,m} : m<\omega\rangle\in\mathbb{P}^*_\emptyset(\bar{t},(K,\Sigma))$ (for $\ell<2$) are such that*

$$s_0\preceq s_1 \quad \textit{and} \quad (\forall^\infty m)(s_{0,m}\in W(m^{s_{0,m}}_{\rm dn},m^{s_{0,m}}_{\rm up},\eta{\restriction}[m^{s_{0,m}}_{\rm dn},m^{s_{0,m}}_{\rm up})))$$

then

$$(\forall^\infty m)(s_{1,m}\in W(m^{s_{1,m}}_{\rm dn},m^{s_{1,m}}_{\rm up},\eta{\restriction}[m^{s_{1,m}}_{\rm dn},m^{s_{1,m}}_{\rm up}))).$$

PROPOSITION 5.3.4. *Suppose that (K,Σ) is a creating pair, $\mathcal{C}(\mathbf{nor})$ is one of the norm conditions introduced in 1.1.10 (i.e. it is one of $(\mathrm{s}\infty)$, (∞), $(\mathrm{w}\infty)$ or (f) for some fast function f) and $\bar{t}\in\mathrm{PC}_{\mathcal{C}(\mathbf{nor})}(K,\Sigma)$.*

1. *$(\mathbb{P}^*_{\mathcal{C}(\mathbf{nor})}(\bar{t},(K,\Sigma)),\preceq_{\mathcal{C}(\mathbf{nor})})$ is a countably closed partial ordering.*
2. *Assume CH. Further suppose that if $\mathcal{C}(\mathbf{nor})\in\{(\mathrm{s}\infty),(\mathrm{w}\infty)\}$ then (K,Σ) is growing. Let $W:\omega\times\omega\times\omega^{\underline{\omega}}\longrightarrow\mathcal{P}(K)$ be a $\bar{t}$–system which is regular if $\mathcal{C}(\mathbf{nor})=(f)$ (for some fast function f). Then there exists a quasi-W-generic Γ in $\mathbb{P}^*_{\mathcal{C}(\mathbf{nor})}(\bar{t},(K,\Sigma))$.*

PROOF. We will show this for $\mathcal{C}(\mathbf{nor})=(\infty)$. In other instances the proof is similar and requires very small changes only.

1) It should be clear that $\preceq$ is a partial order on $\mathbb{P}^*_\infty(\bar{t},(K,\Sigma))$. To show that it is countably closed suppose that $\bar{s}_n = \langle s_{n,m} : m < \omega\rangle \in \mathbb{P}^*_\infty(\bar{t},(K,\Sigma))$ are such that $\bar{s}_n \preceq \bar{s}_{n+1}$ for all $n \in \omega$. Choose an increasing sequence $m_0 < m_1 < \ldots < \omega$ such that $(\forall i \in \omega)(\exists m \in \omega)(m_i = m^{s_{i,m}}_{\mathrm{dn}})$ and for each $n \leq i < \omega$:

if $m < \omega$, $m_i \leq m^{s_{n,m}}_{\mathrm{dn}}$ then $\mathbf{nor}[s_{n,m}] \geq i$ and

if $m < \omega$, $m_i \leq m^{s_{i,m}}_{\mathrm{dn}}$ then $(\exists m' < m'' < \omega)(s_{i,m} \in \Sigma(s_{n,m'},\ldots,s_{n,m''-1}))$.

Now choose $\bar{s} = \langle s_m : m < \omega\rangle \in \mathrm{PC}(K,\Sigma)$ such that

if $m \in \omega$, $m^{s_{0,m}}_{\mathrm{dn}} < m_1$ then $s_m = s_{0,m}$,

if $m \in \omega$, $m^{s_{k+1,m}}_{\mathrm{dn}} \in [m_{k+1}, m_{k+2})$ then $s_{k+1,m} = s_{m^*}$ for some $m^* \in \omega$.

Clearly the choice is possible (and uniquely determined) and, by the niceness, we are sure that $\bar{s} \in \mathrm{PC}(K,\Sigma)$. Moreover, by the choice of m_i's, we have that $\bar{s} \in \mathrm{PC}_\infty(K,\Sigma)$ and so $\bar{s} \in \mathbb{P}^*_\infty(\bar{t},(K,\Sigma))$. Plainly, $\bar{s}_n \preceq \bar{s}$ for all $n \in \omega$.

2) First note that if $\bar{s} \in \mathbb{P}^*_\infty(\bar{t},(K,\Sigma))$, $\eta \in \omega^\omega$ then there is $\bar{s}^* = \langle s^*_m : m < \omega\rangle \in \mathbb{P}^*_\infty(\bar{t},(K,\Sigma))$ such that $\bar{s} \leq \bar{s}^*$ and $(\forall^\infty m)(s^*_m \in W(m^{s^*_m}_{\mathrm{dn}}, m^{s^*_m}_{\mathrm{up}}, \eta{\restriction}[m^{s^*_m}_{\mathrm{dn}}, m^{s^*_m}_{\mathrm{up}})))$ (by 5.3.1(1d), remember that the weight of W is unbounded and non-decreasing).

Using this remark, (1) above, and the assumption of CH we may build a $\preceq$-increasing sequence $\langle \bar{s}_\alpha : \alpha < \omega_1\rangle \subseteq \mathbb{P}^*_\infty(\bar{t},(K,\Sigma))$ such that

$$(\forall \eta \in \omega^\omega)(\exists \alpha < \omega_1)(\forall^\infty n)(s_{\alpha,n} \in W(m^{s_{\alpha,n}}_{\mathrm{dn}}, m^{s_{\alpha,n}}_{\mathrm{up}}, \eta{\restriction}[m^{s_{\alpha,n}}_{\mathrm{dn}}, m^{s_{\alpha,n}}_{\mathrm{up}}))).$$

This sequence gives a quasi-W-generic $\Gamma = \{\bar{s}_\alpha : \alpha < \omega_1\}$.

Note that proving (2) for $\mathcal{C}(\mathbf{nor}) \in \{(\mathrm{s}\infty),(\mathrm{w}\infty)\}$ we have to assume something about the creating pair (K,Σ). The assumption that it is growing is the most natural one (in our context). It allows us to obtain the respective version of the first sentence of the proof of (2) for (∞). Similarly, if $\mathcal{C}(\mathbf{nor})$ is (f) then we need too assume something about the weight of the system W. The assumption that W is regular is much more than really needed. □

REMARK 5.3.5. If W_i are $\bar{t}$–systems (for $i \in \omega_1$) then we may construct in a similar way (under CH) Γ which is quasi-W_i-generic for all $i < \omega_1$.

DEFINITION 5.3.6. Let (K,Σ) be a creating pair, $\bar{t} \in \mathrm{PC}_{\mathcal{C}(\mathbf{nor})}(K,\Sigma)$ (where $\mathcal{C}(\mathbf{nor})$ is a norm condition). Suppose that $W : \omega \times \omega \times \omega^{\underline{\omega}} \longrightarrow \mathcal{P}(K)$ is a $\bar{t}$–system for (K,Σ), and $\Gamma \subseteq \mathbb{P}^*_{\mathcal{C}(\mathbf{nor})}(\bar{t},(K,\Sigma))$ is quasi-W-generic. We say that a proper forcing notion $\mathbb{P}$ is (Γ,W)*–genericity preserving* (or Γ*–genericity preserving* if W is clear) if $\Vdash_{\mathbb{P}}$ " Γ is quasi-W-generic".

REMARK 5.3.7.
1. Note that if $\mathbb{P}$ is a proper forcing notion and $\Gamma \subseteq \mathbb{P}^*_{\mathcal{C}(\mathbf{nor})}(\bar{t},(K,\Sigma))$ is quasi-W-generic then

$$\Vdash_{\mathbb{P}} \text{“}(\Gamma,\preceq) \text{ is directed and countably closed”}.$$

Which may fail after the forcing is condition 5.3.1(3b), so the real meaning of 5.3.6 is that this condition is preserved.
2. The composition of Γ–genericity preserving forcing notions is clearly Γ–genericity preserving. To handle the limit stages in countable support iterations we use the main result of [**Sh:f**, Ch XVIII, §3], see 5.3.12 below.

DEFINITION 5.3.8. Let (K,Σ) be a creating pair, $\bar{t} \in \mathrm{PC}(K,\Sigma)$. We say that a $\bar{t}$–system $W : \omega \times \omega \times \omega^{\underline{\omega}} \longrightarrow \mathcal{P}(K)$ is

1. *Cohen–sensitive* if for all sufficiently large $m' \leq m'' < \omega$

$$(\forall s \in \Sigma(t_{m'},\ldots,t_{m''}))(\exists \sigma : [m^s_{\mathrm{dn}}, m^s_{\mathrm{up}}) \longrightarrow \omega)(s \notin W(m^s_{\mathrm{dn}}, m^s_{\mathrm{up}}, \sigma)),$$

2. *directed* if for every $m' \leq m'' < \omega$, $\sigma_0, \ldots, \sigma_m : [m^{t_{m'}}_{\rm dn}, m^{t_{m''}}_{\rm up}) \longrightarrow \omega$ $(m < \omega)$ there is $\sigma : [m^{t_{m'}}_{\rm dn}, m^{t_{m''}}_{\rm up}) \longrightarrow \omega$ such that
$$W(m^{t_{m'}}_{\rm dn}, m^{t_{m''}}_{\rm up}, \sigma) \subseteq \bigcap_{\ell \leq m} W(m^{t_{m'}}_{\rm dn}, m^{t_{m''}}_{\rm up}, \sigma_\ell).$$

PROPOSITION 5.3.9. *Suppose that (K,Σ) is a creating pair, $\bar{t} \in {\rm PC}_{\mathcal{C}(\mathbf{nor})}(K,\Sigma)$, $W : \omega \times \omega \times \omega^{\underline{\omega}} \longrightarrow \mathcal{P}(K)$ is a $\bar{t}$–system and $\Gamma \subseteq \mathbb{P}^*_{\mathcal{C}(\mathbf{nor})}(\bar{t},(K,\Sigma))$ is quasi-W-generic. Let $\mathbb{P}$ be a proper forcing notion.*

1. *If W is Cohen–sensitive and $\mathbb{P}$ is Γ–genericity preserving then $\mathbb{P}$ does not add Cohen reals.*
2. *If W is directed, (K,Σ) is simple and $\mathbb{P}$ is ω^ω–bounding then $\mathbb{P}$ is Γ–genericity preserving.*

PROOF. 1) Note that if $\eta \in \omega^\omega$ is a Cohen real over $\mathbf{V}$, W is Cohen–sensitive and $\bar{s} = \langle s_m : m < \omega\rangle \in {\rm PC}(K,\Sigma) \cap \mathbf{V}$ then
$$(\exists^\infty m)(s_m \notin W(m^{s_m}_{\rm dn}, m^{s_m}_{\rm up}, \eta{\restriction}[m^{s_m}_{\rm dn}, m^{s_m}_{\rm up}))).$$
2) Suppose that $\dot{\eta}$ is a $\mathbb{P}$–name for a real in ω^ω, $p \in \mathbb{P}$. As $\mathbb{P}$ is ω^ω–bounding we find a function $\eta \in \omega^\omega$ and a condition $q \geq p$ such that $q \Vdash_{\mathbb{P}} (\forall n < \omega)(\dot{\eta}(n) < \eta(n))$. Since W is directed we can find $\eta^* \in \omega^\omega$ such that for each $m < \omega$
$$W(m^{t_m}_{\rm dn}, m^{t_m}_{\rm up}, \eta^*{\restriction}[m^{t_m}_{\rm dn}, m^{t_m}_{\rm up})) \subseteq \bigcap\{W(m^{t_m}_{\rm dn}, m^{t_m}_{\rm up}, \sigma) : \sigma \in \prod_{m^{t_m}_{\rm dn} \leq k < m^{t_m}_{\rm up}} \eta(k)\}.$$
Next, as Γ is quasi-W-generic, we find $\bar{s} = \langle s_m : m < \omega\rangle \in \Gamma$ such that
$$(\forall^\infty m)(s_m \in W(m^{s_m}_{\rm dn}, m^{s_m}_{\rm up}, \eta^*{\restriction}[m^{s_m}_{\rm dn}, m^{s_m}_{\rm up}))).$$
We finish noting that $(\forall m < \omega)(s_m \in \Sigma(t_m))$, as (K,Σ) is simple. □

DEFINITION 5.3.10. Suppose that (K_ℓ,Σ_ℓ) are simple creating pairs and $\bar{t}_\ell = \langle t_{\ell,m} : m < \omega\rangle \in {\rm PC}(K_\ell,\Sigma_\ell)$ (for $\ell < 2$) are such that $(\forall m < \omega)(m^{t_{0,m}}_{\rm dn} = m^{t_{1,m}}_{\rm dn})$.

1. Let $h_0, h_1 \in \omega^\omega$. We say that $\bar{t}_1$–systems $W_0, W_1 : \omega \times \omega \times \omega^{\underline{\omega}} \longrightarrow \mathcal{P}(K_1)$ (for (K_1,Σ_1)) are $(\bar{t}_0,h_0,h_1)$*–coherent* if there are functions ρ_0,ρ_1 (called $(\bar{t}_0,h_0,h_1)$*–coherence witnesses*) such that
   (a) $\rho_0,\rho_1 : \omega^{\underline{\omega}} \longrightarrow \bigcup\{\Sigma_0(t_{0,m}) : m < \omega\}$,
   (b) if $\sigma : [m^{t_{1,m}}_{\rm dn}, m^{t_{1,m}}_{\rm up}) \longrightarrow \omega$, $m < \omega$ then $\rho_\ell(\sigma) \in \Sigma_0(t_{0,m})$ and $\mathbf{nor}[\rho_\ell(\sigma)] \leq h_\ell(m^{t_{0,m}}_{\rm dn})$ (for $\ell = 0,1$),
   (c) for sufficiently large $m < \omega$, for every $\sigma_0 : [m^{t_{1,m}}_{\rm dn}, m^{t_{1,m}}_{\rm up}) \longrightarrow \omega$, there is $\sigma_0^* : [m^{t_{1,m}}_{\rm dn}, m^{t_{1,m}}_{\rm up}) \longrightarrow \omega$ such that
   $$W_1(m^{t_{1,m}}_{\rm dn}, m^{t_{1,m}}_{\rm up}, \sigma_1) \subseteq W_0(m^{t_{1,m}}_{\rm dn}, m^{t_{1,m}}_{\rm up}, \sigma_0)$$
   whenever $\sigma_1 : [m^{t_{1,m}}_{\rm dn}, m^{t_{1,m}}_{\rm up}) \longrightarrow \omega$ is such that $\rho_0(\sigma_0^*) \in \Sigma_0(\rho_1(\sigma_1))$; the sequence σ_0^*, as well as $\rho_0(\sigma_0^*)$, will be called *a ρ_0–cover for σ_0*,
   (d) for each $s \in \Sigma_0(t_{0,m})$, $m < \omega$, if $\mathbf{nor}[s] \leq h_1(m^{t_{0,m}}_{\rm dn})$ then there is $\sigma : [m^{t_{1,m}}_{\rm dn}, m^{t_{1,m}}_{\rm up}) \longrightarrow \omega$ such that $\rho_1(\sigma) = s$.
2. Suppose that $\bar{\mathcal{F}} = (\mathcal{F}, <^*_{\mathcal{F}})$ is a $\bar{t}_0$–good partial order (see 5.2.8). We say that a family $\langle W^k_h : h \in \mathcal{F}, k \in \omega\rangle$ of $\bar{t}_1$–systems for (K_1,Σ_1) is $(\bar{t}_0,\bar{\mathcal{F}})$*–coherent* if for every $k \in \omega$ and $h_0 \in \mathcal{F}$ there is $h_1 \in \mathcal{F}$ such that $h_0 <^*_{\mathcal{F}} h_1$ and the systems $W^k_{h_0}, W^{k+1}_{h_1}$ are $(\bar{t}_0,h_0,h_1)$–coherent.

THEOREM 5.3.11. *Let $(K_0,\Sigma_0),(K_1,\Sigma_1)$ be simple creating pairs and let*

$$\bar{t}_0\in \mathrm{PC}_\infty(K_0,\Sigma_0),\qquad \bar{t}_1\in \mathrm{PC}_{\mathcal{C}(\mathbf{nor})}(K_1,\Sigma_1)$$

be such that $(\forall m<\omega)(m^{t_{0,m}}_{\rm dn}=m^{t_{1,m}}_{\rm dn})$. Assume that $\bar{\mathcal{F}}=(\mathcal{F},<^*_{\mathcal{F}})$ is a $\bar{t}_0$–good partial order and $\langle W^k_h: h\in\mathcal{F},k\in\omega\rangle$ is a $(\bar{t}_0,\bar{\mathcal{F}})$–coherent family of $\bar{t}_1$–systems for (K_1,Σ_1). Further suppose that $\Gamma\subseteq\mathbb{P}^*_{\mathcal{C}(\mathbf{nor})}(\bar{t}_1,(K_1,\Sigma_1))$ is quasi-W^k_h-generic for all $h\in\mathcal{F}$, $k\in\omega$. Then *every $(\bar{t}_0,\bar{\mathcal{F}})$–bounding proper forcing notion is (Γ,W^k_h)–genericity preserving for all $h\in\mathcal{F}$, $k\in\omega$.*

PROOF. Let $\mathbb{P}$ be a proper $(\bar{t}_0,\bar{\mathcal{F}})$–bounding forcing notion. We have to show that for all $h\in\mathcal{F}$, $k\in\omega$

$$\Vdash_{\mathbb{P}} \text{“}\Gamma \text{ is quasi-}W^k_h\text{-generic”}.$$

For this suppose that $\dot{\eta}$ is a $\mathbb{P}$–name for a real in ω^ω, $p\in\mathbb{P}$, $k\in\omega$ and $h_0\in\mathcal{F}$. Take $h_1\in\mathcal{F}$ such that $h_0<^*_{\mathcal{F}}h_1$ and the systems $W^k_{h_0}$, $W^{k+1}_{h_1}$ are $(\bar{t}_0,h_0,h_1)$–coherent and let functions $\rho_0,\rho_1:\omega^{\underline{\omega}}\longrightarrow\bigcup\{\Sigma_0(t_{0,m}):m\in\omega\}$ witness this fact. Let $\dot{\bar{s}}=\langle\dot{s}_m:m<\omega\rangle$ be a $\mathbb{P}$–name for an element of $U_{h_0}(\bar{t}_0)$ such that for some $N\in\omega$

$$p\Vdash_{\mathbb{P}} \text{“ }(\forall m\geq N)(\dot{s}_m \text{ is a } \rho_0\text{–cover for } \dot{\eta}{\restriction}[m^{t_{1,m}}_{\rm dn},m^{t_{1,m}}_{\rm up}))\text{ ”}\qquad\text{(see 5.3.10(1c))}$$

(remember clause 5.3.10(1b)). Now, as $\mathbb{P}$ is $(\bar{t}_0,\bar{\mathcal{F}})$–bounding and $h_0<^*_{\mathcal{F}}h_1$, we find a condition $q\geq p$ and $\bar{s}^*=\langle s^*_m:m<\omega\rangle\in U_{h_1}(\bar{t}_0)$ such that $q\Vdash_{\mathbb{P}}\bar{s}^*\leq\dot{\bar{s}}$. Since (K_0,Σ_0) is simple this means that $q\Vdash_{\mathbb{P}}(\forall m\in\omega)(\dot{s}_m\in\Sigma_0(s^*_m))$. Let $\eta\in\omega^\omega$ be such that for each $m\in\omega$ we have $\rho_1(\eta{\restriction}[m^{t_{1,m}}_{\rm dn},m^{t_{1,m}}_{\rm up}))=s^*_m$ (remember $\mathbf{nor}[s^*_m]\leq h_1(m^{s^*_m}_{\rm dn})$; see clause 5.3.10(1d)). We know that Γ is quasi-$W^{k+1}_{h_1}$-generic, so there is $\bar{s}=\langle s_m:m<\omega\rangle\in\Gamma$ such that

$$(\forall^\infty m)(s_m\in W^{k+1}_{h_1}(m^{s_m}_{\rm dn},m^{s_m}_{\rm up},\eta{\restriction}[m^{s_m}_{\rm dn},m^{s_m}_{\rm up}))).$$

But (K_1,Σ_1) is simple too, so $m^{s_m}_{\rm dn}=m^{s^*_m}_{\rm dn}$, $m^{s_m}_{\rm up}=m^{s^*_m}_{\rm up}$. Thus we may apply 5.3.10(1c) and conclude that for sufficiently large m

$$q\Vdash_{\mathbb{P}} s_m\in W^{k+1}_{h_1}(m^{s_m}_{\rm dn},m^{s_m}_{\rm up},\eta{\restriction}[m^{s_m}_{\rm dn},m^{s_m}_{\rm up}))\subseteq W^k_{h_0}(m^{s_m}_{\rm dn},m^{s_m}_{\rm up},\dot{\eta}{\restriction}[m^{s_m}_{\rm dn},m^{s_m}_{\rm up})),$$

so we are done. □

THEOREM 5.3.12. *Suppose that $(K,\Sigma)\in\mathcal{H}(\aleph_1)$ is a creating pair and $\mathcal{C}(\mathbf{nor})$ is a norm condition. Assume that*

$\bar{t}=\langle t_k:k<\omega\rangle\in\mathrm{PC}_{\mathcal{C}(\mathbf{nor})}(K,\Sigma)$,
$W:\omega\times\omega\times\omega^{\underline{\omega}}\longrightarrow\mathcal{P}(K)$ *is a $\bar{t}$–system and*
$\Gamma\subseteq\mathbb{P}^*_{\mathcal{C}(\mathbf{nor})}(\bar{t},(K,\Sigma))$ *is a quasi-W-generic for (K,Σ).*

Let $\langle\mathbb{P}_\alpha,\dot{\mathbb{Q}}_\alpha:\alpha<\beta\rangle$ be a countable support iteration of proper forcing notions such that for each $\alpha<\beta$:

$$\Vdash_{\mathbb{P}_\alpha}\text{“ }\dot{\mathbb{Q}}_\alpha \textit{ is } \Gamma\textit{–genericity preserving ”}.$$

Then $\mathbb{P}_\beta$ is Γ–genericity preserving.

PROOF. We will use the preservation theorem [**Sh:f**, Ch XVIII, 3.6] and therefore we will follow the notation and terminology of [**Sh:f**, Ch XVIII, §3], checking all necessary details. First we have to define our context $(\bar{R}^{\Gamma,W},S^{\Gamma,W},\mathbf{g}^{\Gamma,W})$ (see [**Sh:f**, Ch XVIII, 3.1]).

For each $m \in \{m^{t_k}_{\rm dn} : k \in \omega\}$ we fix a mapping

$$\psi^m : \omega \stackrel{\rm onto}{\longrightarrow} \{s \in K : (\exists k \leq \ell < \omega)(s \in \Sigma(t_k, \dots, t_\ell) \quad \& \quad m^{t_k}_{\rm dn} = m)\}.$$

Next, for $\eta \in \omega^\omega$ we let $\psi(\eta) = \langle \psi^{m_0}(\eta(0)), \psi^{m_1}(\eta(1)), \dots \rangle$, where $m^{t_0}_{\rm dn} = m_0 < m_1 < m_2 < \dots < \omega$ are chosen in such a way that $m^{\psi^{m_k}(\eta(k))}_{\rm up} = m_{k+1}$. Note that $\psi(\eta) \in \mathbb{P}^*_\emptyset(\bar{t}, (K, \Sigma))$, though it does not have to be in $\mathbb{P}^*_{\mathcal{C}(\mathbf{nor})}(\bar{t}, (K, \Sigma))$ (we do not control the norms).

Now we choose $(\bar{R}^{\Gamma,W}, S^{\Gamma,W}, \mathbf{g}^{\Gamma,W})$ such that:

- $S^{\Gamma,W}$ is, in the ground model $\mathbf{V}$, the collection of all intersections $N \cap \mathcal{H}(\aleph_1)$, where N is a countable elementary submodel of $(\mathcal{H}(\chi), \in, <^*_\chi)$; so $S^{\Gamma,W} \subseteq ([\mathcal{H}(\aleph_1)^{\mathbf{V}}]^{\leq \aleph_0})^{\mathbf{V}}$ is stationary (we could replace $S^{\Gamma,W}$ by any stationary subset),
- for each $a \in S^{\Gamma,W}$ we let $d[a] = c[a] = \omega$ (so $d'[a] = c'[a] = \omega$),
- $\alpha^* = \omega$,
- for $n < \alpha^*$ and $\eta, g \in \omega^\omega$ we let
  $\eta \; R_n \; g$ if and only if
  if $\psi(g) = \langle s^g_m : m < \omega \rangle$ and $m \in \omega$ is such that $m^{s^g_m}_{\rm up} \geq n$ then
  $$s^g_m \in W(m^{s^g_m}_{\rm dn}, m^{s^g_m}_{\rm up}, \eta{\restriction}[m^{s^g_m}_{\rm dn}, m^{s^g_m}_{\rm up})),$$
- $\bar{R}^{\Gamma,W}$ is a three place relation such that $(\eta, n, g) \in \bar{R}^{\Gamma,W}$ if and only if $\eta, g \in \omega^\omega$, $n \in \omega$ and $\eta \; R_n \; g$
  (note: this is a definition of a relation, not a fixed object from $\mathbf{V}$),
- $\mathbf{g}^{\Gamma,W} = \langle \mathbf{g}_a : a \in S^{\Gamma,W} \rangle \subseteq \omega^\omega$ is such that for every $a, a' \in S^{\Gamma,W}$:
  (α) $\psi(\mathbf{g}_a) = \langle \mathbf{g}^\psi_{a,m} : m < \omega \rangle \in \Gamma$,
  (β) $(\forall \eta \in a \cap \omega^\omega)(\forall^\infty m)(\mathbf{g}^\psi_{a,m} \in W(m^{\mathbf{g}^\psi_{a,m}}_{\rm dn}, m^{\mathbf{g}^\psi_{a,m}}_{\rm up}, \eta{\restriction}[m^{\mathbf{g}^\psi_{a,m}}_{\rm dn}, m^{\mathbf{g}^\psi_{a,m}}_{\rm up})))$,
  (γ) if $a' \in a \cap S^{\Gamma,W}$, $\bar{s} \in a \cap \Gamma$ then $\psi(\mathbf{g}_{a'}) \preceq \psi(\mathbf{g}_a)$ and $\bar{s} \preceq \psi(\mathbf{g}_a)$.

Note that we may choose $\mathbf{g}_a$ by $\in$–induction for $a \in S^{\Gamma,W}$ considering all $a' \in a \cap S^{\Gamma,W}$, $\bar{s} \in a \cap \Gamma$ and $\eta \in a \cap \omega^\omega$. So, before we choose $\mathbf{g}_a$, we first take $\bar{s}_\eta = \langle s_{\eta,m} : m < \omega \rangle \in \Gamma$ for $\eta \in a \cap \omega^\omega$ such that

$$(\beta^*) \qquad (\forall^\infty m)(s_{\eta,m} \in W(m^{s_{\eta,m}}_{\rm dn}, m^{s_{\eta,m}}_{\rm up}, \eta{\restriction}[m^{s_{\eta,m}}_{\rm dn}, m^{s_{\eta,m}}_{\rm up})))$$

(possible by 5.3.1(3b)). Next, as $(\Gamma, \preceq)$ is directed and countably closed (by 5.3.1(3a)) we may find $\bar{s}_0 \in \Gamma$ such that

$$(\forall \eta \in a \cap \omega^\omega)(\forall a' \in a \cap S^{\Gamma,W})(\forall \bar{s} \in a \cap \Gamma)(\psi(\mathbf{g}_{a'}) \preceq \bar{s}_0 \;\&\; \bar{s} \preceq \bar{s}_0 \;\&\; \bar{s}_\eta \preceq \bar{s}_0).$$

Let $\mathbf{g}_a \in \omega^\omega$ be such that $\psi(\mathbf{g}_a) = \bar{s}_0$. It is easy to check that $\mathbf{g}_a$ is as required (in (α)–(γ) above; for (β) we use 5.3.3(2)).

CLAIM 5.3.12.1. 1. *$(\bar{R}^{\Gamma,W}, S^{\Gamma,W}, \mathbf{g}^{\Gamma,W})$ covers in $\mathbf{V}$ (see* [**Sh:f**, Ch XVIII, 3.2]*), i.e.:*
if $x \in \mathbf{V}$ then there is a countable elementary submodel N of $(\mathcal{H}(\chi), \in, <^*_\chi)$ such that $a \stackrel{\rm def}{=} N \cap \mathcal{H}(\aleph_1)^{\mathbf{V}} \in S^{\Gamma,W}$, $(\bar{R}^{\Gamma,W}, S^{\Gamma,W}, \mathbf{g}^{\Gamma,W})$, $x \in N$ and $(\forall \eta \in N \cap \omega^\omega)(\exists n < \omega)(\eta \; R_n \; \mathbf{g}_a)$.

2. *Let $\mathbb{P}$ be a proper forcing notion. Then the following conditions are equivalent:*
$(\oplus)_1$ $\Vdash_{\mathbb{P}}$ *"$(\bar{R}^{\Gamma,W}, S^{\Gamma,W}, \mathbf{g}^{\Gamma,W})$ covers",*
$(\oplus)_2$ *$\mathbb{P}$ is Γ-genericity preserving,*

$(\oplus)_3$ if $p \in \mathbb{P}$ *and* N *is a countable elementary submodel of* $(\mathcal{H}(\chi), \in, <^*_\chi)$ *such that* $p, \mathbb{P}, S^{\Gamma,W}, \mathbf{g}^{\Gamma,W} \in N$, $a \stackrel{\rm def}{=} N \cap \mathcal{H}(\aleph_1) \in S^{\Gamma,W}$ then *there is an* $(N,\mathbb{P})$*–generic condition* $q \in \mathbb{P}$ *stronger than* p *and such that*

$$q \Vdash_{\mathbb{P}} \text{“}(\forall \eta \in \omega^\omega \cap N[\Gamma_{\mathbb{P}}])(\exists n < \omega)(\eta\ R_n\ \mathbf{g}_a)\text{”}.$$

Proof of the claim: 1) By the choice of $(\bar{R}^{\Gamma,W}, S^{\Gamma,W}, \mathbf{g}^{\Gamma,W})$ (see condition (β) of the choice of $\mathbf{g}^{\Gamma,W}$).

2) Assume $(\oplus)_1$. Let $\dot{\eta}$ be a $\mathbb{P}$-name for a real in ω^ω, $p \in \mathbb{P}$. By the assumption we find $q \geq p$, $a \in S^{\Gamma,W}$ and a $\mathbb{P}$-name $\dot{N}$ for an elementary submodel such that

$$q \Vdash_{\mathbb{P}} \text{“}\dot{N} \cap \mathcal{H}(\aleph_1)^{\mathbf{V}} = a \quad \& \quad \dot{\eta} \in \dot{N} \quad \& \quad (\forall \eta \in \omega^\omega \cap \dot{N})(\exists n < \omega)(\eta\ R_n\ \mathbf{g}_a)\text{”}.$$

But, as $\psi(\mathbf{g}_a) \in \Gamma$, this is enough to conclude $(\oplus)_2$ (see the definitions of R_n, $\bar{R}^{\Gamma,W}$).

Now, suppose that $(\oplus)_2$ holds true. Let N, p be as in the assumptions of $(\oplus)_3$ (so $a \stackrel{\rm def}{=} N \cap \mathcal{H}(\aleph_1) \in S^{\Gamma,W}$). Let $q \in \mathbb{P}$ be any $(N,\mathbb{P})$–generic condition stronger than p. Then, by $(\oplus)_2$, the condition q forces in $\mathbb{P}$ that

$$(\forall \eta \in \omega^\omega \cap N[\Gamma_{\mathbb{P}}])(\exists \langle s_m : m < \omega\rangle \in \Gamma \cap N)(\forall^\infty m)(s_m \in W(m^{s_m}_{\rm dn}, m^{s_m}_{\rm up}, \eta{\restriction}[m^{s_m}_{\rm dn}, m^{s_m}_{\rm up})))$$

(note that $\mathrm{rng}(\mathbf{g}^{\Gamma,W}) \in N$ is a cofinal subset of Γ). But now, using 5.3.3(2) and clause (γ) of the choice of $\mathbf{g}^{\Gamma,W}$, we conclude

$$q \Vdash_{\mathbb{P}} (\forall \eta \in \omega^\omega \cap N[\Gamma_{\mathbb{P}}])(\exists n < \omega)(\eta\ R_n\ \mathbf{g}_a).$$

The implication $(\oplus)_3 \Rightarrow (\oplus)_1$ is straightforward.

CLAIM 5.3.12.2. *Suppose that* $\mathbb{P}$ *is a proper forcing notion such that*

$$\Vdash_{\mathbb{P}} \text{“}(\bar{R}^{\Gamma,W}, S^{\Gamma,W}, \mathbf{g}^{\Gamma,W}) \text{ covers”}.$$

Then:

1. *If* $\dot{\mathbb{Q}}$ *is a* $\mathbb{P}$*-name for a proper* Γ*-genericity preserving forcing notion then*
$$\Vdash_{\mathbb{P}} \text{“}\dot{\mathbb{Q}} \text{ is } (\bar{R}^{\Gamma,W}, S^{\Gamma,W}, \mathbf{g}^{\Gamma,W})\text{-preserving (for Possibility } A^*)\text{”}$$
(see [**Sh:f**, Ch XVIII, 3.4]*).*
2. $\Vdash_{\mathbb{P}}$ “$(\bar{R}^{\Gamma,W}, S^{\Gamma,W}, \mathbf{g}^{\Gamma,W})$ *strongly covers in the sense of Possibility* A^*” *(see* [**Sh:f**, Ch XVIII, 3.3]*).*

Proof of the claim: 1) We have to show that the following condition holds true in $\mathbf{V}^{\mathbb{P}}$:

$(*)$ Assume
- (i) χ_1 is large enough, $\chi > 2^{\chi_1}$,
- (ii) N is a countable elementary submodel of $(\mathcal{H}(\chi), \in, <^*_\chi)$, $a \stackrel{\rm def}{=} N \cap \mathcal{H}(\aleph_1)^{\mathbf{V}} \in S^{\Gamma,W}$, and $\dot{\mathbb{Q}}, S^{\Gamma,W}, \mathbf{g}^{\Gamma,W}, \chi_1, \ldots \in N$,
- (iii) $(\forall \eta \in \omega^\omega \cap N)(\exists n < \omega)(\eta\ R_n\ \mathbf{g}_a)$,
- (iv) $\dot{\eta}_0 \in N$ is a $\dot{\mathbb{Q}}$-name for a real in ω^ω,
- (v) $\eta^*_0 \in \omega^\omega$,
- (vi) $p, p_n \in \dot{\mathbb{Q}} \cap N$ are such that $p \leq_{\dot{\mathbb{Q}}} p_n \leq_{\dot{\mathbb{Q}}} p_{n+1}$ for all $n \in \omega$,
- (vii) $\eta^*_0, \langle p_n : n < \omega\rangle \in N$,
- (viii) $(\forall x \in \omega)(\forall^\infty n)(p_n \Vdash_{\dot{\mathbb{Q}}} \dot{\eta}_0(x) = \eta^*_0(x))$,
- (ix) $n_0 < \omega$ is such that $\eta^*_0\ R_{n_0}\ \mathbf{g}_a$,

(xi) there is a countable elementary submodel N_1 of $(\mathcal{H}(\chi_1), \in, <^*_{\chi_1})$ such that $\dot{\mathbb{Q}}, S^{\Gamma,W}, \mathbf{g}^{\Gamma,W} \in N_1 \in N$, and for every open dense subset $\mathcal{I}$ of $\dot{\mathbb{Q}}$, $\mathcal{I} \in N_1$ for some $n \in \omega$ we have $p_n \in \mathcal{I} \cap N_1$ (i.e. $\langle p_n : n \in \omega\rangle$ is a generic sequence over N_1).

Then there is an $(N, \dot{\mathbb{Q}})$–generic condition $q \in \dot{\mathbb{Q}}$ stronger than p and such that

(a) $q \Vdash_{\dot{\mathbb{Q}}}$ "$\dot{\eta}_0\ R_{n_0}\ \mathbf{g}_a$" and

(b) $q \Vdash_{\dot{\mathbb{Q}}}$ "$(\forall \eta \in \omega^\omega \cap N[\Gamma_{\dot{\mathbb{Q}}}])(\exists n < \omega)(\eta\ R_n\ \mathbf{g}_a)$".

So suppose that $\chi_1, \chi, N, N_1, a, \dot{\eta}_0, \eta_0^*, n_0, p$ and p_n (for $n \in \omega$) are as in the assumptions of $(*)$. Passing to a subsequence (in N) we may assume that $p_n \Vdash_{\dot{\mathbb{Q}}}$ "$\dot{\eta}_0{\restriction}n = \eta_0^*{\restriction}n$". Remember that we work in $\mathbf{V}^{\mathbb{P}}$.

So, as $(\bar{R}^{\Gamma,W}, S^{\Gamma,W}, \mathbf{g}^{\Gamma,W})$ covers and $N \prec (\mathcal{H}(\chi), \in, <^*_\chi)$, we find a countable elementary submodel N_2 of $(\mathcal{H}(\chi_1), \in, <^*_{\chi_1})$ such that

$$\dot{\mathbb{Q}}, S^{\Gamma,W}, \mathbf{g}^{\Gamma,W}, \dot{\eta}_0, \eta_0^*, \langle p_n : n < \omega\rangle, N_1, \ldots \in N_2 \in N, \quad a_2 \stackrel{\rm def}{=} N_2 \cap \mathcal{H}(\aleph_1)^{\mathbf{V}} \in S^{\Gamma,W}$$

$$\text{and} \qquad (\forall \eta \in \omega^\omega \cap N_2)(\exists n < \omega)(\eta\ R_n\ \mathbf{g}_{a_2}).$$

By the choice of $\mathbf{g}^{\Gamma,W}$ we know that $\psi(\mathbf{g}_{a_2}) \preceq \psi(\mathbf{g}_a)$ (as $a_2 \in a$) and hence we find $m^* \in [n_0, \omega)$ such that

if $m < \omega$, $m_{\rm up}^{\mathbf{g}^\psi_{a,m}} \geq m^*$ then for some $m' \leq m'' < \omega$ we have

$$\mathbf{g}^\psi_{a,m} \in \Sigma(\mathbf{g}^\psi_{a_2,m'}, \ldots, \mathbf{g}^\psi_{a_2,m''}).$$

Now, working in N, we inductively choose sequences $\langle n_\ell : \ell < \omega\rangle$, $\langle k_\ell : \ell < \omega\rangle$, $\langle m_\ell : \ell < \omega\rangle$, $\langle q_\ell : \ell < \omega\rangle$ and $\langle \sigma_\ell : \ell < \omega\rangle$, all from N.

STEP $\ell = 0$.

The n_0 is given already. Let m_0 be the first such that $m^* \leq m_{\rm dn}^{\mathbf{g}^\psi_{a_2,m_0}}$ and let $k_0 = n_0$, $q_0 = p_{n_0}$, $\sigma_0 = \eta_0^*{\restriction}m_{\rm up}^{\mathbf{g}^\psi_{a_2,m_0}}$.

STEP $\ell + 1$.

Suppose we have defined $n_\ell, k_\ell, m_\ell, q_\ell, \sigma_\ell$. We let $n_{\ell+1} = m_{\rm up}^{\mathbf{g}^\psi_{a_2,m_\ell}}$ and we choose an $(N_2, \dot{\mathbb{Q}})$–generic condition $q_{\ell+1} \geq p_{n_{\ell+1}}$, integers $k_{\ell+1} \in [n_{\ell+1}, \omega)$ and $m_{\ell+1} \in (m_\ell, \omega)$, and a finite function $\sigma_{\ell+1} : [n_{\ell+1}, m_{\rm up}^{\mathbf{g}^\psi_{a_2,m_{\ell+1}}}) \longrightarrow \omega$ such that:

(α) $q_{\ell+1} \Vdash_{\dot{\mathbb{Q}}}$ "$(\forall \eta \in \omega^\omega \cap N_2[\Gamma_{\dot{\mathbb{Q}}}])(\exists n < \omega)(\eta\ R_n\ \mathbf{g}_{a_2})$",

(β) $q_{\ell+1} \Vdash_{\dot{\mathbb{Q}}}$ "$\dot{\eta}_0\ R_{k_{\ell+1}}\ \mathbf{g}_{a_2}$",

(γ) $m_{\ell+1}$ is the first such that $k_{\ell+1} \leq m_{\rm dn}^{\mathbf{g}^\psi_{a_2,m_{\ell+1}}}$,

(δ) $q_{\ell+1} \Vdash_{\dot{\mathbb{Q}}}$ "$\dot{\eta}_0{\restriction}[n_{\ell+1}, m_{\rm up}^{\mathbf{g}^\psi_{a_2,m_{\ell+1}}}) = \sigma_{\ell+1}$"

(possible as, in $\mathbf{V}^{\mathbb{P}}$, $\dot{\mathbb{Q}}$ is a proper Γ-genericity preserving forcing notion, remember 5.3.12.1(2)). Note that all parameters needed for the construction are in N. After it we have

$$n_0 = k_0 \leq m_{\rm dn}^{\mathbf{g}^\psi_{a_2,m_0}} < m_{\rm up}^{\mathbf{g}^\psi_{a_2,m_0}} = n_1 \leq k_1 \leq m_{\rm dn}^{\mathbf{g}^\psi_{a_2,m_1}} < m_{\rm up}^{\mathbf{g}^\psi_{a_2,m_1}} = n_2 \leq k_2 \leq \ldots$$

and $\sigma_0 : [0, n_1) \longrightarrow \omega$, $\sigma_{\ell+1} : [n_{\ell+1}, n_{\ell+2}) \longrightarrow \omega$. Let $\eta \stackrel{\rm def}{=} \sigma_0{}^\frown\sigma_1{}^\frown\sigma_2{}^\frown\ldots \in N \cap \omega^\omega$. By **(iii)**, we find $\ell > 0$ such that $\eta\ R_{n_\ell}\ \mathbf{g}_a$. We claim that $q_\ell \Vdash_{\dot{\mathbb{Q}}}$ "$\dot{\eta}_0\ R_{n_0}\ \mathbf{g}_a$". If not, then we find a condition $q \geq q_\ell$ and $m < \omega$ such that $m_{\rm up}^{\mathbf{g}^\psi_{a,m}} \geq n_0$ and

$$q \Vdash_{\dot{\mathbb{Q}}} \text{“}\mathbf{g}^\psi_{a,m} \notin W(m_{\rm dn}^{\mathbf{g}^\psi_{a,m}}, m_{\rm up}^{\mathbf{g}^\psi_{a,m}}, \dot{\eta}_0{\restriction}[m_{\rm dn}^{\mathbf{g}^\psi_{a,m}}, m_{\rm up}^{\mathbf{g}^\psi_{a,m}}))\text{”}.$$

Let $n = n_W(\mathbf{g}^\psi_{a,m})$ (see 5.3.1(1b)) and let $m' < \omega$ be such that $m_{\rm dn}^{\mathbf{g}^\psi_{a_2,m'}} \leq m_{\rm dn}^{t_n} < m_{\rm up}^{t_n} \leq m_{\rm up}^{\mathbf{g}^\psi_{a_2,m'}}$. Consider the following three possibilities.

CASE 1: $m_{\rm dn}^{\mathbf{g}^\psi_{a_2,m'}} \geq k_\ell$.

By the choice of m_0 and m^* (remember $\ell > 0$, so $m_{\rm up}^{\mathbf{g}^\psi_{a,m}} > m_{\rm dn}^{\mathbf{g}^\psi_{a_2,m'}} \geq k_1 > m^*$) we know that for some $m'' \leq m''' < \omega$

$$\mathbf{g}^\psi_{a,m} \in \Sigma(\mathbf{g}^\psi_{a_2,m''}, \ldots, \mathbf{g}^\psi_{a_2,m'''})$$

and by the choice of k_ℓ we know that

$$q_\ell \Vdash_{\dot{\mathbb{Q}}} \text{“}\mathbf{g}^\psi_{a_2,m'} \in W(m_{\rm dn}^{\mathbf{g}^\psi_{a_2,m'}}, m_{\rm up}^{\mathbf{g}^\psi_{a_2,m'}}, \dot{\eta}_0{\restriction}[m_{\rm dn}^{\mathbf{g}^\psi_{a_2,m'}}, m_{\rm up}^{\mathbf{g}^\psi_{a_2,m'}}))\text{”}.$$

By 5.3.1(1c) we conclude that

$$q_\ell \Vdash_{\dot{\mathbb{Q}}} \text{“}\mathbf{g}^\psi_{a,m} \in W(m_{\rm dn}^{\mathbf{g}^\psi_{a,m}}, m_{\rm up}^{\mathbf{g}^\psi_{a,m}}, \dot{\eta}_0{\restriction}[m_{\rm dn}^{\mathbf{g}^\psi_{a,m}}, m_{\rm up}^{\mathbf{g}^\psi_{a,m}}))\text{”}$$

(remember the choice of m', note $m'' \leq m' \leq m'''$), a contradiction.

CASE 2: $m_{\rm dn}^{\mathbf{g}^\psi_{a_2,m'}} < n_\ell$.

Then, by the choice of n_ℓ (remember $\ell > 0$), we have $m_{\rm up}^{\mathbf{g}^\psi_{a_2,m'}} \leq n_\ell$ (and $m' \leq m_\ell$). As $q_\ell \Vdash_{\dot{\mathbb{Q}}}$“$\dot{\eta}_0{\restriction}n_\ell = \eta_0^*{\restriction}n_\ell$” and $\eta_0^*\ R_{n_0}\ \mathbf{g}_a$ we immediately get a contradiction (remember 5.3.1(1b) and the choice of m'). So we are left with the following possibility.

CASE 3: $n_\ell \leq m_{\rm dn}^{\mathbf{g}^\psi_{a_2,m'}} < k_\ell$ (so $m' < m_\ell$).
Now the choice of η, n_ℓ and clause (δ) of the choice of q_ℓ work: we know that

$$m_{\rm up}^{\mathbf{g}^\psi_{a_2,m'}} \leq m_{\rm dn}^{\mathbf{g}^\psi_{a_2,m_\ell}} < m_{\rm up}^{\mathbf{g}^\psi_{a_2,m_\ell}} = n_{\ell+1},$$
$$q_\ell \Vdash_{\dot{\mathbb{Q}}} \text{“ } \dot{\eta}_0{\restriction}[n_\ell, n_{\ell+1}) = \eta{\restriction}[n_\ell, n_{\ell+1}) \text{ ”} \qquad \text{and} \qquad \eta\ R_{n_\ell}\ \mathbf{g}_a.$$

Consequently we get a contradiction like in the previous cases.

Now, choosing an $(N, \dot{\mathbb{Q}})$–generic condition $q \geq q_\ell$ such that

$$q \Vdash_{\dot{\mathbb{Q}}} \text{“}(\forall \eta \in \omega^\omega \cap N[\Gamma_{\dot{\mathbb{Q}}}])(\exists n < \omega)(\eta\ R_n\ \mathbf{g}_a)\text{”}$$

(possible by 5.3.12.1(2)) we finish.

2) Work in $\mathbf{V}^{\mathbb{P}}$. We know that $(\bar{R}^{\Gamma,W}, S^{\Gamma,W}, \mathbf{g}^{\Gamma,W})$ covers. Clearly each R_n (for $n < \omega$) is (a definition of) a closed relation on ω^ω. So what is left are the following two requirements:

$\otimes$ if $a_1, a_2 \in S^{\Gamma,W}$, $a_1 \in a_2$ then for every $\eta \in \omega^\omega$ we have

$$(\exists n < \omega)(\eta\ R_n\ \mathbf{g}_{a_1}) \quad \Rightarrow \quad (\exists n < \omega)(\eta\ R_n\ \mathbf{g}_{a_2});$$

$\oplus_1$ if $\dot{\mathbb{Q}}$, $\dot{\eta}_0$, p, N, N_1, G_1, n_0 are such that (in $\mathbf{V}^{\mathbb{P}}$):

(a) $\dot{\mathbb{Q}}$ is a proper forcing notion,

(b) $N \prec (\mathcal{H}(\chi), \in, <^*_\chi)$ is countable, $a \stackrel{\rm def}{=} N \cap \mathcal{H}(\aleph_1)^{\mathbf{V}} \in S^{\Gamma,W}$,

$$(\forall \eta \in \omega^\omega \cap N)(\exists n < \omega)(\eta\ R_n\ \mathbf{g}_a),$$

$\dot{\mathbb{Q}}, S^{\Gamma,W}, \mathbf{g}^{\Gamma,W}, \chi_1, \ldots \in N$, $p \in \dot{\mathbb{Q}} \cap N$,

(c) $\dot{\eta}_0 \in N$ is a $\dot{\mathbb{Q}}$–name for a function in ω^ω,

(d) $\chi_1 < \chi$ (χ_1 large enough), $N_1 \in N$, $N_1 \prec (\mathcal{H}(\chi_1), \in, <^*_{\chi_1})$ is countable, $\dot{\mathbb{Q}}, p, S^{\Gamma,W}, \mathbf{g}^{\Gamma,W}, \dot{\eta}_0, \ldots \in N_1$, and $G_1 \in N$ is a $\dot{\mathbb{Q}}$–generic filter over N_1, $p \in G_1$,

(e) $\dot{\eta}_0[G_1]\ R_{n_0}\ \mathbf{g}_a$,

then for every $y \in N \cap \mathcal{H}(\chi_1)$ there are N_2, G_2 satisfying the parallel of clause (d) and such that $y \in N_2$ and $\dot{\eta}_0[G_2]\ R_{n_0}\ \mathbf{g}_a$.

Now, concerning $\otimes$, look at our choice of $\mathbf{g}_a$'s: by (γ) we know that

$$a_1 \in a_2 \quad \Rightarrow \quad \psi(\mathbf{g}_{a_1}) \preceq \psi(\mathbf{g}_{a_2}).$$

Thus we may use 5.3.3(2) and the definition of $\bar{R}^{\Gamma,W}$ to get $\otimes$.
To show $\oplus_1$ we proceed similarly as in the proof of (1) above. So suppose that $\dot{\mathbb{Q}}$, $\dot{\eta}_0$, p, N, N_1, G_1, n_0 are as in the assumptions of $\oplus_1$ and $y \in N \cap \mathcal{H}(\chi_1)$. We work in the universe $\mathbf{V}^{\mathbb{P}}$. Choose a $\leq_{\dot{\mathbb{Q}}}$–increasing sequence $\langle p_n : n \in \omega \rangle \in N$, $\dot{\mathbb{Q}}$–generic over N_1, such that $\{p_n : n \in \omega\} \subseteq G_1$ and p_n decides the value of $\dot{\eta}_0 {\restriction} n$. Let $N_2 \in N$ be a countable elementary submodel of $(\mathcal{H}(\chi_1), \in, <^*_{\chi_1})$ such that $N_1, y, \ldots \in N_2$ and then choose a countable $N_2^+ \prec (\mathcal{H}(\chi_1), \in, <^*_{\chi_1})$ such that $N_2 \in N_2^+ \in N$, $a_2 \stackrel{\rm def}{=} N_2^+ \cap \mathcal{H}(\aleph_1) \in S^{\Gamma,W}$ and

$$(\forall \eta \in N_2^+ \cap \omega^\omega)(\exists n \in \omega)(\eta\ R_n\ \mathbf{g}_{a_2})$$

(remember that $(\bar{R}^{\Gamma,W}, S^{\Gamma,W}, \mathbf{g}^{\Gamma,W})$ covers in $\mathbf{V}^{\mathbb{P}}$). Let $m^* \geq n_0$ be such that

$$(\forall m \in \omega)\big(m_{\rm up}^{\mathbf{g}^\psi_{a,m}} \geq m^* \ \Rightarrow\ (\exists m' \leq m'' < \omega)(\mathbf{g}^\psi_{a,m} \in \Sigma(\mathbf{g}^\psi_{a_2,m'}, \ldots, \mathbf{g}^\psi_{a_2,m''}))\big).$$

Next, working in N, construct inductively sequences $\langle n_\ell : \ell < \omega \rangle$, $\langle \sigma_\ell : \ell < \omega \rangle$, $\langle G^\ell : \ell < \omega \rangle$ (all from N) such that

(α) G^ℓ is $\dot{\mathbb{Q}}$–generic over N_2, $p_{n_\ell} \in G^\ell \in N_2^+$ (so $\dot{\eta}_0[G_\ell] \in N_2^+$),

(β) $\dot{\eta}_0[G^\ell]\ R_{n_{\ell+1}}\ \mathbf{g}_{a_2}$, $n_{\ell+1} > n_\ell + m^*$ and $n_{\ell+1} = m_{\rm dn}^{\mathbf{g}^\psi_{a_2,k}}$ for some $k \in \omega$,

(γ) $\sigma_\ell = \dot{\eta}_0[G^\ell]{\restriction}[n_\ell, n_{\ell+1})$.

Finally let $\eta = \dot{\eta}_0[G_1]{\restriction} n_0{}^\frown\sigma_0{}^\frown\sigma_1{}^\frown\sigma_2 \ldots \in \omega^\omega \cap N$ and let $\ell > 0$ be such that $\eta\ R_{n_\ell}\ \mathbf{g}_a$. As in (1), one shows now that $\dot{\eta}_0[G^\ell]\ R_{n_0}\ \mathbf{g}_a$.

CLAIM 5.3.12.3. *The forcing notion $\mathbb{P}_\beta$ is $(\bar{R}^{\Gamma,W}, S^{\Gamma,W}, \mathbf{g}^{\Gamma,W})$–preserving and hence*

$$\Vdash_{\mathbb{P}_\beta} \text{“}(\bar{R}^{\Gamma,W}, S^{\Gamma,W}, \mathbf{g}^{\Gamma,W}) \text{ covers”}.$$

Proof of the claim: Due to 5.3.12.2, we may apply [**Sh:f**, Ch XVIII, 3.6(1)] to get the conclusion.

Putting together 5.3.12.3 and 5.3.12.1(2) we finish the proof of the theorem. □

5.4. Examples

EXAMPLE 5.4.1. Let $F \in \omega^\omega$ be strictly increasing.
There are increasing functions $f^F = f, g^F = g \in \omega^\omega$, and $(K^F_{5.4.1}, \Sigma^F_{5.4.1}) = (K_{5.4.1}, \Sigma_{5.4.1})$, $(K^{\ell,F}_{5.4.1}, \Sigma^{\ell,F}_{5.4.1}) = (K^\ell_{5.4.1}, \Sigma^\ell_{5.4.1})$, $\bar{t}^F = \bar{t}$, $\bar{t}^F_\ell = \bar{t}_\ell$, $\bar{\mathcal{F}}^F_\ell = \bar{\mathcal{F}}_\ell = (\mathcal{F}_\ell, <^*_\ell)$ (for $\ell < \omega$) and $\langle W^k_{\ell,h} : h \in \mathcal{F}_\ell,\ k, \ell < \omega\rangle$ such that for every $\ell < \omega$:

1. $(K_{5.4.1}, \Sigma_{5.4.1})$, $(K^\ell_{5.4.1}, \Sigma^\ell_{5.4.1})$ are simple, strongly finitary and forgetful creating pairs for $\mathbf{H}_{5.4.1}$, $\mathbf{H}^\ell_{5.4.1}$, respectively;
2. $\bar{t} \in \mathrm{PC}_\infty(K_{5.4.1}, \Sigma_{5.4.1})$, $\bar{t}_\ell \in \mathrm{PC}_\infty(K^\ell_{5.4.1}, \Sigma^\ell_{5.4.1})$;
3. $\bar{\mathcal{F}}_\ell$ is a countable $\bar{t}_\ell$–good partial order;
4. $\bar{t}_\ell$ is $(2^g, h)$–additive for each $h \in \mathcal{F}_\ell$;
5. $\langle W^k_{\ell,h} : h \in \mathcal{F}_\ell,\ k \in \omega\rangle$ is a $(\bar{t}_\ell, \bar{\mathcal{F}}_\ell)$–coherent sequence of regular $\bar{t}$–systems;
6. each $W^k_{\ell,h}$ (for $h \in \mathcal{F}_\ell$, $k \in \omega$) is Cohen sensitive;
7. $(\forall h \in \mathcal{F}_\ell)(\forall^\infty m)(|V^m_h(\bar{t}_\ell)| < f(m))$;
8. $(\forall m \in \omega)(g(m+1) = F(f(m)))$.

Moreover, the sequence $\langle W^k_{\ell,h} : h \in \mathcal{F}_\ell,\ k, \ell < \omega\rangle$ has the following property:

$(\circledast)_{5.4.1}$ if $\Gamma \subseteq \mathbb{P}^*_\infty(K_{5.4.1}, \Sigma_{5.4.1})$ is quasi-$W^k_{\ell,h}$-generic for every $k, \ell < \omega$, $h \in \mathcal{F}_\ell$ and $\mathbb{B}$ is a measure algebra (i.e. adding a number of random reals) then $\mathbb{B}$ is $(\Gamma, W^k_{\ell,h})$–genericity preserving for all $k, \ell < \omega$, $h \in \mathcal{F}_\ell$.

CONSTRUCTION. Let $F \in \omega^\omega$ be a strictly increasing function. We inductively define $f, g \in \omega^\omega$, $\langle n^*_i, \ell^*_i, k^*_i : i < \omega\rangle \subseteq \omega$ and a function $\psi : \omega \times \omega \longrightarrow \omega$ such that:

(α) $g(0) = F(1)$, $n^*_0 > 2^{2g(0)}$ satisfies $2^{2g(0)+1} < \frac{(n^*_0)^{n^*_0}}{(n^*_0)!}$,

(β) ℓ^*_i is such that $n^*_i < \ell^*_i$ and

$$(\ell^*_i \cdot n^*_i)^{n^*_i} \cdot \frac{(\ell^*_i \cdot n^*_i - n^*_i)!}{(\ell^*_i \cdot n^*_i)!} \leq 2,$$

(γ) $k^*_i = n^*_i(\ell^*_i)^{n^*_i} + n^*_i + 1$,

(δ_0) $\psi(i, 0) = 2^{g(i)(i+1)} \cdot (n^*_i)^{k^*_i}$,

(δ_1) $\psi(i, \ell+1) = [2^{2g(i)(i+1)+\psi(i,\ell)} \cdot (\psi(i,\ell)!)]^{g(i)+1} \cdot (n^*_i)^{k^*_i}$,

(ϵ) $f(i) = 2^{\psi(i,i)}$, $g(i+1) = F(f(i))$,

(ζ) n^*_{i+1} is such that $2^{2g(i+1)} \cdot k^*_i < n^*_{i+1}$ and

$$2^{2g(i+1)(i+1)^2+1} < \frac{(n^*_{i+1})^{n^*_{i+1}}}{(n^*_{i+1})!}.$$

Why is the choice possible? For clauses (α), (ζ) remember that $\lim\limits_{n\to\infty} \frac{n^n}{n!} = \infty$. For clause ($\beta$) note that

$$N^{n^*_i} \cdot \frac{(N - n^*_i)!}{N!} = \frac{N^{n^*_i}}{N \cdot (N-1) \cdot \ldots \cdot (N - (n^*_i - 1))} \stackrel{N\to\infty}{\longrightarrow} 1.$$

Now, we define $\mathbf{H}_{5.4.1}, \mathbf{H}^\ell_{5.4.1}$. For $i \in \omega$ we let

$$\mathbf{H}_{5.4.1}(i) = \{e \subseteq \mathcal{P}([(n^*_i, k^*_i)]^{n^*_i}) : \bigcup e = (n^*_i, k^*_i)\ \&\ (\forall u, u' \in e)(u \neq u' \ \Rightarrow\ u \cap u' = \emptyset)\},$$

$$\mathbf{H}^\ell_{5.4.1}(i) = \{(j, x) : j < \psi(i, \ell)\ \&\ x : (n^*_i, k^*_i) \longrightarrow n^*_i\}.$$

A creature $t \in \mathrm{CR}[\mathbf{H}_{5.4.1}]$ is in $K_{5.4.1}$ if for some $m \in \omega$, $B \subseteq \mathcal{P}([(n^*_m, k^*_m)]^{n^*_m})$:

- $\mathbf{val}[t] = \{\langle u, w\rangle \in \prod\limits_{i<m} \mathbf{H}_{5.4.1}(i) \times \prod\limits_{i\leq m} \mathbf{H}_{5.4.1}(i) : u \lhd w \ \&\ B \subseteq w(m)\}$,
- $\mathbf{nor}[t] = \frac{\log_2\left(1+\frac{k^*_m - n^*_m - 1}{n^*_m} - |B|\right)}{\log_2(\ell^*_m)} = \frac{\log_2\left(1+(\ell^*_m)^{n^*_m} - |B|\right)}{\log_2(\ell^*_m)}$,
- $\mathbf{dis}[t] = B$.

The composition operation $\Sigma_{5.4.1}$ on $K_{5.4.1}$ is given by

$$\Sigma_{5.4.1}(t) = \{s \in K_{5.4.1} : m^s_{\rm dn} = m^t_{\rm dn} \ \&\ \mathbf{dis}[t] \subseteq \mathbf{dis}[s]\} \qquad \text{for } t \in K_{5.4.1}.$$

Now we define $(K^\ell_{5.4.1}, \Sigma^\ell_{5.4.1})$ for $\ell < \omega$. The family $K^\ell_{5.4.1}$ consists of all creatures $t \in \mathrm{CR}[\mathbf{H}^\ell_{5.4.1}]$ such that for some $m \in \omega$ and a nonempty set $C \subseteq \mathbf{H}^\ell_{5.4.1}(m)$ such that $(\forall (j,f),(j'f') \in C)(j = j' \ \Rightarrow f = f')$ we have:

- $\mathbf{val}[t] = \{\langle u, w\rangle \in \prod\limits_{i<m} \mathbf{H}^\ell_{5.4.1}(i) \times \prod\limits_{i\leq m} \mathbf{H}^\ell_{5.4.1}(i) : u \lhd w \ \&\ w(m) \in C\}$,
- $\mathbf{nor}[t] = \frac{\log_2(|C|)}{g(m)}$,
- $\mathbf{dis}[t] = C$.

The composition operation $\Sigma^\ell_{5.4.1}$ on $K^\ell_{5.4.1}$ is such that for $t \in K^\ell_{5.4.1}$:

$$\Sigma^\ell_{5.4.1}(t) = \{s \in K^\ell_{5.4.1} : m^s_{\rm dn} = m^t_{\rm dn} \ \&\ \mathbf{dis}[s] \hookrightarrow \mathbf{dis}[t]\},$$

where $\mathbf{dis}[s] \hookrightarrow \mathbf{dis}[t]$ means that there is an embedding

$$\mathbf{i} : \{j < \psi(m,\ell) : (\exists x)((j,x) \in \mathbf{dis}[s])\} \xrightarrow{1-1} \{j < \psi(m,\ell) : (\exists x)((j,x) \in \mathbf{dis}[t])\}$$

such that $(\forall j \in \mathrm{dom}(\mathbf{i}))(\forall x)((j,x) \in \mathbf{dis}[s] \ \Rightarrow\ (\mathbf{i}(j),x) \in \mathbf{dis}[t])$. Later we may identify elements $s_0, s_1 \in \Sigma^\ell_{5.4.1}(t)$ such that $\mathbf{dis}[s_0] \hookrightarrow \mathbf{dis}[s_1]$ and $\mathbf{dis}[s_1] \hookrightarrow \mathbf{dis}[s_0]$. Therefore we may think that we have the following inequality:

$$|\Sigma^\ell_{5.4.1}(t)| \leq 2^{\mathbf{dis}[t]}.$$

It should be clear that $(K_{5.4.1}, \Sigma_{5.4.1})$, $(K^\ell_{5.4.1}, \Sigma^\ell_{5.4.1})$ are strongly finitary, simple and forgetful creating pairs. Now we have to define $\bar{t}$, $\bar{t}_\ell$. The first is the minimal member of $\mathrm{PC}(K_{5.4.1}, \Sigma_{5.4.1})$:

$$\bar{t} = \langle t_m : m < \omega\rangle \quad \text{ is such that } \quad (\forall m \in \omega)(m^{t_m}_{\rm dn} = m \ \ \&\ \ \mathbf{dis}[t_m] = \emptyset).$$

Next, for each $m \in \omega$ (and $\ell \in \omega$), we choose $t_{\ell,m} \in (K^\ell_{5.4.1}, \Sigma^\ell_{5.4.1})$ such that $m^{t_{\ell,m}}_{\rm dn} = m$ and

$$(\forall x : (n^*_m, k^*_m) \longrightarrow n^*_m)\big(|\{j < \psi(m,\ell) : (j,x) \in \mathbf{dis}[t_{\ell,m}]\}| = \frac{\psi(m,\ell)}{(n^*_m)^{k^*_m - n^*_m - 1}}\big).$$

Then we let $\bar{t}_\ell = \langle t_{\ell,m} : m < \omega\rangle$. Note that

$$\mathbf{nor}[t_m] = \frac{\log_2\left(1 + (\ell^*_m)^{n^*_m}\right)}{\log_2(\ell^*_m)} \longrightarrow \infty \quad \text{ and } \quad \mathbf{nor}[t_{\ell,m}] = \frac{\log_2(\psi(m,\ell))}{g(m)} \longrightarrow \infty$$

(when m goes to ∞, ℓ is fixed). Moreover, if n is such that

$$2^{g(m)(n+1)} < \frac{\psi(m,\ell)}{(n^*_m)^{k^*_m - n^*_m - 1}} \qquad \text{(where } m,\ell < \omega)$$

then the creature $t_{\ell,m}$ is $(2^{g(m)}, n)$–additive. Why? Note that if $\mathbf{nor}[s_i] \leq n$, $s_i \in \Sigma^\ell_{5.4.1}(t_{\ell,m})$ then $\sum\limits_{i<2^{g(m)}} |\mathbf{dis}[s_i]| \leq 2^{g(m)(n+1)}$ and it is smaller than $\frac{\psi(m,\ell)}{(n^*_m)^{k^*_m - n^*_m - 1}}$, which is the number of repetitions of each function from $(n^*_m)^{(n^*_m, k^*_m)}$ in $\mathbf{dis}[t_{\ell,m}]$.

For each $\ell \in \omega$ we choose a countable $\bar{t}_\ell$–good partial order $\bar{\mathcal{F}}_\ell = (\mathcal{F}_\ell, <^*_\ell)$ such that for every $h \in \mathcal{F}_\ell$:

(i) $g(m)(h(m)+1) < \log_2\left(\frac{\psi(m,\ell)}{(n^*_m)^{k^*_m-n^*_m-1}}\right)$ for all $m\in\omega$,

(ii) there is $h^*\in\mathcal{F}_\ell$ such that

$$h <^*_\ell h^* \quad \text{and} \quad (\forall m\in\omega)(h^*(m)\leq h(m)+m),$$

(iii) there is a function $h^\otimes_\ell\in\mathcal{F}_{\ell+1}$ such that

$$h^\otimes_\ell(m)\geq g(m)(m+1)+\psi(m,\ell)+\log_2(\psi(m,\ell)!) \qquad \text{(for all } m\in\omega).$$

There should be no problems in carrying the construction of the $\mathcal{F}_\ell$. Note that we may do this inductively, building a linear order (and so it will be isomorphic to rationals). The clause (iii) is not an obstacle (in the presence of (i)) as $\psi(m,\cdot)$ is increasing fast enough:

$$\begin{aligned}&(n^*_m+1)\log_2(n^*_m)+g(m)(g(m)(m+1)+\psi(m,\ell)+\log_2(\psi(m,\ell)!)+2)<\\&(n^*_m+1)\log_2(n^*_m)+(g(m)+1)(2g(m)(m+1)+\psi(m,\ell)+\log_2(\psi(m,\ell)!))=\\&\log_2\left(\frac{\psi(m,\ell+1)}{(n^*_m)^{k^*_m-n^*_m-1}}\right).\end{aligned}$$

Note that the clause (i) and the previous remark imply that $\bar{t}_\ell$ is $(2^g,h)$–additive for each $h\in\mathcal{F}_\ell$. Moreover, by the choice of the function f we have that for every $\ell\leq m<\omega$ and $h\in\mathcal{F}_\ell$

$$|V^m_h(\bar{t}_\ell)|<|\Sigma(t_{\ell,m})|\leq 2^{|\mathbf{dis}[t_{\ell,m}]|}=2^{\psi(m,\ell)}\leq 2^{\psi(m,m)}=f(m).$$

Finally, we are going to define $\bar{t}$–systems $W^k_{\ell,h}$ for $k,\ell\in\omega$ and $h\in\mathcal{F}_\ell$. First, for each $\ell\in\omega$, $h\in\mathcal{F}_\ell$ we fix a function $\rho^\ell_h:\omega^{\underline{\omega}}\longrightarrow\bigcup\limits_{m\in\omega}V^m_h(\bar{t}_\ell)$ such that for $m\in\omega$:

$$\rho^\ell_h{\restriction}\omega^{[m,m+1)}:\omega^{[m,m+1)}\stackrel{\text{onto}}{\longrightarrow}V^m_h(\bar{t}_\ell).$$

Next, for $m\in\omega$ and $\sigma:[m,m+1)\longrightarrow\omega$ (and $\ell,k\in\omega$, $h\in\mathcal{F}_\ell$) we let:
if $m\leq k$ then $W^k_{\ell,h}(m,m+1,\sigma)=\Sigma_{5.4.1}(t_m)$,
if $m>k$ then

$$\begin{aligned}W^k_{\ell,h}(m,m+1,\sigma)=\ &\{s\in\Sigma_{5.4.1}(t_m):\ \text{for some } u\in\mathbf{dis}[s] \text{ we have}\\&\quad|\{(j,x)\in\mathbf{dis}[\rho^\ell_h(\sigma)]:x[u]=n^*_m\}|<\tfrac{|\mathbf{dis}[\rho^\ell_h(\sigma)]|}{2^{g(m)(m+1)(k+1)}}\}\end{aligned}$$

(in all other instances we let $W^k_{\ell,h}(m',m'',\sigma)=\emptyset$).

CLAIM 5.4.1.1. *For each $k,\ell\in\omega$ and $h\in\mathcal{F}_\ell$, the function $W^k_{\ell,h}$ is a Cohen–sensitive regular $\bar{t}$–system.*

Proof of the claim: First we have to check that $W^k_{\ell,h}$ is a $\bar{t}$–system. Immediately by its definition we have that 5.3.1(1a-c) are satisfied (remember $(K_{5.4.1},\Sigma_{5.4.1})$ is simple; see 5.3.3(1)). What might be problematic is 5.3.1(1d). So suppose that $k,\ell,m\in\omega$, $m>k$ (otherwise trivial), $h\in\mathcal{F}_\ell$, $\sigma:[m,m+1)\longrightarrow\omega$, $s\in\Sigma_{5.4.1}(t_m)$, $\mathbf{nor}[s]>1$. The last means that

$$\frac{k^*_m-n^*_m-1}{n^*_m}-|\mathbf{dis}[s]|>\ell^*_m-1.$$

Let $N=\ell^*_m\cdot n^*_m$. Choose a set $X\subseteq(n^*_m,k^*_m)$ such that $|X|=N$ and $X\cap\bigcup\mathbf{dis}[s]=\emptyset$. Note that for each $(j,x)\in\mathbf{dis}[\rho^\ell_h(\sigma)]$ we have

$$\frac{|\{u\in[X]^{n^*_m}:x[u]=n^*_m\}|}{|[X]^{n^*_m}|}=\frac{|x^{-1}[\{0\}]\cap X|\cdot\ldots\cdot|x^{-1}[\{n^*_m-1\}]\cap X|}{\binom{N}{n^*_m}}\leq$$

$$\left(\sum_{i<n_m^*}\frac{|x^{-1}[\{i\}]\cap X|}{n_m^*}\right)^{n_m^*}\cdot\frac{(n_m^*)!\cdot(N-n_m^*)!}{N!}=N^{n_m^*}\cdot\frac{(N-n_m^*)!}{N!}\cdot\frac{(n_m^*)!}{(n_m^*)^{n_m^*}}.$$

Now look at clause (β) of the choice of ℓ_m^*. It implies that $N^{n_m^*}\cdot\frac{(N-n_m^*)!}{N!}\leq 2$ and hence

$$\frac{|\{u\in[X]^{n_m^*}:x[u]=n_m^*\}|}{|[X]^{n_m^*}|}\leq 2\frac{(n_m^*)!}{(n_m^*)^{n_m^*}}<\frac{1}{2^{2g(m)m^2}}\leq\frac{1}{2^{g(m)(m+1)(k+1)}}$$

(the second inequality follows from clause (ζ) of the choice of n_m^*, remember $m>k$). Consequently, applying the Fubini theorem, we find $u_0\in[X]^{n_m^*}$ such that

$$\frac{|\{(j,x)\in\mathbf{dis}[\rho_h^\ell(\sigma)]:x[u_0]=n_m^*\}|}{|\mathbf{dis}[\rho_h^\ell(\sigma)]|}<\frac{1}{2^{g(m)(m+1)(k+1)}}.$$

Thus, choosing $s^*\in\Sigma_{5.4.1}(s)$ such that $\mathbf{dis}[s^*]=\mathbf{dis}[s]\cup\{u_0\}$ we will have $\mathbf{nor}[s^*]\geq\mathbf{nor}[s]-1$ and $s^*\in W_{\ell,h}^k(m,m+1,\sigma)$. This shows that each $W_{\ell,h}^k$ is a regular $\bar{t}$–system.

Finally we show that $W_{\ell,h}^k$ is Cohen–sensitive. Suppose that $s\in\Sigma_{5.4.1}(t_m)$, $m>k$. Choose a function $x^*:(n_m^*,k_m^*)\longrightarrow n_m^*$ such that $(\forall u\in\mathbf{dis}[s])(x^*[u]=n_m^*)$. Next take $\sigma:[m,m+1)\longrightarrow\omega$ such that $\mathbf{dis}[\rho_h^\ell(\sigma)]=\{(j_0,f^*)\}$ for some $j_0<\psi(m,\ell)$. It should be clear that $s\notin W_{\ell,h}^k(m,m+1,\sigma)$.

CLAIM 5.4.1.2. *For each $\ell\in\omega$, the sequence $\langle W_{\ell,h}^k:h\in\mathcal{F}_\ell,\ k\in\omega\rangle$ is $(\bar{t}_\ell,\bar{\mathcal{F}}_\ell)$–coherent.*

Proof of the claim: Let $h_0\in\mathcal{F}_\ell$, $k\in\omega$. By the demand (ii) of the choice of $\mathcal{F}_\ell$ we find $h_1\in\mathcal{F}_\ell$ such that

$$h_0<_\ell^* h_1\quad\text{ and }\quad(\forall m\in\omega)(h_1(m)\leq h_0(m)+m).$$

We want to show that the systems W_{ℓ,h_0}^k and W_{ℓ,h_1}^{k+1} are $(\bar{t}_\ell,h_0,h_1)$–coherent and that this is witnessed by the functions $\rho_{h_0}^\ell,\rho_{h_1}^\ell$. Clearly these functions satisfy the demands (1a), (1b) and (1d) of 5.3.10, so what we have to check is 5.3.10(1c) only.

Suppose that $m>k+1$, $\sigma_0:[m,m+1)\longrightarrow\omega$. Look at the creature $s=\rho_{h_0}^\ell(\sigma_0)$. We know that $\mathbf{nor}[s]\leq h_0(m)$ and hence $|\mathbf{dis}[s]|\leq 2^{g(m)h_0(m)}$. Since $2^{g(m)h_0(m)}<\frac{\psi(m,\ell)}{(n_m^*)^{k_m^*-n_m^*-1}}$ (remember clause (i) of the choice of $\mathcal{F}_\ell$) we find a creature $s^*\in\Sigma_{5.4.1}^\ell(t_{\ell,m})$ such that $\mathbf{dis}[s]\hookrightarrow\mathbf{dis}[s^*]$, $\mathbf{nor}[s^*]=h_0(m)$ (i.e. $|\mathbf{dis}[s^*]|=2^{g(m)h_0(m)}$) and for each function $x^*:(n_m^*,k_m^*)\longrightarrow n_m^*$ we have

$$\frac{|\{(j,x)\in\mathbf{dis}[s]:x=x^*\}|}{|\mathbf{dis}[s]|}\leq 2\frac{|\{(j,x)\in\mathbf{dis}[s^*]:x=x^*\}|}{|\mathbf{dis}[s^*]|}.$$

How? We just "repeat" each (j,x) from $\mathbf{dis}[s]$ successively, till we get the required size. We have enough space for this as the number of the required repetitions for each function from (n_m^*,k_m^*) to n_m^* is less than $2^{g(m)h_0(m)}$.
Take $\sigma_0^*:[m,m+1)\longrightarrow\omega$ such that $\rho_{h_0}^\ell(\sigma_0^*)=s^*$. We want to show that this σ_0^* is a $\rho_{h_0}^\ell$–cover for σ_0. So suppose that $\sigma_1:[m,m+1)\longrightarrow\omega$ is such that $\rho_{h_0}^\ell(\sigma_0^*)\in\Sigma_{5.4.1}^\ell(\rho_{h_1}^\ell(\sigma_1))$. Let $t\in W_{\ell,h_1}^{k+1}(m,m+1,\sigma_1)$. This means that we can find $u\in\mathbf{dis}[t]$ such that

$$|\{(j,x)\in\mathbf{dis}[\rho_{h_1}^\ell(\sigma_1)]:x[u]=n_m^*\}|<\frac{|\mathbf{dis}[\rho_{h_1}^\ell(\sigma_1)]|}{2^{g(m)(m+1)(k+2)}}\leq\frac{2^{g(m)h_1(m)}}{2^{g(m)(m+1)(k+2)}}.$$

Hence, remembering that $\mathbf{dis}[s^*] \hookrightarrow \mathbf{dis}[\rho^\ell_{h_1}(\sigma_1)]$ and the choice of h_1, we get

$$|\{(j,x) \in \mathbf{dis}[s^*] : x[u] = n^*_m\}| <$$
$$< \frac{2^{g(m)h_0(m)}}{2^{g(m)(m+1)(k+1)}} \cdot \frac{2^{mg(m)}}{2^{g(m)(m+1)}} = \frac{1}{2^{g(m)}} \cdot \frac{|\mathbf{dis}[s^*]|}{2^{g(m)(m+1)(k+1)}}.$$

But we are interested in s. By the choice of s^* we have

$$\frac{|\{(j,x) \in \mathbf{dis}[s] \colon x[u]=n^*_m\}|}{|\mathbf{dis}[s]|} \le 2\frac{|\{(j,x) \in \mathbf{dis}[s^*] \colon x[u]=n^*_m\}|}{|\mathbf{dis}[s^*]|} < \frac{1}{2^{g(m)(m+1)(k+1)}}$$

and therefore $t \in W^k_{\ell,h_0}(m, m+1, \sigma_0)$. Thus we have proved

$$W^{k+1}_{\ell,h_1}(m, m+1, \sigma_1) \subseteq W^k_{\ell,h_0}(m, m+1, \sigma_0)$$

whenever $\rho^\ell_{h_0}(\sigma^*_0) \in \Sigma^\ell_{5.4.1}(\rho^\ell_{h_1}(\sigma_1))$. This finishes the claim.

CLAIM 5.4.1.3. *Suppose that $\Gamma \subseteq \mathbb{P}^*_\infty(K_{5.4.1}, \Sigma_{5.4.1})$ is quasi-$W^k_{\ell,h}$-generic for all $k, \ell \in \omega$, $h \in \mathcal{F}_\ell$. Let $\mathbb{B}$ be a measure algebra. Then*

$$\Vdash_{\mathbb{B}} \text{“}\Gamma \textit{ is quasi-}W^k_{\ell,h}\textit{-generic for all } k, \ell \in \omega,\ h \in \mathcal{F}_\ell\text{”}.$$

Proof of the claim: Let μ be a σ-additive measure on the complete Boolean algebra $\mathbb{B}$. Let $k, \ell \in \omega$, $h \in \mathcal{F}_\ell$. Suppose that $\dot\eta$ is a $\mathbb{B}$–name for a real in ω^ω, $b \in \mathbb{B}^+$ (i.e. $\mu(b) > 0$). To simplify notation let us define, for $m \in \omega$,

$$K_m = |V^m_h(\bar t_\ell)|, \quad M_m = 2^{g(m)(m+1)} \cdot K_m, \quad N_m = \psi(m, \ell)!$$

and let $h^\otimes_\ell \in \mathcal{F}_{\ell+1}$ be the function given by the clause (iii) of the choice of $\bar{\mathcal{F}}_{\ell+1}$.

Fix $m \in \omega$ for a moment.

For $s \in V^m_h(\bar t_\ell)$ choose a creature $t(s) \in \Sigma^{\ell+1}_{5.4.1}(t_{\ell+1,m})$ such that

- $|\mathbf{dis}[t(s)]| = N_m$,
- for each $x^* : (n^*_m, k^*_m) \longrightarrow n^*_m$

$$\frac{|\{(j,x) \in \mathbf{dis}[s] : x = x^*\}|}{|\mathbf{dis}[s]|} = \frac{|\{(j,x) \in \mathbf{dis}[t(s)] : x = x^*\}|}{|\mathbf{dis}[t(s)]|}$$

(possible be the choice of N_m and the fact that each x^* is repeated more than N_m times in $\mathbf{dis}[t_{\ell+1,m}]$). Further, we choose integers $g(m,s) < M_m$ for $s \in V^m_h(\bar t_\ell)$ such that

$$\sum_{s \in V^m_h(\bar t_\ell)} g(m,s) = M_m \quad \text{and} \quad \left|\frac{\mu([\![\rho^\ell_h(\dot\eta{\restriction}[m, m+1)) = s]\!]_{\mathbb{B}} \cdot b)}{\mu(b)} - \frac{g(m,s)}{M_m}\right| \le \frac{1}{M_m}$$

(where $[\![\cdot]\!]_{\mathbb{B}}$ stands for the Boolean value). Take a creature $t^*_m \in \Sigma^{\ell+1}_{5.4.1}(t_{\ell+1,m})$ such that for some sequence $\langle A_s : s \in V^m_h(\bar t_\ell)\rangle$ of disjoint subsets of $\psi(m, \ell+1)$ we have

- $|A_s| = g(m,s) \cdot N_m$,
- $\bigcup\{A_s : s \in V^m_h(\bar t_\ell)\} = \{j < \psi(m, \ell+1) : (\exists x)((j,x) \in \mathbf{dis}[t^*_m])\}$,
- for some bijection $\pi_s : g(m,s) \times \mathbf{dis}[t(s)] \longrightarrow A_s$ we have

$$(\forall k < g(m,s))(\forall (j,x) \in \mathbf{dis}[t(s)])\big((\pi_s(k,(j,x)), x) \in \mathbf{dis}[t^*_m]\big).$$

Why is the choice of the t^*_m possible? Note that our requirements imply that

$$|\mathbf{dis}[t^*_m]| = M_m \cdot N_m = 2^{g(m)(m+1)} \cdot |V^m_h(\bar t_\ell)| \cdot \psi(m,\ell)! < \frac{\psi(m, \ell+1)}{(n^*_m)^{k^*_m - n^*_m - 1}}$$

and the last number says how often each function is repeated in $\mathbf{dis}[t_{\ell+1,m}]$. Moreover

$$\mathbf{nor}[t^*_m] = \frac{g(m)(m+1)+\log_2(|V^m_h(\bar{t}_\ell)|)+\log_2(\psi(m,\ell)!)}{g(m)} < \frac{g(m)(m+1)+\psi(m,\ell)+\log_2(\psi(m,\ell)!)}{g(m)} \leq \frac{h^\otimes_\ell(m)}{g(m)} < h^\otimes_\ell(m).$$

Thus $t^*_m \in V^m_{h^\otimes_\ell}(\bar{t}_{\ell+1})$.

Let $\eta \in \omega^\omega$ be such that $\rho^{\ell+1}_{h^\otimes_\ell}(\eta\restriction[m,m+1)) = t^*_m$. Since Γ is quasi-$W^{k+1}_{\ell+1,h^\otimes_\ell}$-generic, we find $\bar{s} = \langle s_m : m < \omega\rangle \in \Gamma$ such that for some $m^* > k+4$

$$(\forall m \geq m^*)(s_m \in W^{k+1}_{\ell+1,h^\otimes_\ell}(m,m+1,\eta\restriction[m,m+1)).$$

Fix $m \geq m^*$ for a moment. We know that for some $u \in \mathbf{dis}[s_m]$ we have

$$|\{(j,x) \in \mathbf{dis}[t^*_m] : x[u] = n^*_m\}| < \frac{|\mathbf{dis}[t^*_m]|}{2^{g(m)(m+1)(k+2)}}.$$

For $s \in V^m_h(\bar{t}_\ell)$ let

$$Y_m(s) \stackrel{\text{def}}{=} \frac{|\{(j,x) \in \mathbf{dis}[s] : x[u] = n^*_m\}|}{|\mathbf{dis}[s]|} = \frac{|\{(j,x) \in \mathbf{dis}[t(s)] : x[u] = n^*_m\}|}{|\mathbf{dis}[t(s)]|}$$

and note that $\sum\limits_s Y_m(s)\cdot g(m,s)\cdot N_m = |\{(j,x) \in \mathbf{dis}[t^*_m] : x[u] = n^*_m\}|$. Let

$$\mathcal{X}_m = \{s \in V^m_h(\bar{t}_\ell) : Y_m(s) \geq \frac{1}{2^{g(m)(m+1)(k+1)}}\}$$

(so $\rho^\ell_h(\sigma) \notin \mathcal{X}_m$ implies $s_m \in W^k_{\ell,h}(m,m+1,\sigma)$). Note that

$$\sum_{s\in\mathcal{X}_m} \frac{g(m,s)\cdot N_m}{2^{g(m)(m+1)(k+1)}} \leq \sum_{s\in\mathcal{X}_m} Y_m(s)\cdot g(m,s)\cdot N_m \leq \frac{M_m\cdot N_m}{2^{g(m)(m+1)(k+2)}}$$

and therefore $\sum\limits_{s\in\mathcal{X}_m} \frac{g(m,s)}{M_m} \leq \frac{1}{2^{g(m)(m+1)}}$. Hence

$$\sum_{s\in\mathcal{X}_m} \frac{\mu([\![\rho^\ell_h(\dot{\eta}\restriction[m,m+1)) = s]\!]_{\mathbb{B}}\cdot b)}{\mu(b)} \leq \sum_{s\in\mathcal{X}_m}\left(\frac{g(m,s)}{M_m} + \frac{1}{M_m}\right) \leq \frac{1}{2^{g(m)m+g(m)-1}}.$$

Let $b_m = \sum^{\mathbb{B}}_{s\in\mathcal{X}_m} [\![\rho^\ell_h(\dot{\eta}\restriction[m,m+1)) = s]\!]_{\mathbb{B}}\cdot b$. By the above estimations we have $\mu(b_m) \leq \frac{\mu(b)}{2^{m+4}}$ (remember $m \geq m^* > k+4 \geq 4$). Look at the condition $b^* = b - (\sum^{\mathbb{B}}_{m\geq m^*} b_m)$. Clearly $\mu(b^*) > 0$ and

$$b^* \Vdash_{\mathbb{B}} (\forall m \geq m^*)(\rho^\ell_h(\dot{\eta}\restriction[m,m+1) \notin \mathcal{X}_m))$$

and therefore

$$b^* \Vdash_{\mathbb{B}} (\forall m \geq m^*)(s_m \in W^k_{\ell,h}(m,m+1,\dot{\eta}\restriction[m,m+1))).$$

This finishes the proof of the claim and thus checking that the construction is as required. □

CONCLUSION 5.4.2. Let $F \in \omega^\omega$ be strictly increasing and f, g, $(K_{5.4.1},\Sigma_{5.4.1})$, $(K^\ell_{5.4.1},\Sigma^\ell_{5.4.1})$, $\bar{t}$, $\bar{t}_\ell$, $\bar{\mathcal{F}}_\ell = (\mathcal{F}_\ell,<^*_\ell)$ (for $\ell < \omega$) and $\langle W^k_{\ell,h} : h \in \mathcal{F}_\ell, k,\ell < \omega\rangle$ be given by 5.4.1 (for F). Suppose that $\Gamma \subseteq \mathbb{P}^*_\infty(K_{5.4.1},\Sigma_{5.4.1})$ is quasi-$W^k_{\ell,h}$-generic for every $k,\ell < \omega$, $h \in \mathcal{F}_\ell$ (exists e.g. under CH, see 5.3.4(2), 5.3.5).

1. Countable support iterations of proper forcing notions which are $(\Gamma, W^k_{\ell,h})$–genericity preserving for all $k, \ell \in \omega$, $h \in \mathcal{F}_\ell$ is $(\Gamma, W^k_{\ell,h})$–genericity preserving (for $k, \ell \in \omega$, $h \in \mathcal{F}$) and hence does not add Cohen reals.
2. Every (f,g)-bounding proper forcing notion (this includes proper forcing notions with the Laver property) and random real forcing are $(\Gamma, W^k_{\ell,h})$–genericity preserving (for $k, \ell \in \omega$, $h \in \mathcal{F}_\ell$),
3. Assume CH. Then any countable support iteration of proper forcing notions of one of the following types:

   (f,g)–bounding, Laver property, random forcing

   does not add Cohen reals.

PROOF. By 5.4.1, 5.3.9(1), 5.3.11, 5.2.7 and 5.3.12. □

EXAMPLE 5.4.3. We define a strictly increasing function $F \in \omega^\omega$ and a creating pair $(K_{5.4.3}, \Sigma_{5.4.3})$ which: captures singletons, is strongly finitary, reducible, forgetful, simple, essentially f^F–big and h–limited for some function $h : \omega \times \omega \longrightarrow \omega$ such that $(\forall^\infty n)(\prod\limits_{m<n} h(m,m) < g^F(n) < f^F(n))$, where f^F, g^F are given by 5.4.1 for F.

CONSTRUCTION. For $N < \omega$ we define a nice pre-norm H_N on $\mathcal{P}(\mathcal{P}(N) \setminus \{N\})$ by:

$$H_N(A) = \log_2(1 + \max\{k < \omega : (\forall x \in [N]^k)(\exists a \in A)(x \subseteq a)\})$$

(for $A \subseteq \mathcal{P}(N) \setminus \{N\}$). Note that if $B \subseteq C \subseteq \mathcal{P}(N) \setminus \{N\}$, $H_N(C) > 1$ then:

1. $\max\{k < \omega : (\forall x \in [N]^k)(\exists a \in C)(x \subseteq a)\} \geq 2$,
2. $H_N(B) \leq H_N(C)$,
3. $\max\{H_N(B), H_N(C \setminus B)\} \geq H_N(C) - 1$.

For 3) above note that if it fails then we find $k_0 \in \omega$ such that

$$1 + \max\{H_N(B), H_N(C \setminus B)\} \leq \log_2(2k_0) < H_N(C).$$

By the first inequality we find $x_0, x_1 \in [N]^{k_0}$ such that

$$(\forall a \in B)(x_0 \not\subseteq a) \quad \text{and} \quad (\forall a \in C \setminus B)(x_1 \not\subseteq a).$$

But the second inequality implies that there is $a \in C$ such that $x_0 \cup x_1 \subseteq a$, what gives a contradiction.
As clearly $H_N(\{a\}) = 0$ for $a \in \mathcal{P}(N) \setminus \{N\}$, H_N is really a nice pre–norm.

Let $F \in \omega^\omega$ be defined by $F(m) = 2^{(m+1)^3 \cdot 2^{2^m}}$ (for $m \in \omega$) and let f^F, g^F be from 5.4.1 (for F). To simplify notation let $M_n = 2^{f^F(n)}$, $N_n = (n+1)^2 \cdot 2^{M_n}$ (for $n \in \omega$).

Let $\mathbf{H} : \omega \longrightarrow \mathcal{P}([\omega]^{<\omega})$ be given by $\mathbf{H}(n) = [N_n]^{2^{M_n}}$.

Let $K_{5.4.3}$ consist of creatures $t \in \mathrm{CR}[\mathbf{H}]$ such that, letting $m = m^t_{\mathrm{dn}}$,

(a) $\mathbf{dis}[t]$ is a subset of $\mathbf{H}(m)$,
(b) $\mathbf{val}[t] = \{\langle w, u\rangle \in \prod\limits_{i<m} \mathbf{H}(i) \times \prod\limits_{i\leq m} \mathbf{H}(i) : w \lhd u \ \&\ u(m) \in \mathbf{dis}[t]\}$,
(c) $\mathbf{nor}[t] = \frac{H_{N_m}(\mathbf{dis}[t])}{(m+1)\cdot f^F(m)}$.

The composition operation $\Sigma_{5.4.3}$ is the trivial one: it gives a nonempty result for singletons only and then $\Sigma_{5.4.3}(t) = \{s \in K_{5.4.3} : m^s_{\mathrm{dn}} = m^t_{\mathrm{dn}} \ \&\ \mathbf{dis}[s] \subseteq \mathbf{dis}[t]\}$. Now, we have to check that $(K_{5.4.3}, \Sigma_{5.4.3})$ has the required properties. Clearly it is a strongly finitary, simple and forgetful creating pair. Plainly, it captures singletons. Note that if $t_m \in K_{5.4.3}$ (for $m \in \omega$) is such that $\mathbf{dis}[t_m] = \mathbf{H}(m)$ then

$$\mathbf{nor}[t_m] = \frac{H_{N_m}(\mathbf{H}(m))}{(m+1)\cdot f^F(m)} = \frac{\log_2(1+2^{M_m})}{(m+1)\cdot f^F(m)} > \frac{2^{f^F(m)}}{(m+1)\cdot f^F(m)} \stackrel{m\to\infty}{\longrightarrow} \infty.$$

Consequently the forcing notions $\mathbb{Q}^*_\infty(K_{5.4.3},\Sigma_{5.4.3})$, $\mathbb{Q}^*_{\mathrm{w}\infty}(K_{5.4.3},\Sigma_{5.4.3})$ etc will be non-trivial.
To verify that $(K_{5.4.3},\Sigma_{5.4.3})$ is reducible use the fact that H_N is a nice pre-norm on $\mathcal{P}(\mathcal{P}(N)\setminus\{N\})$: if $a \in C \in \mathcal{P}(\mathcal{P}(N)\setminus\{N\})$, $H_N(C) > 1$ then $H_N(C) - 1 \leq H_N(C\setminus\{a\})$. For similar reasons $(K_{5.4.3},\Sigma_{5.4.3})$ is essentially f^F–big, remember that we divide the respective value of H_{N_m} by $(m+1)\cdot f^F(m)$ (where $m = m^t_{\mathrm{dn}}$). Finally, let $h(m,k) = 2^{N_m}$ for $m,k \in \omega$. Then $|\mathbf{H}(m)| = \binom{N_m}{2^{M_m}} < 2^{N_m}$ for all $m \in \omega$, so $(K_{5.4.3},\Sigma_{5.4.3})$ is h–limited (actually much more). Moreover, for every $m \in \omega$:

$$g^F(m+1) = F(f^F(m)) > 2^{(m+1)^3\cdot 2^{2^{f^F(m)}}} = 2^{(m+1)^3\cdot 2^{M_m}} > \prod_{k\leq m} 2^{N_k} = \prod_{k\leq m} h(k,k).$$

Thus $(K_{5.4.3},\Sigma_{5.4.3})$ is as required. □

CONCLUSION 5.4.4. The forcing notion $\mathbb{Q}^*_{\mathrm{w}\infty}(K_{5.4.3},\Sigma_{5.4.3})$:

1. is proper and ω^ω–bounding,
2. preserves non-meager sets,
3. is (f^F,g^F)–bounding,
4. makes the ground model reals have measure zero.

Moreover, assuming CH, countable support iterations of $\mathbb{Q}^*_{\mathrm{w}\infty}(K_{5.4.3},\Sigma_{5.4.3})$ with Laver's forcing notion, Miller's forcing notion and random forcing do not add Cohen reals.

PROOF. The first required property is a consequence of 2.1.12 and 3.1.3. The second follows from 3.2.6 and the third property is a consequence of 5.1.10. The "moreover" part holds true by 5.4.2.

Let $\mathcal{X} = \prod\limits_{m\in\omega} N_m$ be equipped with the product measure. Of course, this space is measure isomorphic to the reals, so what we have to show is the following claim.

CLAIM 5.4.4.1. $\Vdash_{\mathbb{Q}^*_{\mathrm{w}\infty}(K_{5.4.3},\Sigma_{5.4.3})}$ "*$\mathbf{V}\cap\mathcal{X}$ is a null set*".

Proof of the claim: Let $\dot{W}$ be the $\mathbb{Q}^*_{\mathrm{w}\infty}(K_{5.4.3},\Sigma_{5.4.3})$–name for the generic real (see 1.1.13) and let $\dot{A}$ be a $\mathbb{Q}^*_{\mathrm{w}\infty}(K_{5.4.3},\Sigma_{5.4.3})$–name for a subset of $\mathcal{X}$ such that

$$\Vdash_{\mathbb{Q}^*_{\mathrm{w}\infty}(K_{5.4.3},\Sigma_{5.4.3})} \text{“}\dot{A} = \{x\in\mathcal{X} : (\exists^\infty n)(x(n)\in\dot{W}(n))\}\text{”}.$$

Note that $\Vdash$" $(\forall m\in\omega)(\dot{W}(m)\in[N_m]^{2^{M_m}})$" and $\frac{2^{M_m}}{N_m} = \frac{1}{(m+1)^2}$. Consequently, $\Vdash$" $\dot{A}$ is a null subset of $\mathcal{X}$". By the definition of $(K_{5.4.3},\Sigma_{5.4.3})$ (remember the definition of H_N) one easily shows that $\Vdash$" $\mathbf{V}\cap\mathcal{X}\subseteq\dot{A}$", what finishes the proof of the claim and the conclusion. □

One may consider tree versions of 5.4.3.

EXAMPLE 5.4.5. Let $F\in\omega^\omega$, $h:\omega\times\omega\longrightarrow\omega$ and $\mathbf{H}:\omega\longrightarrow\mathcal{P}([\omega]^{<\omega})$ be as defined in 5.4.3. There are finitary tree–creating pairs $(K^\ell_{5.4.5},\Sigma^\ell_{5.4.5})$ (for $\ell<3$) for $\mathbf{H}$ such that

(a) $(K^0_{5.4.5}, \Sigma^0_{5.4.5})$ is local, reducible, h–limited and essentially f^F–big (where f^F is given by 5.4.1),
(b) $(K^1_{5.4.5}, \Sigma^1_{5.4.5})$ is t-omittory, reducible and essentially f^F–big,
(c) $(K^2_{5.4.5}, \Sigma^2_{5.4.5})$ is reducible, h–limited and essentially f^F–big,
(d) the forcing notions $\mathbb{Q}^{\rm tree}_0(K^0_{5.4.5}, \Sigma^0_{5.4.5})$, $\mathbb{Q}^{\rm tree}_1(K^1_{5.4.5}, \Sigma^1_{5.4.5})$ and $\mathbb{Q}^{\rm tree}_1(K^2_{5.4.5}, \Sigma^2_{5.4.5})$ are equivalent,
(e) the forcing notion $\mathbb{Q}^{\rm tree}_1(K^0_{5.4.5}, \Sigma^0_{5.4.5})$ is proper, ω^ω-bounding, makes ground model reals meager and null (even more), is (f^F, g^+)–bounding, where $g^+(n) = \prod\limits_{i<n} |\mathbf{H}(i)|$,
(f) the forcing notions $\mathbb{Q}^{\rm tree}_1(K^1_{5.4.5}, \Sigma^1_{5.4.5})$ and $\mathbb{Q}^{\rm tree}_1(K^2_{5.4.5}, \Sigma^2_{5.4.5})$ are proper, ω^ω–bounding, preserve non-meager sets, make ground model reals null and are (f^F, g^F)–bounding.

CONSTRUCTION. Let $F, N_m, M_m, \mathbf{H}, h$ be as in 5.4.3.
A tree creature $t \in {\rm TCR}_\eta[\mathbf{H}]$ is in $K^0_{5.4.5}$ if

- $\mathbf{dis}[t] \subseteq \mathbf{H}(\ell g(\eta))$,
- $\mathbf{val}[t] = \{\langle \eta, \nu\rangle : \eta \lhd \nu \ \&\ \ell g(\nu) = \ell g(\eta) + 1 \ \&\ \nu(\ell g(\eta)) \in \mathbf{dis}[t]\}$,
- $\mathbf{nor}[t] = \frac{H_{N_{\ell g(\eta)}}(\mathbf{dis}[t])}{(\ell g(\eta)+1)\cdot f^F(\ell g(\eta))}$.

The tree composition $\Sigma^0_{5.4.5}$ on $K^0_{5.4.5}$ is trivial: $\Sigma^0_{5.4.5}(t) = \{s \in K^0_{5.4.5} : \mathbf{val}[s] \subseteq \mathbf{val}[t]\}$.
The family $K^1_{5.4.5}$ consists of these tree–creatures $t \in {\rm TCR}[\mathbf{H}]$ that for some $\eta \trianglelefteq \eta^* \in \bigcup\limits_{n<\omega} \prod\limits_{i<n} \mathbf{H}(i)$ we have

- $\mathbf{dis}[t] \subseteq \mathbf{H}(\ell g(\eta^*))$,
- $\mathbf{val}[t] = \{\langle \eta, \nu\rangle : \eta^* \lhd \nu \ \&\ \ell g(\nu) = \ell g(\eta^*) + 1 \ \&\ \nu(\ell g(\eta^*)) \in \mathbf{dis}[t]\}$,
- $\mathbf{nor}[t] = \frac{H_{N_{\ell g(\eta^*)}}(\mathbf{dis}[t])}{(\ell g(\eta^*)+1)\cdot f^F(\ell g(\eta^*))}$.

The tree composition $\Sigma^1_{5.4.5}$ is such that

$$\Sigma^1_{5.4.5}(t_\nu : \nu \in \hat{T}) = \{t \in K^1_{5.4.5} : {\rm dom}(\mathbf{val}[t]) = \{{\rm root}(T)\} \ \&\ {\rm rng}(\mathbf{val}[t]) \subseteq \max(T)\}.$$

In a similar manner we define $(K^2_{5.4.5}, \Sigma^2_{5.4.5})$. A tree creature $t \in {\rm TCR}_\eta[\mathbf{H}]$ is in $K^2_{5.4.5}$ if

- $\mathbf{dis}[t] = (A_t, \langle \nu^t_x : x \in A_t\rangle)$, where $A_t \subseteq \mathbf{H}(\ell g(\eta))$ and $\nu^t_x \in \bigcup\limits_{n<\omega} \prod\limits_{i<n} \mathbf{H}(i)$ (for $x \in A_t$) are such that $\eta \lhd \eta^\frown\langle x\rangle \trianglelefteq \nu^t_x$,
- $\mathbf{val}[t] = \{\langle \eta, \nu^t_x\rangle : x \in A_t\}$,
- $\mathbf{nor}[t] = \frac{H_{N_{\ell g(\eta)}}(A_t)}{(\ell g(\eta)+1)\cdot f^F(\ell g(\eta))}$,

and the tree composition $\Sigma^2_{5.4.5}$ is defined like $\Sigma^1_{5.4.5}$.
Checking that $(K^\ell_{5.4.5}, \Sigma^\ell_{5.4.5})$ have the desired properties is straightforward and similar to 5.4.3 (remember 5.1.5, 5.1.8, 3.2.2, 3.2.8). □

COROLLARY 5.4.6. *Let $F \in \omega^\omega$ be an increasing function, $f, \mathbf{H}$ be as defined in 2.4.6 (for F) and let $F^*(n) = F(\varphi_{\mathbf{H}}(n))$. Then the forcing notion $\mathbb{Q}^*_f(K_{2.4.6}, \Sigma_{2.4.6})$ (defined as in 2.4.6 for F) is proper, ω^ω-bounding, $(F^*, \varphi_{\mathbf{H}})$-bounding and makes ground model reals meager.*

PROOF. By 3.1.2, 5.1.12 and 3.2.8(2). □

CHAPTER 6

Playing with ultrafilters

This chapter originated in the following question of Matet and Pawlikowski. Are the cardinals $\mathfrak{m}_1$, $\mathfrak{m}_2$ equal, where

$\mathfrak{m}_1$ is the least cardinality of a set $Z \subseteq \bigcup_{E\subseteq\omega} \omega^E$ such that

(i) $(\exists F \in \omega^\omega)(\forall f \in Z)(\forall n \in \mathrm{dom}(f))(f(n) < F(n))$,

(ii) the family $\{\mathrm{dom}(f) : f \in Z\}$ has the finite intersection property, and

(iii) $(\forall x \in \prod_{n\in\omega} [\omega]^{\leq n})(\exists f \in Z)(\forall^\infty n \in \mathrm{dom}(f))(f(n) \notin x(n))$;

$\mathfrak{m}_2$ is defined in a similar manner, but **(iii)** is replaced by

(iii)$^-$ $(\forall g \in \omega^\omega)(\exists f \in Z)(\forall^\infty n \in \mathrm{dom}(f))(f(n) \neq g(n))$?

It was known that $\mathfrak{m}_2 \leq \lambda \leq \mathfrak{m}_1$, where λ is the least size of a basis of an ideal on ω which is not a weak q-point (see Matet Pawlikowski [**MaPa98**]). We answer the Matet – Pawlikowski question in 6.4.6, showing that it is consistent that $\lambda = \aleph_1$ (and so $\mathfrak{m}_2 = \aleph_1$) and $\mathfrak{m}_1 = \aleph_2$. On the way to this result we have to deal with preserving some special ultrafilters on ω. The technology developed in the previous section is very useful for this (both to describe the required properties and to preserve them at limit stages of countable support iterations).

In the first part of the chapter we present the framework: ultrafilters generated by quasi-W-generic Γ. Then we introduce several properties of ultrafilters and discuss relations between them. The third section shows how forcing notions constructed according to our schema may preserve some special ultrafilters. Finally, in the last part, we apply all these tools to answer the Matet – Pawlikowski question.

6.1. Generating an ultrafilter

DEFINITION 6.1.1. We say that a creating pair (K, Σ) *generates an ultrafilter* if

$(\otimes_{6.1.1})$ for every $k < \omega$ there is $k^* < \omega$ such that
if $t \in K$, $\mathbf{nor}[t] \geq k^*$ and $c : [m^t_{\mathrm{dn}}, m^t_{\mathrm{up}}) \longrightarrow 2$,
then for some $s \in \Sigma(t)$ and $i^* < 2$ we have $\mathbf{nor}[s] \geq k$ and

$$[u \in \mathbf{basis}(s)\ \&\ v \in \mathrm{pos}(u, s)\ \&\ m^t_{\mathrm{dn}} \leq m < m^t_{\mathrm{up}}\ \&\ v(m) \neq 0] \quad \Rightarrow \quad c(m) = i^*$$

and if $\mathbf{nor}[t] > 0$, $u \in \mathbf{basis}(t)$ then there is $v \in \mathrm{pos}(u, t)$ such that for some $m \in [m^t_{\mathrm{dn}}, m^t_{\mathrm{up}})$ we have $v(m) \neq 0$.

PROPOSITION 6.1.2. *Assume that (K, Σ) is an omittory and monotonic (see 5.2.3) creating pair which generates an ultrafilter. Then*

$$\Vdash_{\mathbb{Q}^*_{\mathrm{s}\infty}(K,\Sigma)} \text{“}\{m : \dot{W}(m) \neq 0\}\ \textit{induces an ultrafilter on the algebra}\ \mathcal{P}(\omega)^{\mathbf{V}}/\mathrm{Fin}\text{”}.$$

PROOF. For $k \in \omega$, let $g(k)$ be the k^* given by $(\otimes_{6.1.1})$ (for k). Let $\dot{Z}$ be a $\mathbb{Q}^*_{s\infty}(K,\Sigma)$-name such that

$$\Vdash_{\mathbb{Q}^*_{s\infty}(K,\Sigma)} \dot{Z} = \{m \in \omega : \dot{W}(m) \neq 0\}.$$

Clearly $\Vdash_{\mathbb{Q}^*_{s\infty}(K,\Sigma)} \dot{Z} \in [\omega]^\omega$ (by the second requirement of 6.1.1). We have to show that for each $A \in [\omega]^\omega$

$$\Vdash_{\mathbb{Q}^*_{s\infty}(K,\Sigma)} \text{"either } \dot{Z} \cap A \text{ or } \dot{Z} \setminus A \text{ is finite"}.$$

So suppose that $p \in \mathbb{Q}^*_{s\infty}(K,\Sigma)$, $A \in [\omega]^\omega$ are such that

$$p \Vdash_{\mathbb{Q}^*_{s\infty}(K,\Sigma)} \text{"both } \dot{Z} \cap A \text{ and } \dot{Z} \setminus A \text{ are infinite"}.$$

Since (K,Σ) is omittory we may assume that $p = (w^p, t_0, t_1, \dots)$ where $\mathbf{nor}[t_\ell] \geq g(\ell + m^{t_\ell}_{\rm dn})$ (see 2.1.3.(2)). Let $c_\ell : [m^{t_\ell}_{\rm dn}, m^{t_\ell}_{\rm up}) \longrightarrow 2$ be the characteristic function of $A \cap [m^{t_\ell}_{\rm dn}, m^{t_\ell}_{\rm up})$. Applying the condition $(\otimes_{6.1.1})$ and the choice of g for each $\ell < \omega$ we find $s'_\ell \in \Sigma(t_\ell)$ such that $\mathbf{nor}[s'_\ell] \geq \ell + m^{s'_\ell}_{\rm dn}$ and for some $i^\ell < 2$ we have

$(\otimes^*)$ *if* $u \in \mathbf{basis}(s'_\ell)$, $v \in \mathrm{pos}(u, s'_\ell)$, $m^{s'_\ell}_{\rm dn} \leq m < m^{s'_\ell}_{\rm up}$ and $v(m) \neq 0$ *then* $c_\ell(m) = i^\ell$.

Now choose $i^* < 2$ and an increasing sequence $\langle \ell_k : k < \omega \rangle \subseteq \omega$ such that $i^{\ell_k} = i^*$ (for $k < \omega$) and take $s_0 = s'_{\ell_0} \restriction [m^{s'_0}_{\rm dn}, m^{s'_{\ell_0}}_{\rm up})$, $s_{k+1} = s'_{\ell_{k+1}} \restriction [m^{s'_{\ell_k}}_{\rm up}, m^{s'_{\ell_{k+1}}}_{\rm up})$. Once again, since (K,Σ) is omittory we get $q \stackrel{\rm def}{=} (w^p, s_0, s_1, s_2, \dots) \in \mathbb{Q}^*_{s\infty}(K,\Sigma)$ and it is stronger than p. Suppose $i^* = 0$. We claim that in this case $q \Vdash_{\mathbb{Q}^*_{s\infty}(K,\Sigma)} \dot{Z} \cap A \subseteq m^{s_0}_{\rm dn}$ (contradicting the choice of p, A). If not then we find $k < \omega$, $w \in \mathrm{pos}(w^p, s_0, \dots, s_k)$ and $n \in [m^{s_0}_{\rm dn}, m^{s_k}_{\rm up}) \cap A$ such that $w(n) \neq 0$. By smoothness we may additionally demand that, if $k > 0$ then $n \in [m^{s_k}_{\rm dn}, m^{s_k}_{\rm up}) = [m^{s'_{\ell_{k-1}}}_{\rm up}, m^{s'_{\ell_k}}_{\rm up})$. By the smoothness and monotonicity of (K,Σ) we have

$$w \in \mathrm{pos}(w \restriction m^{s_k}_{\rm dn}, s_k) = \{v : \langle w \restriction m^{s_k}_{\rm dn}, v \rangle \in \mathbf{val}[s_k]\}.$$

Now, by the choice of s_k we may conclude that

$$w \restriction [m^{s_k}_{\rm dn}, m^{s'_{\ell_k}}_{\rm dn}) = \mathbf{0}_{[m^{s_k}_{\rm dn}, m^{s'_{\ell_k}}_{\rm dn})} \quad \text{and thus } n \in [m^{s'_{\ell_k}}_{\rm dn}, m^{s'_{\ell_k}}_{\rm up}).$$

Moreover $\langle w \restriction m^{s'_{\ell_k}}_{\rm dn}, w \rangle \in \mathbf{val}[s'_{\ell_k}]$ so we may apply $(\otimes^*)$ (to ℓ_k) and conclude that $c_{\ell_k}(n) = i^* = 0$, a contradiction. Similarly one shows $q \Vdash \dot{Z} \setminus A \subseteq m^{s_0}_{\rm dn}$ if $i^* = 1$. □

DEFINITION 6.1.3. Suppose that (K,Σ) is a creating pair, $\bar{t} \in \mathrm{PC}_{\mathcal{C}(\mathbf{nor})}(K,\Sigma)$ and $W : \omega \times \omega \times \omega^{\underline{\omega}} \longrightarrow \mathcal{P}(K)$ is a $\bar{t}$–system (see 5.3.1). Let $\Gamma \subseteq \mathbb{P}^*_{\mathcal{C}(\mathbf{nor})}(K,\Sigma)$ be quasi-W-generic.

1. We define $\mathcal{D}(\Gamma)$ as the family of all sets $A \subseteq \omega$ such that: for some $\bar{s} = \langle s_n : n < \omega \rangle \in \Gamma$ and $N < \omega$, for every $w \in \mathbf{basis}(s_0)$ and $u \in \mathrm{pos}(w, s_0, \dots, s_N)$ we have

$$(\forall m > N)(\forall v \in \mathrm{pos}(u, s_{N+1}, \dots, s_m))(\{k \in [\ell g(u), \ell g(v)) : v(k) \neq 0\} \subseteq A).$$

2. We say that Γ *generates a filter (an ultrafilter, respectively) on* ω if $\mathcal{D}(\Gamma)$ is a filter (an ultrafilter, resp.).

REMARK 6.1.4. Note the close relation of 6.1.3 and 6.1.2. Below it becomes even closer.

DEFINITION 6.1.5. A creating pair (K, Σ) is *interesting* if for each creature $t \in K$ such that $\mathbf{nor}[t] > 0$ and every $u \in \mathbf{basis}(t)$ we have

$$|\{m \in [m^t_{\rm dn}, m^t_{\rm up}) : (\exists v \in \mathrm{pos}(u,t))(v(m) \neq 0)\}| > 1.$$

PROPOSITION 6.1.6. *Let (K, Σ) be a forgetful creating pair, $\bar{t} \in \mathrm{PC}_\infty(K, \Sigma)$ and $W : \omega \times \omega \times \omega^{\underline{\omega}} \longrightarrow \mathcal{P}(K)$ be a $\bar{t}$-system.*

1. *If $\Gamma \subseteq \mathbb{P}^*_\infty(\bar{t}, (K, \Sigma))$ is quasi-W-generic then $\mathcal{D}(\Gamma)$ is a filter on ω containing all co-finite sets. If, additionally, (K, Σ) is interesting and condensed (see 4.3.1(3)) and $A \subseteq \omega$ is such that*

$$(\forall^\infty n)(|A \cap [m^{t_n}_{\rm dn}, m^{t_n}_{\rm up})| \leq 1)$$

   then $A \notin \mathcal{D}(\Gamma)$.
2. *If Γ is quasi-W-generic in $\mathbb{P}^*_\infty(\bar{t}, (K, \Sigma))$ then $\mathcal{D}(\Gamma)$ is a p-point (see 6.2.1).*
3. *Assume CH. Suppose that, additionally, (K, Σ) is finitary, omittory, monotonic and generates an ultrafilter. Then there exists a quasi-W-generic $\Gamma \subseteq \mathbb{P}^*_\infty(K, \Sigma)$ such that $\mathcal{D}(\Gamma)$ is an ultrafilter on ω.*

PROOF. 1), 2) Should be obvious.
3) Modify the proof of 5.3.4(2), noting that if $\bar{s} \in \mathrm{PC}_\infty(K, \Sigma)$, $A \subseteq \omega$ then there is $\bar{s}^* = \langle s^*_n : n < \omega \rangle \in \mathrm{PC}_\infty(K, \Sigma)$ such that $\bar{s} \leq \bar{s}^*$ and either

$$(\forall u \in \mathbf{basis}(s^*_0))(\forall n < \omega)(\forall v \in \mathrm{pos}(u, s^*_0, \ldots, s^*_n))(\{k \in [\ell g(u), \ell g(v)) : v(k) \neq 0\} \subseteq A)$$

or a similar requirement with $\omega \setminus A$ instead of A holds. (Compare the proof of 6.1.2, remember (K, Σ) is forgetful.) □

CONCLUSION 6.1.7. Suppose that (K, Σ) is a forgetful and monotonic creating pair, $\bar{t} \in \mathrm{PC}_\infty(K, \Sigma)$ and W is a $\bar{t}$-system. Assume that $\Gamma \subseteq \mathbb{P}^*_\infty(\bar{t}, (K, \Sigma))$ is quasi-W-generic and $\mathcal{D}(\Gamma)$ is an ultrafilter. Let δ be a limit ordinal and $\langle \mathbb{P}_\alpha, \dot{\mathbb{Q}}_\alpha : \alpha < \delta \rangle$ be a countable support iteration of proper forcing notions such that for each $\alpha < \delta$:

(a) $\Vdash_{\mathbb{P}_\alpha}$ "Γ is quasi-W-generic",
(b) $\Vdash_{\mathbb{P}_\alpha}$ "$\mathcal{D}(\Gamma)$ is an ultrafilter".

Then

1. $\Vdash_{\mathbb{P}_\delta}$ "Γ is quasi-W-generic",
2. $\Vdash_{\mathbb{P}_\delta}$ "$\mathcal{D}(\Gamma)$ is an ultrafilter".

PROOF. 1) It follows from 5.3.12.
2) Since, by 6.1.6(2), $\mathcal{D}(\Gamma)$ is a p-point we may use [**Sh:f**, Ch VI, 5.2] (another presentation of this result might be found in [**BaJu95**, 6.2]). □

REMARK 6.1.8.
1. If $\bar{t} = \langle t_n : n < \omega \rangle \in \mathrm{PC}_\infty(K, \Sigma)$, W is a $\bar{t}$–system and Γ is quasi-W-generic generating a filter then we make think of $\mathcal{D}(\Gamma)$ as a filter on $\bigcup\limits_{i \in \omega} \{i\} \times (m^{t_i}_{\rm up} - m^{t_i}_{\rm dn})$ (just putting the intervals $[m^{t_i}_{\rm dn}, m^{t_i}_{\rm up})$ vertically). This will be our approach in the further part, where we will consider ultrafilters on $\bigcup\limits_{i \in \omega} \{i\} \times (i+1)$.
2. We may treat $\mathcal{D}(\Gamma)$ as a canonical filter on ω with a property described by Γ (or, more accurately, by the $\bar{t}$–system W). This is the way we are going to use 6.1.7 later: it will allow us to claim that the additional property of an ultrafilter is preserved at limit stages of an iteration.

6.2. Between Ramsey and p-points

Here we recall some definitions of special properties of ultrafilters on ω and we introduce more of them. Then we comment on relations between these notions.

DEFINITION 6.2.1. Let $\mathcal{D}$ be a filter on ω. We say that:

1. $\mathcal{D}$ *is Ramsey* if for each colouring $F : [\omega]^2 \longrightarrow 2$ there is a set $A \in \mathcal{D}$ homogeneous for F.
2. $\mathcal{D}$ *is a p-point* if for every partition $\langle A_n : n \in \omega \rangle$ of ω into sets from the dual ideal (i.e. $\omega \setminus A_n \in \mathcal{D}$) we find a set $A \in \mathcal{D}$ with
$$(\forall n \in \omega)(|A_n \cap A| < \omega).$$
3. $\mathcal{D}$ *is a q-point* if for every partition $\langle A_n : n \in \omega \rangle$ of ω into finite sets there is a set $A \in \mathcal{D}$ with
$$(\forall n \in \omega)(|A_n \cap A| \leq 1).$$
4. $\mathcal{D}$ *is a weak q-point* if for each set $B \subseteq \omega$ such that $\omega \setminus B \notin \mathcal{D}$ and a partition $\langle A_n : n \in \omega \rangle$ of B into finite sets there is a set $A \subseteq B$ such that
$$\omega \setminus A \notin \mathcal{D} \qquad \text{and} \qquad (\forall n \in \omega)(|A_n \cap A| \leq 1).$$

REMARK 6.2.2. Clearly, if $\mathcal{D}$ is an ultrafilter on ω which is a weak q-point then $\mathcal{D}$ is a q-point. (So the two notions coincide for ultrafilters).

DEFINITION 6.2.3.
1. For a filter $\mathcal{D}$ on ω let $G^R(\mathcal{D})$ be the game of two players, I and II, in which Player I in his n^{th} move plays a set $A_n \in \mathcal{D}$ and Player II answers choosing a point $k_n \in A_n$. Thus a result of a play is a pair of sequences $\langle \langle A_n : n \in \omega \rangle, \langle k_n : n \in \omega \rangle \rangle$ such that $k_n \in A_n \in \mathcal{D}$.
Player I wins the play of the game $G^R(\mathcal{D})$ if and only if the result $\langle \langle A_n : n \in \omega \rangle, \langle k_n : n \in \omega \rangle \rangle$ satisfies: $\{k_n : n \in \omega\} \notin \mathcal{D}$.
2. Similarly we define the game $G^p(\mathcal{D})$ allowing the second player to play finite sets $a_n \subseteq A_n$ (instead of points $k_n \in A_n$).
Player I wins if $\bigcup\limits_{n \in \omega} a_n \notin \mathcal{D}$

REMARK 6.2.4. Let us recall that if $\mathcal{D}$ is an ultrafilter on ω then the following conditions are equivalent (see [**Sh:f**, Ch VI, 5.6] or [**BaJu95**, 4.5]):

(a) $\mathcal{D}$ is Ramsey,
(b) $\mathcal{D}$ is both a p-point and a q-point,
(c) Player I does not have a winning strategy in the game $G^R(\mathcal{D})$.

Similarly, an ultrafilter $\mathcal{D}$ is a p-point if and only if Player I does not have a winning strategy in $G^p(\mathcal{D})$.

As we are interested in ultrafilters which are not q-points (see the discussion of the Matet — Pawlikowski problem at the beginning of this chapter) it is natural to fix a partition of ω which witnesses this. Thus, after renaming, we may consider ultrafilters on $\bigcup\limits_{i \in \omega} \{i\} \times (i+1)$ instead (compare with the last remark of the previous section).

DEFINITION 6.2.5. Let $\mathcal{D}$ be a filter on $\bigcup\limits_{i<\omega} \{i\} \times (i+1)$.

1. We say that the filter $\mathcal{D}$ is *interesting* if for each function $h \in \prod\limits_{i \in \omega} (i+1)$ the set $\{(i, h(i)) : i \in \omega\}$ is not in $\mathcal{D}$.

2. Let $G^{sR}(\mathcal{D})$ be the game of two players in which Player I in his n^{th} move plays a set $A_n \in \mathcal{D}$ and the second player answers choosing an integer i_n and a set $a_n \in [A_n \cap \{i_n\} \times (i_n+1)]^{\leq n}$.
Finally, Player I wins the play if $\bigcup\limits_{n\in\omega} a_n \notin \mathcal{D}$.
3. The game $G^{aR}(\mathcal{D})$ is a modification of $G^{sR}(\mathcal{D})$ such that now, the first Player (in his n^{th} move) chooses a set $A_n \in \mathcal{D}$, $L_n < \omega$ and a function
$$f_n : \bigcup_{i<\omega} [\{i\} \times (i+1)]^{\leq n} \longrightarrow L_n.$$
Player II answers playing $i_n \in \omega$ and a set $a_n \in [A_n \cap \{i_n\} \times (i_n+1)]^{\leq n}$ homogeneous for f_n (i.e. such that $f_n{\restriction}[a_n]^k$ is constant for $k \leq n$).
Player I wins the play if $\bigcup\limits_{n\in\omega} a_n \notin \mathcal{D}$.
4. We say that the filter $\mathcal{D}$ is *semi–Ramsey* if the first player has no winning strategy in the game $G^{sR}(\mathcal{D})$.
5. The filter $\mathcal{D}$ is *almost Ramsey* if it is semi–Ramsey and for every colouring
$$f : \bigcup_{i<\omega} [\{i\} \times (i+1)]^{\leq n} \longrightarrow L, \qquad (n, L < \omega)$$
there is a set $A \in \mathcal{D}$ which is almost homogeneous for f in the following sense:
$$(\forall i \in \omega)(f{\restriction}[A \cap (\{i\} \times (i+1))]^k \quad \text{is constant for each } k \leq n).$$

PROPOSITION 6.2.6. *Suppose $\mathcal{D}$ is a non-principal ultrafilter on $\bigcup\limits_{i<\omega} \{i\} \times (i+1)$.*

1. *If $\mathcal{D}$ is interesting then it is not a q-point.*
2. *If $\mathcal{D}$ is Ramsey then $\mathcal{D}$ is almost Ramsey.*
3. *If $\mathcal{D}$ is semi–Ramsey then it is a p-point.*
4. *If $\mathcal{D}$ is semi–Ramsey then*
$$\mathcal{D}^* \stackrel{\text{def}}{=} \{A \subseteq \omega : \bigcup_{i\in A} \{i\} \times (i+1) \in \mathcal{D}\}$$
is a Ramsey ultrafilter.

PROOF. Compare the games and definitions. □

THEOREM 6.2.7. *Assume CH. There exists an ultrafilter $\mathcal{D}$ on $\bigcup\limits_{i<\omega} \{i\} \times (i+1)$ which is semi–Ramsey but not almost Ramsey.*

PROOF. For $i \in \omega$ let k_i be the integer part of the square root of $i+1$. Choose partitions $\langle e_i^m : m < k_i \rangle$ of $\{i\} \times (i+1)$ such that $(\forall m < k_i)(|e_i^m| \geq k_i)$. Let $f : \bigcup\limits_{i\in\omega} [\{i\} \times (i+1)]^2 \longrightarrow 2$ be such that
$$f((i,\ell_0),(i,\ell_1)) = 1 \quad \text{if and only if} \quad (\forall m < k_i)((i,\ell_0) \in e_i^m \Leftrightarrow (i,\ell_1) \in e_i^m).$$
Assuming CH, we will construct a semi–Ramsey ultrafilter containing no almost homogeneous set for f. To this end we choose an enumeration $\{\varphi_\alpha : \alpha < \omega_1\}$ of all functions from ω to $[\bigcup\limits_{i\in\omega} \{i\} \times (i+1)]^\omega$. By induction on $\alpha < \omega_1$ define sequences $\langle i_n^\alpha : n < \omega \rangle$ and $\langle a_n^\alpha : n < \omega \rangle$ such that for $\alpha < \beta < \omega_1$:

(a) $i_0^\alpha < i_1^\alpha < i_2^\alpha < \ldots < \omega$,
(b) $a_n^\alpha \in [\{i_n^\alpha\} \times (i_n^\alpha + 1)]^{\leq n}$ (for $n \in \omega$),

(c) either $(\exists m \in \omega)(\forall n \in \omega)(a^\alpha_n \cap \varphi_\alpha(m) = \emptyset)$ or $(\forall n \in \omega)(a^\alpha_{n+1} \subseteq \varphi_\alpha(i^\alpha_n))$,

(d) $\lim\limits_{n\to\infty} |\{m < k_{i^\alpha_n} : e^m_{i^\alpha_n} \cap a^\alpha_n \neq \emptyset\}| = \infty$,
$\lim\limits_{n\to\infty} \min\{|a^\alpha_n \cap e^m_{i^\alpha_n}| : m < k_{i^\alpha_n} \ \&\ a^\alpha_n \cap e^m_{i^\alpha_n} \neq \emptyset\} = \infty$,

(e) $|\bigcup\limits_{n\in\omega} a^\beta_n \setminus \bigcup\limits_{n\in\omega} a^\alpha_n| < \omega$.

There should be no problems with carrying out the construction. Let $A_\alpha = \bigcup\limits_{n\in\omega} a^\alpha_n$. Clearly the sequence $\langle A_\alpha : \alpha < \omega_1\rangle$ generates a non-principal ultrafilter $\mathcal{D}$ on $\bigcup\limits_{i\in\omega} \{i\}\times(i+1)$ (remember that the constant functions are among the φ_α's). By the demand (d), no set A_α is almost homogeneous for f, so $\mathcal{D}$ is not almost Ramsey. To show that $\mathcal{D}$ is semi–Ramsey suppose that σ is a winning strategy for the first player in the game $G^{sR}(\mathcal{D})$. Then σ is a function defined on finite sequences $\bar{x} = \langle (i_0, a_0), \dots, (i_{n-1}, a_{n-1})\rangle$ such that $i_0 < \dots < i_{n_1}$ and

$$(\forall \ell < n)(a_\ell \in [\{i_\ell\} \times (i_\ell + 1)]^{\leq \ell})$$

and with values in $\mathcal{D}$. For $j < \omega$ put

$$\varphi(j) = \bigcap \{\sigma((i_0, a_0), \dots, (i_{n-1}, a_{n-1})) : \ i_0 < \dots < i_{n-1} \leq j \text{ and } a_\ell \in [\{i_\ell\} \times (i_\ell + 1)]^{\leq \ell}\}.$$

Thus $\varphi : \omega \longrightarrow \mathcal{D}$, so for some $\alpha < \omega_1$ we have $\varphi = \varphi_\alpha$. But now look at the sequence $\bar{a} = \langle (i^\alpha_0, a^\alpha_0), (i^\alpha_1, a^\alpha_1), (i^\alpha_2, a^\alpha_2), \dots\rangle$. Since $A_\alpha \in \mathcal{D}$ and $\varphi_\alpha(m) \in \mathcal{D}$ for all $m \in \omega$ we necessarily have

$$(\forall n \in \omega)(a^\alpha_{n+1} \subseteq \varphi_\alpha(i^\alpha_n) \subseteq \sigma((i^\alpha_0, a^\alpha_0), \dots, (i^\alpha_n, a^\alpha_n))).$$

This means that the sequence $\bar{a}$ is a result of a legal play of the second player against the strategy σ. Hence $A_\alpha = \bigcup\limits_{n\in\omega} a^\alpha_n \notin \mathcal{D}$, a contradiction. □

THEOREM 6.2.8. *Suppose that $\mathcal{D}$ is a semi–Ramsey ultrafilter on $\bigcup\limits_{i<\omega} \{i\}\times(i+1)$. Then the following conditions are equivalent.*

(a) *$\mathcal{D}$ is almost Ramsey,*

(b) *the first player has no winning strategy in the game $G^{aR}(\mathcal{D})$,*

(c) *for each $m, L \in \omega$ and a colouring $f : \bigcup\limits_{i<\omega} [\{i\} \times (i+1)]^{\leq m} \longrightarrow L$, the first player has no winning strategy in the following modification $G^{sR}_f(\mathcal{D})$ of the game $G^{sR}(\mathcal{D})$: rules are like in $G^{sR}(\mathcal{D})$ but the sets a_n chosen by the second player have to be homogeneous for f.*

PROOF. The implications (b) $\Rightarrow$ (c) $\Rightarrow$ (a) are immediate by the definitions.

The implication (a) $\Rightarrow$ (b) is easy too: suppose that σ is a strategy for the first player in the game $G^{aR}(\mathcal{D})$. Let σ^* be a strategy for Player I in $G^{sR}(\mathcal{D})$ such that if $\sigma((i_0, a_0), \dots, (i_{n-1}, a_{n-1})) = (f_n, A_n)$ then $\sigma^*((i_0, a_0), \dots, (i_{n-1}, a_{n-1})) \in \mathcal{D}$ is an almost f_n–homogeneous subset of A_n (exists by the assumption **(a)**). Now, σ^* cannot be the winning strategy for Player I as $\mathcal{D}$ is semi–Ramsey. But then the play witnessing this shows that σ is not winning in $G^{aR}(\mathcal{D})$. □

THEOREM 6.2.9. *Assume that $\mathcal{D}$ is a semi–Ramsey ultrafilter on $\bigcup\limits_{i<\omega} \{i\}\times(i+1)$. Suppose that $\mathbb{P}$ is a proper ω^ω–bounding forcing notion such that*

$$\Vdash_{\mathbb{P}} \text{“ } \mathcal{D} \text{ generates an ultrafilter ”}.$$

Then

$$\Vdash_{\mathbb{P}} \text{“}\,\mathcal{D} \textit{ generates a semi–Ramsey ultrafilter }\text{”}.$$

PROOF. This is very similar to [**Sh:f**, Ch VI, 5.1]. We know that, in $\mathbf{V}^{\mathbb{P}}$, $\mathcal{D}$ generates an ultrafilter. What we have to show is that Player I has no winning strategy in the game $G^{sR}(\mathcal{D})$ (in $\mathbf{V}^{\mathbb{P}}$). So suppose that $\sigma \in \mathbf{V}^{\mathbb{P}}$ is a winning strategy of the first player in $G^{sR}(\mathcal{D})$. We may assume that the values of σ are elements of $\mathcal{D}$ (so from the ground model). But now, as $\mathbb{P}$ is proper and ω^{ω}–bounding, we find a function $\sigma^{+} \in \mathbf{V}$ such that $\mathrm{dom}(\sigma^{+}) = \mathrm{dom}(\sigma)$, $\mathrm{rng}(\sigma^{+}) \subseteq [\mathcal{D}]^{<\omega}$ and $\sigma(\bar{a}) \in \sigma^{+}(\bar{a})$ for all $\bar{a} \in \mathrm{dom}(\sigma)$. Letting $\sigma^{*}(\bar{a}) = \bigcap \sigma^{+}(\bar{a})$ for $\bar{a} \in \mathrm{dom}(\sigma)$ we will get, in $\mathbf{V}$, a winning strategy for Player I in $G^{sR}(\mathcal{D})$, a contradiction. □

REMARK 6.2.10. One may note that we did not mention anything about the existence of almost Ramsey ultrafilters. Of course it is done like 6.2.7, under CH. However we want to have an explicit representation of the ultrafilter as $\mathcal{D}(\Gamma)$ for some quasi-generic Γ. This will give us the preservation of the “colouring” part of the definition of almost Ramsey ultrafilters at limit stages. As the representation is very specific we postpone it for a moment and we will present this in Examples (see 6.4.1, 6.4.2).

6.3. Preserving ultrafilters

In this section we show when forcing notions of the type $\mathbb{Q}_1^{\mathrm{tree}}$ preserve ultrafilters introduced in the previous part. The key property of a tree–creating pair needed for this is formulated in the following definition.

DEFINITION 6.3.1. Let $\mathcal{D}$ be a filter on ω. We say that a tree creating pair (K,Σ) *is of the* $\mathbf{UP}(\mathcal{D})^{\mathrm{tree}}$*–type* if the following condition is satisfied:

$(\circledast)^{\mathrm{tree}}_{\mathbf{UP}(\mathcal{D})}$ Assume that $1 \leq m < \omega$, $p \in \mathbb{Q}^{\mathrm{tree}}_{\emptyset}(K,\Sigma)$, $\mathbf{nor}[t^p_\nu] > m+1$ for each $\nu \in T^p$, and $F_0, F_1, \ldots$ are fronts of T^p such that

$$(\forall n \in \omega)(\forall \nu \in F_{n+1})(\exists \eta \in F_n)(\eta \vartriangleleft \nu).$$

Further suppose that $u_n \subseteq F_n$ (for $n \in \omega$) are sets such that there is **no** system $\langle s_\nu : \nu \in \hat{T}\rangle \subseteq K$ with:

(a) $T \subseteq \{\nu \in T^p : (\exists \eta \in F_n)(\nu \trianglelefteq \eta)\}$ is a (well founded) quasi tree with $\max(T) \subseteq u_n$ and $\mathrm{root}(T) = \mathrm{root}(T^p)$,

(b) for each $\nu \in \hat{T}$:
$\mathrm{root}(s_\nu) = \nu$, $\mathrm{pos}(s_\nu) = \mathrm{succ}_T(\nu)$, $\mathbf{nor}[s_\nu] \geq m$ and
$s_\nu \in \Sigma(t^p_\eta : \eta \in \hat{S}_\nu)$ for some (well founded) quasi tree $S_\nu \subseteq T^p$.

Then there is a condition $q \in \mathbb{Q}^{\mathrm{tree}}_{\emptyset}(K,\Sigma)$ such that

(α) $p \leq q$,

(β) the set

$$Z \stackrel{\mathrm{def}}{=} \{n \in \omega : u_n \cap \mathrm{dcl}(T^q) = \emptyset\}$$

is not in the ideal $\mathcal{D}^c$ dual to $\mathcal{D}$,

(γ) $(\forall \nu \in T^q)(\mathbf{nor}[t^q_\nu] \geq \min\{\mathbf{nor}[t^p_\eta] - m : \nu \trianglelefteq \eta \in T^p\})$.

If we may additionally demand that the condition $q \in \mathbb{Q}^{\mathrm{tree}}_{\emptyset}(K,\Sigma)$ above satisfies

(δ) $\mathrm{root}(q) = \mathrm{root}(p)$ and $(\forall n \in Z)(F_n \cap T^q$ is a front of $T^q)$

then we say that (K,Σ) is of the $\mathbf{sUP}(\mathcal{D})^{\rm tree}$–type.
If $\mathcal{D}$ is the filter of all co-finite subsets of ω then we say that (K,Σ) is of the $\mathbf{UP}^{\rm tree}$–type ($\mathbf{sUP}^{\rm tree}$–type, respectively) instead of $\mathbf{UP}(\mathcal{D})^{\rm tree}$–type ($\mathbf{sUP}(\mathcal{D})^{\rm tree}$–type, resp.).

THEOREM 6.3.2. *Let $\mathcal{D}$ be a Ramsey ultrafilter on ω. Suppose that (K,Σ) is a finitary 2–big tree-creating pair of the $\mathbf{UP}(\mathcal{D})^{\rm tree}$–type. Then:*

$$\Vdash_{\mathbb{Q}_1^{\rm tree}(K,\Sigma)} \text{“ } \mathcal{D} \text{ generates an ultrafilter on } \omega \text{ ”.}$$

Consequently, $\Vdash_{\mathbb{Q}_1^{\rm tree}(K,\Sigma)}$ *“ $\mathcal{D}$ generates a Ramsey ultrafilter on ω ”.*

PROOF. Let $\dot{X}$ be a $\mathbb{Q}_1^{\rm tree}(K,\Sigma)$–name for a subset of ω, $p_0 \in \mathbb{Q}_1^{\rm tree}(K,\Sigma)$. Consider the following strategy for Player I in the game $G^R(\mathcal{D})$:

in the $n^{\rm th}$ move he chooses a condition $p_{n+1} \in \mathbb{Q}_1^{\rm tree}(K,\Sigma)$ such that $p_n \leq^1_n p_{n+1}$ and $p_{n+1} \Vdash k_n \in \dot{X}$ (where k_n is the last point played so far by Player II). Then he plays the set

$$B(p_{n+1},n+1) \stackrel{\rm def}{=} \{k\in\omega : (\exists q\in \mathbb{Q}_1^{\rm tree}(K,\Sigma))(p_{n+1} \leq^1_{n+1} q \ \ \& \ \ q\Vdash k\in\dot{X})\}.$$

As $\mathcal{D}$ is Ramsey, this strategy cannot be the winning one for Player I and therefore there is a play (determined by $k_0,k_1,k_2,\ldots$) according to this strategy in which Player I looses. This means that one of the following two possibilities holds:

CASE A *In the course of the play all sets $B(p_{n+1},n+1)$ are in the ultrafilter $\mathcal{D}$ and $\{k_0,k_1,\ldots\}\in\mathcal{D}$.*
In this situation we look at the sequence $\langle p_n : n\in\omega\rangle$. By 1.3.11 it has the limit $p^* = \lim\limits_{n\in\omega} p_n$. Clearly $p^*\Vdash \{k_0,k_1,\ldots\}\subseteq\dot{X}$ and we are done.

CASE B *In the course of the play it occurs that for some $n\in\omega$ the set $B(p_{n+1},n+1)$ is not in the ultrafilter.*
Take $q\in\mathbb{Q}_1^{\rm tree}(K,\Sigma)$, $p_{n+1}\leq^1_{n+1} q$ and fronts F^*, F_k of T^q such that the condition $q^{[\eta]}$ decides the truth value of "$k\in\dot{X}$" for each $\eta\in F_k$ and $k\in\omega$ and

$$(\forall\nu\in F^*)(\forall\eta\in T^q)(\nu\trianglelefteq\eta \ \ \Rightarrow\ \ \mathbf{nor}[t^q_\eta] > n+3),$$

$$(\forall k\in\omega)(\forall\nu\in F_k)(\forall\eta\in T^q)(\nu\trianglelefteq\eta \quad\Rightarrow\quad \mathbf{nor}[t^q_\eta] > 2^k+n+3)$$

(possible by 2.3.7(2) and 2.3.11). Of course we may assume that

$$(\forall k\in\omega)(\forall\nu\in F_{k+1})(\exists\eta\in F_k)(\eta\vartriangleleft\nu)$$

and that the fronts F_k are "above" F^* and F^* is "above" $F^1_{n+1}(q)$. Further let

$$u_k = \{\eta\in F_k : q^{[\eta]}\Vdash_{\mathbb{Q}_1^{\rm tree}(K,\Sigma)} k\in\dot{X}\}.$$

Look at the set $C = \omega\setminus B(p_{n+1},n+1)\in\mathcal{D}$. If $k\in C$ then necessarily for some $\rho\in F^*$ there is **no** $\langle s_\nu : \nu\in\hat{T}\rangle\subseteq K$ with:

1. $T\subseteq\{\nu\in T^{q^{[\rho]}} : (\exists\eta\in F_k)(\nu\trianglelefteq\eta)\}$ is a (well founded) quasi tree with $\max(T)\subseteq u_k$ and $\mathrm{root}(T)=\rho$
2. for each $\nu\in\hat{T}$:
   $\mathrm{root}(s_\nu)=\nu$, $\mathrm{pos}(s_\nu)=\mathrm{succ}_T(\nu)$, $\mathbf{nor}[s_\nu]\geq n+2$ and $s_\nu\in\Sigma(t^q_\eta : \eta\in S_\nu)$ for some (well founded) quasi tree $S_\nu\subseteq T^q$.

(Otherwise we could build a condition $\leq^1_{n+1}$–stronger than q (and thus than p_{n+1}) and forcing that $k \in \dot{X}$, contradicting $k \notin B(p_{n+1}, n+1)$.) As F^* is finite we have one $\rho \in F^*$ for which the above holds for all $k \in C_0$ for some $C_0 \in \mathcal{D}$, $C_0 \subseteq C$.

But now we may apply the fact that (K, Σ) is of the $\mathbf{UP}(\mathcal{D})^{\rm tree}$–type (see 6.3.1, remember the choice of F^*): we get $q^* \in \mathbb{Q}^{\rm tree}_\emptyset(K,\Sigma)$ such that $q^{[\rho]} \leq q^*$, $Z \stackrel{\rm def}{=} \{k \in C_0 : u_k \cap {\rm dcl}(T^{q^*}) = \emptyset\} \in \mathcal{D}$ and

$$\mathbf{nor}[t^{q^*}_\nu] \geq \min\{\mathbf{nor}[t^q_\eta] - (n+2) : \nu \trianglelefteq \eta \in T^q\}.$$

Now we easily see that in fact $q^* \in \mathbb{Q}^{\rm tree}_1(K,\Sigma)$ (by the norm requirement and the choice of q and F_k's) and $q^* \Vdash_{\mathbb{Q}^{\rm tree}_1(K,\Sigma)} Z \cap \dot{X} = \emptyset$.

For the "consequently" part, note that, by 3.1.1, the forcing notion $\mathbb{Q}^{\rm tree}_1(K,\Sigma)$ is proper and ω^ω–bounding. Therefore we may apply [**Sh:f**, Ch VI, 5.1]. This finishes the proof of the theorem. □

DEFINITION 6.3.3. We say that a tree creating pair (K, Σ) for $\mathbf{H}$ is *rich* if: for every system $\langle s_\nu : \nu \in \hat{T}\rangle \subseteq K$, $n \in \omega$ and u such that

1. $T \subseteq \bigcup\limits_{k\in\omega} \prod\limits_{m<k} \mathbf{H}(m)$ is a well founded quasi tree, $u \subseteq \max(T)$,
2. ${\rm root}(s_\nu) = \nu$, ${\rm pos}(s_\nu) = {\rm succ}_T(\nu)$, $\mathbf{nor}[s_\nu] > n+3$,
3. there is **no** $\langle s^*_\nu : \nu \in \hat{T}^*\rangle \subseteq K$ such that

$$T^* \subseteq T,\quad \max(T^*) \subseteq u,\quad {\rm root}(T^*) = {\rm root}(T),\quad {\rm root}(s^*_\nu) = \nu,$$
$${\rm pos}(s^*_\nu) = {\rm succ}_{T^*}(\nu),\quad \mathbf{nor}[s^*_\nu] > n+1,\quad \text{and}$$
$$s^*_\nu \in \Sigma(s_\eta : \eta \in \hat{T}_\nu) \text{ for some } T_\nu \subseteq T$$

there is $\langle s^+_\nu : \nu \in \hat{T}^+\rangle \subseteq K$ such that

$$T^+ \subseteq T,\quad \max(T^+) \subseteq \max(T) \setminus u,\quad {\rm root}(T^+) = {\rm root}(T),\quad {\rm root}(s^+_\nu) = \nu,$$
$${\rm pos}(s^+_\nu) = {\rm succ}_{T^+}(\nu),\quad \mathbf{nor}[s^+_\nu] \geq \min\{\mathbf{nor}[s_\eta] : \eta \in T\} - (n+2),$$
$$\text{and}\quad s^+_\nu \in \Sigma(s_\eta : \eta \in \hat{T}_\nu) \text{ for some } T_\nu \subseteq T.$$

THEOREM 6.3.4. *Assume (K, Σ) is a finitary 2-big rich tree–creating pair of the $\mathbf{sUP}^{\rm tree}$–type. Let $\mathcal{D}$ be an almost Ramsey interesting ultrafilter on $\bigcup\limits_{i\in\omega} \{i\} \times (i+1)$. Then*

$$\Vdash_{\mathbb{Q}^{\rm tree}_1(K,\Sigma)} \text{“}\,\mathcal{D} \text{ generates an interesting ultrafilter on } \bigcup_{i\in\omega}\{i\} \times (i+1)\text{ ”}.$$

PROOF. First note that if we show that, in $\mathbf{V}^{\mathbb{Q}^{\rm tree}_1(K,\Sigma)}$, $\mathcal{D}$ generates an ultrafilter then the ultrafilter has to be interesting (remember that $\mathcal{D}$ is interesting).

Let $\dot{X}$ be a $\mathbb{Q}^{\rm tree}_1(K,\Sigma)$–name for a subset of $\bigcup\limits_{i\in\omega}\{i\} \times (i+1)$.

We say that a condition $p \in \mathbb{Q}^{\rm tree}_1(K,\Sigma)$ is *$(\dot{X}, n)$–special* if:

there is a set $C \in \mathcal{D}$ such that:

for every $i \in \omega$ and $a \in [C \cap \{i\} \times (i+1)]^{\leq n}$ there are a condition $p' \geq p$ and a front F of $T^{p'}$ such that

$${\rm root}(p) = {\rm root}(p'),\quad (\forall \nu \in T^{p'})(\mathbf{nor}[t^{p'}_\nu] > n+1),$$

$$(\forall \nu \in T^{p'})(\forall \eta \in F)(\eta \trianglelefteq \nu \ \Rightarrow\ t^p_\nu = t^{p'}_\nu) \quad\text{and}\quad p' \Vdash_{\mathbb{Q}^{\rm tree}_1(K,\Sigma)} a \subseteq \dot{X}.$$

The condition p is *n–special* if it is either $(\dot{X}, n)$–special or $(\omega \setminus \dot{X}, n)$–special.

Note that the part of the definition of special conditions concerning the existence of a front F (of $T^{p'}$) is purely technical and usually easy to get (once we have the rest):

If the condition p is such that $(\forall \nu \in T^p)(\mathbf{nor}[t^p_\nu] > n+1)$ and the values of $\dot{X} \cap \{i\} \times (i+1)$ are decided on some fronts F_i of T^p *then* if we have a condition $p' \geq_0 p$ such that

$$(\forall \nu \in T^{p'})(\mathbf{nor}[t^{p'}_\nu] > n+1)$$

and $p' \Vdash a \subseteq \dot{X}$, then we may find one (weaker than p') which has this property and a front F as there. Moreover, in this situation, if $p \leq_0 p_1$ and p_1 is $(\dot{X}, n)$–special then p is $(\dot{X}, n)$–special.

Note that if $n \geq m$ and p is $(\dot{X}, n)$–special then it is $(\dot{X}, m)$– special.

CLAIM 6.3.4.1. *Let $n < \omega$. Suppose that $p \in \mathbb{Q}^{\rm tree}_1(K, \Sigma)$ is such that*

$$(\forall \nu \in T^p)(\mathbf{nor}[t^p_\nu] > (2^{2n}+1)(n+3))$$

and there are fronts F_i of T^p (for $i \in \omega$) with

$$(\forall i \in \omega)(\forall \nu \in F_i)(p^{[\nu]} \text{ decides } \dot{X} \cap \{i\} \times (i+1)).$$

Then p is n-special.

Proof of the claim: Let $f^+, f^- : \bigcup\limits_{i\in\omega} [\{i\} \times (i+1)]^n \longrightarrow 2$ be such that

$f^+(v) = 1$ if and only if there are $q \geq_0 p$ and a front F of T^q such that $q \Vdash_{\mathbb{Q}^{\rm tree}_1(K,\Sigma)} v \subseteq \dot{X}$ and

$$(\forall \nu \in T^q)(\forall \eta \in F)(\eta \trianglelefteq \nu \ \Rightarrow\ t^q_\nu = t^p_\nu) \ \ \& \ \ (\forall \nu \in T^q)(\mathbf{nor}[t^q_\nu] > n+1),$$

and f^- is defined similarly replacing "$\dot{X}$" by "$\omega \setminus \dot{X}$".

As $\mathcal{D}$ is almost Ramsey and interesting we find $j^+, j^- < 2$ and a set $C \in \mathcal{D}$ such that for each $i \in \omega$:

if $C \cap \{i\} \times (i+1) \neq \emptyset$
then $|C \cap \{i\} \times (i+1)| \geq 2n$, $f^+{\restriction}[C \cap \{i\} \times (i+1)]^n = j^+$ and
$f^-{\restriction}[C \cap \{i\} \times (i+1)]^n = j^-$.

If either $j^+ = 1$ or $j^- = 1$ then plainly p is n-special. So suppose that $j^+ = j^- = 0$ (and we want to get a contradiction).

Take $i \in \omega$ such that $|C \cap \{i\} \times (i+1)| \geq 2n$ (remember the choice of C) and fix $v \in [C \cap \{i\} \times (i+1)]^{2n}$. For each $v_1 \subseteq v$ let

$$u^i_{v_1,v} = \{\nu \in F_i : p^{[\nu]} \Vdash_{\mathbb{Q}^{\rm tree}_1(K,\Sigma)} v \cap \dot{X} = v_1\}.$$

Since (K, Σ) is rich we find v_1 and $\langle s^+_\nu : \nu \in \hat{T}^+\rangle \subseteq K$ such that

$T^+ \subseteq T^p$, $\max(T^+) \subseteq u^i_{v_1,v}$, $\mathrm{root}(T^+) = \mathrm{root}(p)$, $\mathrm{pos}(s^+_\nu) = \mathrm{succ}_{T^+}(\nu)$,
$\mathrm{root}(s^+_\nu) = \nu$, $\mathbf{nor}[s^+_\nu] > n+1$ and $s^+_\nu \in \Sigma(s_\eta : \eta \in \hat{T}_\nu)$ for some $T_\nu \subseteq T^p$.

[How? We try successively each $v_1 \subseteq v$. If we fail with one, we use 6.3.3 to pass to a subtree with the minimum of norms dropping down by at most $n+2$ and we try next candidate. For some $v_1 \subseteq v$ we have to succeed.]

Now look at this v_1 (and suitable $\langle s^+_\nu : \nu \in \hat{T}^+\rangle$). Since $j^+ = 0$ we necessarily have $|v_1| < n$: if not then we may take $v_2 \in [v_1]^n$ and then $f^+(v_2) = 1$ as witnessed by the condition q starting with $\langle s^+_\nu : \nu \in \hat{T}^+\rangle$. Similarly, by $j^- = 0$, we have $|v \setminus v_1| < n$. Together contradiction to $|v| = 2n$. Thus the claim is proved.

Now, let $p \in \mathbb{Q}^{\rm tree}_1(K,\Sigma)$. By 2.3.12 and 2.3.7(2) we find $p_1 \geq p$ and fronts F_n of T^{p_1} such that for $n \in \omega$:

1. $(\forall \nu \in F_{n+1})(\exists \eta \in F_n)(\eta \lhd \nu)$,
2. $(\forall \nu \in F_n)(\forall \eta \in T^{p_1})(\nu \unlhd \eta \quad \Rightarrow \quad \mathbf{nor}[t^{p_1}_\eta] > (2^{2n+1}+5)(n+3))$,
3. $(\forall \nu \in F_n)(p_1^{[\nu]}$ decides $\dot{X} \cap \{n\} \times (n+1))$.

Then, by 6.3.4.1, for each $\nu \in F_n$ the condition $p_1^{[\nu]}$ is n–special. Let

$$u_n = \{\nu \in F_n : p_1^{[\nu]} \text{ is } (\dot{X}, n)\text{–special}\}.$$

Now we consider two cases.

CASE A: *There are $n \in \omega$, $\nu \in F_n$ such that for each $m > n$ there is* **no** *system $\langle s_\eta : \eta \in \hat{T}\rangle \subseteq K$ with:*

$$T \subseteq T^{p_1}, \quad \max(T) \subseteq u_m, \quad \mathrm{root}(T) = \nu, \quad \mathrm{root}(s_\eta) = \eta, \quad \mathrm{pos}(s_\eta) = \mathrm{succ}_T(\eta),$$
$$\mathbf{nor}[s_\eta] \geq (2^{2n}+2)(n+3) \quad \textit{and } s_\eta \in \Sigma(t^{p_1}_\rho : \rho \in \hat{T}_\eta) \quad \textit{for some } T_\eta \subseteq T^{p_1}.$$

Since (K,Σ) is of the $\mathbf{sUP}^{\rm tree}$–type and

$$(\forall \eta \in T^{p_1})(\nu \unlhd \eta \quad \Rightarrow \quad \mathbf{nor}[t^{p_1}_\eta] > (2^{2n+1}+5)(n+3)),$$

we find a condition $q \in \mathbb{Q}^{\rm tree}_1(K,\Sigma)$ such that

(α) $p_1^{[\nu]} \leq^1_0 q$,
(β) $Z \stackrel{\rm def}{=} \{m > n : u_m \cap \mathrm{dcl}(T^q) = \emptyset\} \in [\omega]^\omega$,
(γ) $(\forall \eta \in T^q)(\mathbf{nor}[t^q_\eta] \geq \min\{\mathbf{nor}[t^{p_1}_\rho] : \eta \unlhd \rho \in T^{p_1}\} - (2^{2n}+2)(n+3))$,
(δ) $(\forall m \in Z)(F_m \cap T^q$ is a front of $T^q)$.

Let $m \in Z$ (so then $F_m \cap T^q$ is a front of T^q and $u_m \cap T^q = \emptyset$) and let $\eta \in F_m \cap T^q$. By (γ) and (2) above we know that

$$(\forall \rho \in T^q)(\eta \unlhd \rho \quad \Rightarrow \quad \mathbf{nor}[t^q_\rho] > (2^{2m}+1)(m+3)).$$

Consequently we may use 6.3.4.1 to conclude that $q^{[\eta]}$ is m–special. It cannot be $(\dot{X}, m)$–special as then the condition $p_1^{[\eta]}$ would be $(\dot{X}, m)$–special (compare the remark after the definition of special conditions) contradicting $\eta \notin u_m$. Thus $q^{[\eta]}$ is $(\omega \setminus \dot{X}, m)$–special.

For $m \in Z$ and $\eta \in F_m \cap T^q$ fix a set $C^m_\eta \in \mathcal{D}$ witnessing the fact that " $q^{[\eta]}$ is $(\omega \setminus \dot{X}, m)$–special ". Let $B_m = \bigcap\limits_{\eta \in F_m \cap T^q} C^m_\eta \in \mathcal{D}$ (for $m \in Z$).

Consider the following strategy for Player I in the game $G^{sR}(\mathcal{D})$:

at position number 0:
Player I writes down to the side: $m_0 = \min Z$, $q_0 = q$ and he plays: B_{m_0}.
[Note that he is (trivially) sure that if (i_0, a_0) is an answer of Player II then he may find $q_1 \geq_0 q_0$ such that $q_1 \Vdash a_0 \subseteq \omega \setminus \dot{X}$ and for some front F of q_1, $(\forall \rho \in F)(\forall \eta \in T^{q_1})(\rho \unlhd \eta \Rightarrow t^{q_1}_\eta = t^q_\eta)$.]

at position number $k+1$:
Player I looks at the last move (i_k, a_k) of his opponent. He chooses a condition

$q_{k+1} \geq_k q_k$ and a front F of $T^{q_{k+1}}$ such that

$$(\forall \rho \in F)(\forall \eta \in T^{q_{k+1}})(\rho \trianglelefteq \eta \ \Rightarrow\ t^{q_{k+1}}_\eta = t^q_\eta) \quad \text{and} \quad q_{k+1} \Vdash a_k \subseteq \omega \setminus \dot{X}.$$

Now he takes $m_{k+1} \in Z$ so large that $m_{k+1} > m_k$ and the front $F_{m_{k+1}}$ is "above" both F and $F^1_{k+2}(q_{k+1})$. Finally:
Player I writes down to the side: m_{k+1}, q_{k+1} and he plays: $B_{m_{k+1}}$.
[After this move he is sure that if (i_{k+1}, a_{k+1}) is a legal answer of the second player then he may find a condition $q_{k+2} \geq_{k+1} q_{k+1}$ such that $q_{k+2} \Vdash a_{k+1} \subseteq \omega \setminus \dot{X}$ and for some front F of $T^{q_{k+2}}$, $\rho \trianglelefteq \eta \ \Rightarrow\ t^{q_{k+2}}_\eta = t^q_\eta$ whenever $\rho \in F$, $\eta \in T^{q_{k+2}}$. Why? Remember the choice of $Z, m_{k+1}, B_{m_{k+1}}$ and $F_{m_{k+1}}$; see 2.3.1, clearly $m_{k+1} \geq k+1$.]

The strategy described above cannot be the winning one. Consequently there is a sequence $\langle (i_0, a_0), (i_1, a_1), \ldots \rangle$ such that (i_n, a_n) are legitimative moves of Player II against the strategy and $\bigcup\limits_{k\in\omega} a_k \in \mathcal{D}$. But in this play, Player I constructs (on a side) a sequence $p \leq q = q_0 \leq_0 q_1 \leq_1 q_2 \leq_2 \ldots$ of conditions such that $q_k \Vdash a_k \subseteq \omega \setminus \dot{X}$. Take the limit condition $q_\infty = \lim\limits_{k\in\omega} q_k$; it forces that $\bigcup\limits_{k\in\omega} a_k \subseteq \omega \setminus \dot{X}$, finishing the proof of the theorem in Case A.

CASE B: *Not* CASE A.

Thus for every $\nu \in F_n$, $n \in \omega$ we find $m > n$ and $\langle s_\eta \colon \eta \in \hat{T} \rangle \subseteq K$ such that:

$$T \subseteq T^{p_1},\ \max(T) \subseteq u_m,\ \mathrm{root}(T) = \nu, \quad \mathrm{root}(s_\eta) = \eta,\ \mathrm{pos}(s_\eta) = \mathrm{succ}_T(\eta),$$
$$\mathbf{nor}[s_\eta] \geq (2^{2n} + 2)(n+3) \quad \text{and } s_\eta \in \Sigma(t^{p_1}_\rho : \rho \in \hat{T}_\eta) \quad \text{for some } T_\eta \subseteq T^{p_1}.$$

If $\langle s_\eta : \eta \in \hat{T} \rangle$, $m > n$ are as above then we will say that *$\langle s_\eta : \eta \in \hat{T} \rangle$ is (m,n)–good for ν*. For each $n \in \omega$ and $\nu \in u_n$ fix a set $C^n_\nu \in \mathcal{D}$ witnessing the fact that "the condition $p_1^{[\nu]}$ is $(\dot{X}, n)$–special". Now consider the following strategy for Player I in the game $G^{sR}(\mathcal{D})$:

at position number 0:
For each $\nu \in F_0$, Player I chooses $n(\nu) > 0$ and $\langle s^\nu_\eta : \eta \in \hat{T}_\nu \rangle \subseteq K$ which is $(n(\nu), 0)$–good. He builds a condition q_0 which starts like these sequences. Thus q_0 is such that:

(a) $\mathrm{root}(q_0) = \mathrm{root}(p_1)$,
(b) if $\eta \in T^{p_1}$ is below the front F_0 then $\eta \in T^{q_0}$, $t^{q_0}_\eta = t^{p_1}_\eta$,
(c) if $\eta \in \hat{T}_\nu$ for some $\nu \in F_0$ then $\eta \in T^{q_0}$, $t^{q_0}_\eta = s^\nu_\eta$,
(d) if $\eta \in T^{p_1}$ and there are $\nu \in F_0$ and $\rho \in \max(T_\nu)$ such that $\rho \trianglelefteq \eta$ then $\eta \in T^{q_0}$, $t^{q_0}_\eta = t^{p_1}_\eta$.

Note that $F^*_0 \stackrel{\mathrm{def}}{=} \bigcup\limits_{\nu \in F_0} \max(T_\nu)$ is a front of T^{q_0} above $F^0_2(q_0)$.

Now, Player I writes down to the side: q_0, F^*_0 and he plays:

$$A_0 = \bigcap \{ C^{n(\nu)}_\eta : \nu \in F_0,\ \eta \in \max(T_\nu) \}.$$

[Thus Player I knows that F^*_0 is a front of T^{q_0} above $F^0_2(q_0)$, and for each $\nu \in F^*_0$ the condition $p_1^{[\nu]} = q_0^{[\nu]}$ is $(\dot{X}, 1)$–special and the set A_0 witnesses this fact.]

at the position number $k+1$:
Player I looks at the last move (i_k, a_k) of his opponent. He chooses a condition

$q_{k+1} \geq_k q_k$ and a front F^*_{k+1} of $T^{q_{k+1}}$ such that $q_{k+1} \Vdash a_k \subseteq \dot{X}$, the front F^*_{k+1} is above $F^0_{k+3}(q_{k+1})$ and for each $\nu \in F^*_{k+1}$ the condition $q^{[\nu]}_{k+1} = p^{[\nu]}_1$ is $(\dot{X}, k+2)$–special. How does he find q_{k+1} and F^*_{k+1}? He has q_k and F^*_k and he knows that F^*_k is a front of T^{q_k} above $F^0_{k+2}(q_k)$ and for each $\nu \in F^*_k$, the condition $q^{[\nu]}_k = p^{[\nu]}_1$ is $(\dot{X}, k+1)$-special and the set A_k (played by Player I before) witnesses this fact (this is our inductive hypothesis). Now, as $a_k \subseteq A_k$, for each $\nu \in F^*_k$ the first player may choose a condition $q^+_\nu \geq_0 q^{[\nu]}_k = p^{[\nu]}_1$ which forces that $a_k \subseteq \dot{X}$ and such that $(\forall \eta \in T^{q^+_\nu})(\mathbf{nor}[t^{q^+_\nu}_\eta] > k+1)$ and for some front F^+_ν of q^+_ν, if $\eta \trianglelefteq \rho \in T^{p_1}$, $\eta \in F^+_\nu$ then $\rho \in T^{q^+_\nu}$, $t^{q^+_\nu}_\rho = t^{p_1}_\rho$. We may assume that the fronts F^+_ν are such that $F^+_\nu \subseteq F_{m^+(\nu)}$ for some $m^+(\nu) > k+2$ and F^+_ν is above $F^0_{k+3}(q^+_\nu)$. For each $\rho \in F^+_\nu$, $\nu \in F^*_k$ we may choose $m(\rho) > m^+(\nu)$ and $\langle s^\rho_\eta : \eta \in \hat{T}_\rho \rangle \subseteq K$ such that $\langle s^\rho_\eta : \eta \in \hat{T}_\rho \rangle$ is $(m(\rho), m^+(\nu))$–good. Let

$$F^*_{k+1} = \bigcup \{\max(T_\rho) : (\exists \nu \in F^*_k)(\nu \lhd \rho \in F^+_\nu)\}.$$

The condition q_{k+1} is such that:

(a) below F^*_k it agrees with q_k,

(b) if $\nu \trianglelefteq \eta \lhd \rho \in F^+_\nu$, $\nu \in F^*_k$, $\eta \in T^{q^+_\nu}$ then $\eta \in T^{q_{k+1}}$, $t^{q_{k+1}}_\eta = t^{q^+_\nu}_\eta$,

(c) if $\nu \lhd \rho \trianglelefteq \eta \in \hat{T}_\rho$, $\nu \in F^*_k$, $\rho \in F^+_\nu$ then $\eta \in T^{q_{k+1}}$, $t^{q_{k+1}}_\eta = s^\rho_\eta$,

(d) if $\nu \lhd \rho \lhd \eta_0 \trianglelefteq \eta_1$, $\nu \in F^*_k$, $\rho \in F^+_\nu$, $\eta_0 \in \max(T_\rho)$, $\eta_1 \in T^{p_1}$ then $\eta_1 \in T^{q_{k+1}}$ and $t^{q_{k+1}}_{\eta_1} = t^{p_1}_{\eta_1}$.

Thus $q_{k+1} \geq_k q_k$, F^*_{k+1} is a front of $T^{q_{k+1}}$, and if $\nu \lhd \rho \lhd \eta \in F^*_{k+1}$, $\nu \in F^*_k$, $\rho \in F^+_\nu$ then $q^{[\eta]}_{k+1} = p^{[\eta]}_1$ is $(\dot{X}, m(\rho))$–special, $m(\rho) > m^+(\nu) > k+2$ (so $m(\rho) > k+3$). Now the first player writes down to the side q_{k+1}, F^*_{k+1} and he plays:

$$A_{k+1} = \bigcap \{C^{m(\rho)}_\eta : (\exists \nu \in F^*_k)(\rho \in F^+_\nu \ \ \&\ \ \nu \lhd \rho \lhd \eta \in u_{m(\rho)})\}.$$

[Note that for each $\nu \in F^*_{k+1}$, the set A_{k+1} witnesses that the condition $q^{[\nu]}_{k+1} = p^{[\nu]}_1$ is $(\dot{X}, k+2)$-special; the front F^*_{k+1} is above $F^0_{k+3}(q_{k+1})$.]

The strategy described above cannot be the winning one. Consequently, there is a play according to this strategy in which Player I loses. Thus we have moves $(i_0, a_0), (i_1, a_1), \ldots$ of Player II (legal in this play) for which $\bigcup\limits_{k\in\omega} a_k \in \mathcal{D}$. But in the course of the play the first player constructs conditions $p_1 \leq q_0 \leq_0 q_1 \leq_1 q_2 \leq_2 \cdots$ such that $q_k \Vdash a_k \subseteq \dot{X}$. Then the limit condition $q_\infty = \lim\limits_{k\in\omega} q_k$ forces that $\bigcup\limits_{k\in\omega} a_k \subseteq \dot{X}$. This finishes the proof of the theorem. $\square$

Definition 6.3.5. For $n, k, m \in \omega$ let $R_n(k, m)$ be the smallest integer such that for every colouring $f : [R_n(k,m)]^{\leq n} \longrightarrow k$ there is $a \in [R_n(k,m)]^m$ homogeneous for f (so this is the respective Ramsey number).

Theorem 6.3.6. *Let $\mathcal{D}$ be an almost Ramsey ultrafilter on $\bigcup\limits_{i\in\omega} \{i\} \times (i+1)$. Suppose that (K, Σ) is a finitary 2-big rich tree–creating pair such that*

$$\Vdash_{\mathbb{Q}^{\rm tree}_1(K,\Sigma)} \text{“ } \mathcal{D} \text{ generates an ultrafilter on } \bigcup_{i\in\omega} \{i\} \times (i+1) \text{ ”}.$$

Then $\Vdash_{\mathbb{Q}^{\rm tree}_1(K,\Sigma)}$ “ the ultrafilter $\tilde{\mathcal{D}}$ generated by $\mathcal{D}$ is almost Ramsey ”.

PROOF. Due to 3.1.1(1) we may apply 6.2.9 and get that

$$\Vdash_{\mathbb{Q}_1^{\rm tree}(K,\Sigma)} \text{“ } \mathcal{D} \text{ generates a semi–Ramsey ultrafilter on } \bigcup_{i\in\omega}\{i\}\times(i+1) \text{ ”}.$$

So, what we have to do is to show that, in $\mathbf{V}^{\mathbb{Q}_1^{\rm tree}(K,\Sigma)}$, the ultrafilter $\tilde{\mathcal{D}}$ generated by $\mathcal{D}$ has the colouring property of 6.2.5(5). Suppose that a condition $p\in\mathbb{Q}_1^{\rm tree}(K,\Sigma)$, $n,L<\omega$ and a $\mathbb{Q}_1^{\rm tree}(K,\Sigma)$–name $\dot{\varphi}$ are such that $p\Vdash$“$\dot{\varphi}:\bigcup\limits_{i\in\omega}[\{i\}\times(i+1)]^n\longrightarrow L$”. Note that if $\mathcal{D}$ is not interesting and it is witnessed by $h\in\prod\limits_{i\in\omega}(i+1)$ then the set $\{(i,h(i)):i\in\omega\}$ is almost homogeneous for $\dot{\varphi}$ and it is in $\mathcal{D}$. Consequently, we may assume that $\mathcal{D}$ is interesting.

We say that a condition $q\in\mathbb{Q}_1^{\rm tree}(K,\Sigma)$ is *m-beautiful* (for $m\in\omega$) if:

there is a set $C\in\mathcal{D}$ such that:
for every $i\in\omega$ and $a\in[C\cap\{i\}\times(i+1)]^m$ there are a condition $q'\geq q$ and a front F of $T^{q'}$ such that

$$\operatorname{root}(q)=\operatorname{root}(q'),\quad (\forall\nu\in T^{q'})(\mathbf{nor}[t^{q'}_\nu]>m+1),$$

$$q'\Vdash \text{“}a\text{ is }\dot{\varphi}\text{–homogeneous”}\quad\text{and}\quad (\forall\nu\in T^{q'})(\forall\eta\in F)(\eta\trianglelefteq\nu\ \Rightarrow\ t^q_\nu=t^{q'}_\nu).$$

Clearly, if q is m-beautiful and $k\leq m$ then q is k-beautiful.

Let $R^*(m)=R_n(L,m)$ (see 6.3.5).

CLAIM 6.3.6.1. *Let $m<\omega$, $q\in\mathbb{Q}_1^{\rm tree}(K,\Sigma)$. Assume that*

$$(\forall\nu\in T^q)(\mathbf{nor}[t^q_\nu]>(\binom{R^*(m)}{m}+1)(m+3))$$

and there are fronts F_i of T^q (for $i\in\omega$) such that conditions $q^{[\nu]}$ (for $\nu\in F_i$, $i\in\omega$) decide the value of $\dot{\varphi}{\restriction}[\{i\}\times(i+1)]^n$. Then the condition q is m–beautiful.

Proof of the claim: Look at the following colouring $f:\bigcup\limits_{i\in\omega}[\{i\}\times(i+1)]^m\longrightarrow 2$:

$f(v)=1$ if and only if there are $q'\geq_0 q$ and a front F of $T^{q'}$ such that

$$q'\Vdash_{\mathbb{Q}_1^{\rm tree}(K,\Sigma)}\text{“}v\text{ is homogeneous for }\dot{\varphi}\text{”}\qquad\text{and}$$

$$(\forall\nu\in T^{q'})(\forall\eta\in F)(\eta\trianglelefteq\nu\ \Rightarrow\ t^{q'}_\nu=t^q_\nu)\ \ \&\ \ (\forall\nu\in T^{q'})(\mathbf{nor}[t^{q'}_\nu]>m+1).$$

Since $\mathcal{D}$ is almost Ramsey and interesting we find $j<2$ and a set $C\in\mathcal{D}$ such that for each $i\in\omega$:

if $C\cap\{i\}\times(i+1)\neq\emptyset$
then $|C\cap\{i\}\times(i+1)|\geq R^*(m)$, $\quad f{\restriction}[C\cap\{i\}\times(i+1)]^m=j$.

If $j=1$ then easily the condition q is m-beautiful. Thus we have to exclude the other possibility. So assume $j=0$. Take $i\in\omega$ such that $C\cap\{i\}\times(i+1)\neq\emptyset$ and choose $v\in[C\cap\{i\}\times(i+1)]^{R^*(m)}$. For $v_1\in[v]^m$ put

$$u^i_{v_1,v}=\{\nu\in F_i: q^{[\nu]}\Vdash_{\mathbb{Q}_1^{\rm tree}(K,\Sigma)}\text{“}v_1\text{ is }\dot{\varphi}\text{–homogeneous”}\}.$$

Note that, by the definition of $R^*(m)$, for each $\nu \in F_i$ we find $v_1 \in [v]^m$ such that $\nu \in u^i_{v_1,v}$. Consequently we may apply the assumption that (K, Σ) is rich and we find $v_1 \in [v]^m$ and $\langle s^+_\nu : \nu \in \hat{T}^+ \rangle \subseteq K$ such that

$$T^+ \subseteq T^q,\ \max(T^+) \subseteq u^i_{v_1,v},\ \mathrm{root}(T^+) = \mathrm{root}(q),\quad \mathrm{pos}(s^+_\nu) = \mathrm{succ}_{T^+}(\nu),$$
$$\mathrm{root}(s^+_\nu) = \nu,\ \mathbf{nor}[s^+_\nu] > m+1 \quad \text{and } s^+_\nu \in \Sigma(s_\eta : \eta \in \hat{T}_\nu) \text{ for some } T_\nu \subseteq T^q.$$

[Exactly like in 6.3.4.1.] But with this in hands we easily conclude that $f(v_1) = 1$, contradicting $v_1 \in [C \cap \{i\} \times (i+1)]^m$ and the choice of j, C.

Choose a condition $q \geq p$ such that for some fronts F_m of T^q (for $m \in \omega$) we have

1. $(\forall \nu \in F_{m+1})(\exists \eta \in F_m)(\eta \lhd \nu)$,
2. $(\forall \nu \in F_m)(\forall \eta \in T^q)(\nu \unlhd \eta \ \Rightarrow\ \mathbf{nor}[t^q_\eta] > (\binom{R^*(m)}{m} + 1)(m+3))$,
3. $(\forall \nu \in F_m)(q^{[\nu]}$ decides $\dot{\varphi}{\restriction}[\{m\} \times (m+1)]^n)$

(possible by 2.3.12 and 2.3.7(2)). By 6.3.6.1 we know that for each $\nu \in F_m$, $m \in \omega$ the condition $q^{[\nu]}$ is m–beautiful. So for every $m < \omega$ and $\nu \in F_m$ we may fix a set $C^m_\nu \in \mathcal{D}$ witnessing "$q^{[\nu]}$ is m–beautiful".

Consider the following strategy of the first player in the game $G^{sR}(\mathcal{D})$:

at position number 0:
Player I writes down to the side $q_0 = q$, $F^*_0 = F_0$. He plays $\bigcap\{C^0_\nu : \nu \in F_0\} \in \mathcal{D}$.

arriving at position $k+1$:
Player I has a condition q_k and a front F^*_k of T^{q_k} such that q_k above F^*_k agrees with q. Moreover, the set played by him before witnesses that each $q^{[\nu]}$ is k-beautiful (for $\nu \in F^*_k$). He looks at the last move (i_k, a_k) of his opponent. For each $\nu \in F^*_k$, Player I can find a condition $q_\nu \geq_0 q^{[\nu]}_k = q^{[\nu]}$ such that

(a) $(\forall \eta \in T^{q_\nu})(\mathbf{nor}[t^{q_\nu}_\eta] > k+1)$,
(b) for some fronts F_ν of T^{q_ν}, q_ν above F_ν agrees with q and
(c) $q_\nu \Vdash_{\mathbb{Q}^{\rm tree}_1(K,\Sigma)}$ "a_k is $\dot{\varphi}$–homogeneous".

We may think that for some $m > i_k$ the fronts F_ν are contained in F_m (for $\nu \in F^*_k$). Now let q_{k+1} be such that

(α) below F^*_k it agrees with q_k,
(β) if $\nu \unlhd \eta \lhd \rho \in F_\nu$, $\nu \in F^*_k$, $\eta \in T^{q_\nu}$ then $\eta \in T^{q_{k+1}}$, $t^{q_{k+1}}_\eta = t^{q_\nu}_\eta$,
(γ) if $\nu \lhd \rho \unlhd \eta \in T^{q_\nu}$, $\nu \in F^*_k$, $\rho \in F_\nu$ then $\eta \in T^{q_{k+1}}$, $t^{q_{k+1}}_\eta = t^{q_\nu}_\eta = t^q_\eta$.

Let $F^*_{k+1} = \bigcup\{F_\nu : \nu \in F^*_k\}$. Clearly F^*_{k+1} is a front of $T^{q_{k+1}}$ contained in F_m, $q_{k+1} \geq_k q_k$ and q_{k+1} forces that a_k is $\dot{\varphi}$-homogeneous. Now:
Player I writes down to the side q_{k+1}, F^*_{k+1} and he plays the set

$$\bigcap\{C^m_\eta : \eta \in F^*_{k+1}\} \in \mathcal{D}.$$

[Note that for every $\nu \in F^*_{k+1}$ the condition $q^{[\nu]}_{k+1} = q^{[\nu]}$ is $k+1$-beautiful and the set played by Player I witnesses it.]

This strategy cannot be winning for the first player. Consequently he loses some play according to it. Let $\langle (i_0, a_0), (i_1, a_1), \ldots \rangle$ be the sequence of the respective moves of the second player (so $\bigcup\limits_{k \in \omega} a_k \in \mathcal{D}$) and let $q_0, q_1, q_2, \ldots$ be the sequence of conditions written down by the first player during the play. Let q_∞ be the limit condition $\lim\limits_{k\in\omega} q_k$. Then we have:

$$q_\infty \Vdash_{\mathbb{Q}_1^{\rm tree}(K,\Sigma)} (\forall k\in\omega)(a_k \text{ is } \dot{\varphi}\text{–homogeneous }).$$

This finishes the proof of the theorem. □

Let us finish this section with a theorem showing how several types of forcing notions built according to our scheme may preserve Γ–genericity in the context of ultrafilters. The proof of theorem 6.3.8 below resembles the proof that Blass–Shelah forcing notion preserves p–points (see [**BsSh 242**, 3.3]).

DEFINITION 6.3.7. 1. A creating pair (K,Σ) is *simple except omitting* if it is omittory, $|\mathrm{pos}(u,t)|>1$ whenever $t\in K$, $\mathbf{nor}[t]>0$ and $u\in\mathbf{basis}(t)$, and for every $(t_0,\ldots,t_{n-1})\in\mathrm{PFC}(K,\Sigma)$ and $s\in\Sigma(t_0,\ldots,t_{n-1})$ there is $k<n$ such that $s\in\Sigma(t_k\restriction[m^s_{\rm dn},m^s_{\rm up}))$.

2. Suppose that (K,Σ) is an omittory creating pair, $\bar{t}=\langle t_k: k<\omega\rangle\in \mathrm{PC}_\infty(K,\Sigma)$ and $W:\omega\times\omega\times\omega^{\underline{\omega}}\longrightarrow K$ is a $\bar{t}$–system. We say that W is *omittory compatible* if
   - (α) $k\leq\ell<n<\omega$ and $s\in\Sigma(t_\ell\restriction[m^{t_k}_{\rm dn},m^{t_n}_{\rm dn}))$ imply $n_W(s)=\ell$ (where $n_W(s)$ is as in 5.3.1(1b,c)), and
   - (β) $k\leq\ell<\omega$, $\sigma:[m^{t_k}_{\rm dn},m^{t_\ell}_{\rm up})\longrightarrow\omega$ imply

$$W(m^{t_k}_{\rm dn},m^{t_\ell}_{\rm up},\sigma)=\bigcup\{s\restriction[m^{t_k}_{\rm dn},m^{t_\ell}_{\rm up}): s\in W(m^{t_n}_{\rm dn},m^{t_n}_{\rm up},\sigma\restriction[m^{t_n}_{\rm dn},m^{t_n}_{\rm up})),\ k\leq n\leq\ell\}.$$

THEOREM 6.3.8. *Suppose that a creating pair (K_0,Σ_0) is simple except omitting, forgetful and monotonic. Let $\bar{t}\in\mathrm{PC}_\infty(K,\Sigma)$ and W be an omittory-compatible $\bar{t}$-system. Assume that $\Gamma\subseteq\mathrm{PC}_\infty(K_0,\Sigma_0)$ is quasi-W-generic and generates an ultrafilter.*

1. *If (K_0,Σ_0) is strongly finitary, (K,Σ) is a finitary, monotonic, omittory and omittory–big creating pair then the forcing notion $\mathbb{Q}^*_{\rm s\infty}(K,\Sigma)$ is Γ-genericity preserving.*
2. *If (K,Σ) is a finitary creating pair which captures singletons then the forcing notion $\mathbb{Q}^*_{\rm w\infty}(K,\Sigma)$ is Γ-genericity preserving.*
3. *If (K,Σ) is a finitary t-omittory tree creating pair then the forcing notion $\mathbb{Q}^{\rm tree}_1(K,\Sigma)$ is Γ-genericity preserving.*

PROOF. 1) We have to show that, in $\mathbf{V}^{\mathbb{Q}^*_{\rm s\infty}(K,\Sigma)}$, the demand 5.3.1(3b) is satisfied (as we know that $\mathbb{Q}^*_{\rm s\infty}(K,\Sigma)$ is proper). So suppose that $\dot{\eta}$ is a $\mathbb{Q}^*_{\rm s\infty}(K,\Sigma)$–name for an element of ω^ω and $p\in\mathbb{Q}^*_{\rm s\infty}(K,\Sigma)$. Since (K_0,Σ_0) is strongly finitary we may repeatedly use 2.2.6 and we get a condition $q\geq p$ such that for each $n<\omega$ and $v\in\mathrm{pos}(w^q,t^q_0,\ldots,t^q_{n-1})$ the condition $(v,t^q_n,t^q_{n+1},\ldots)$ decides the value of $W(m^{t_n}_{\rm dn},m^{t_n}_{\rm up},\dot{\eta}\restriction[m^{t_n}_{\rm dn},m^{t_n}_{\rm up}))$, where $\bar{t}=\langle t_0,t_1,t_2,\ldots\rangle$. Fix $v\in\mathrm{pos}(w^q,t^q_0,\ldots,t^q_{n-1})$, $n<\omega$ for a moment. For $k\geq n$ let $v^k=v^\frown\mathbf{0}_{[\ell g(v),m^{t^q_k}_{\rm up})}\in\mathrm{pos}(v,t^q_n,\ldots,t^q_k)$ (remember that (K,Σ) is omittory). Note that, for each $k\geq n$, the condition $(v^k,t^q_{k+1},t^q_{k+2},\ldots)$ decides the value of $W(m^{t_{k+1}}_{\rm dn},m^{t_{k+1}}_{\rm up},\dot{\eta}\restriction[m^{t_{k+1}}_{\rm dn},m^{t_{k+1}}_{\rm up}))$. Thus we find a function $\eta(v)\in\omega^\omega$ such that, for each $\ell\leq n$,

$$(v,t^q_n,t^q_{n+1},\ldots)\Vdash W(m^{t_\ell}_{\rm dn},m^{t_\ell}_{\rm up},\dot{\eta}\restriction[m^{t_\ell}_{\rm dn},m^{t_\ell}_{\rm up}))=W(m^{t_\ell}_{\rm dn},m^{t_\ell}_{\rm up},\eta(v)\restriction[m^{t_\ell}_{\rm dn},m^{t_\ell}_{\rm up}))$$

and for each $k\geq n$ the condition $(v^k,t^q_{k+1},t^q_{k+2},\ldots)$ forces that

$$\text{“ } W(m^{t_{k+1}}_{\rm dn},m^{t_{k+1}}_{\rm up},\dot{\eta}\restriction[m^{t_{k+1}}_{\rm dn},m^{t_{k+1}}_{\rm up}))=W(m^{t_{k+1}}_{\rm dn},m^{t_{k+1}}_{\rm up},\eta(v)\restriction[m^{t_{k+1}}_{\rm dn},m^{t_{k+1}}_{\rm up}))\text{ ”.}$$

Since Γ is quasi-W-generic we find $\bar{s} = \langle s_m : m < \omega \rangle \in \Gamma$ such that for every $n < \omega$ and $v \in \mathrm{pos}(w^q, t^q_0, \ldots, t^q_{n-1})$ we have

$$(\forall^\infty m \in \omega)(s_m \in W(m^{s_m}_{\mathrm{dn}}, m^{s_m}_{\mathrm{up}}, \eta(v){\restriction}[m^{s_m}_{\mathrm{dn}}, m^{s_m}_{\mathrm{up}})))$$

(remember $(\Gamma, \preceq)$ is directed countably closed). Next we choose inductively an increasing sequence $\ell(0) < \ell(1) < \ell(2) < \ldots < \omega$ such that

$$\begin{aligned}&\text{if } v \in \mathrm{pos}(w^q, t^q_0, \ldots, t^q_{n-1}),\ m^{t_n}_{\mathrm{up}} \leq m^{s_{\ell(i)}}_{\mathrm{up}},\ n, i \in \omega\\ &\qquad\text{then } (\forall m \geq \ell(i+1))(s_m \in W(m^{s_m}_{\mathrm{dn}}, m^{s_m}_{\mathrm{up}}, \eta(v){\restriction}[m^{s_m}_{\mathrm{dn}}, m^{s_m}_{\mathrm{up}}))).\end{aligned}$$

For $j < 4$ let $Y_j = \bigcup\limits_{i\in\omega} [m^{s_{\ell(4i+j)}}_{\mathrm{dn}}, m^{s_{\ell(4i+j+1)}}_{\mathrm{dn}})$. Since Γ generates an ultrafilter, exactly one of these sets is in $\mathcal{D}(\Gamma)$. Without loss of generality we may assume that $Y_2 \in \mathcal{D}(\Gamma)$ (otherwise start the sequence of the $\ell(i)$'s from $\ell(1)$, or $\ell(2)$ or $\ell(3)$). This means that we find $\bar{s}^* = \langle s^*_m : m < \omega \rangle \in \Gamma$ such that $\bar{s} \leq \bar{s}^*$ and for sufficiently large m (say $m \geq m^*$), there are $i = i(m)$ and $k = k(m)$ such that

$$m^{s_{\ell(4i+2)}}_{\mathrm{dn}} \leq m^{t_k}_{\mathrm{dn}} < m^{t_k}_{\mathrm{up}} \leq m^{s_{\ell(4i+3)}}_{\mathrm{dn}} \quad \text{and} \quad s^*_m \in \Sigma(t_k \restriction [m^{s^*_m}_{\mathrm{dn}}, m^{s^*_m}_{\mathrm{up}}))$$

(remember (K_0, Σ_0) is simple–except–omitting and monotonic). Let $x(i) \in \omega$ be such that $m^{t_{x(i)}}_{\mathrm{dn}} = m^{s_{\ell(4i+3)}}_{\mathrm{dn}}$ (for $i \in \omega$). Now we define a condition $q^* = (w^{q^*}, t^{q^*}_0, t^{q^*}_1, \ldots)$:

$$w^{q^*} = w^q,\quad t^{q^*}_0 = t^q_{x(0)} \restriction [\ell g(w^q), m^{t^q_{x(0)}}_{\mathrm{up}}),\quad \text{and}\quad t^{q^*}_{i+1} = t^q_{x(i+1)} \restriction [m^{t^q_{x(i)}}_{\mathrm{up}}, m^{t^q_{x(i+1)}}_{\mathrm{up}}).$$

Plainly, this defines a condition in $\mathbb{Q}^*_{\mathrm{s}\infty}(K, \Sigma)$ stronger than q. We claim that

$$q^* \Vdash (\forall m > m^*)(s^*_m \in W(m^{s^*_m}_{\mathrm{dn}}, m^{s^*_m}_{\mathrm{up}}, \dot{\eta}{\restriction}[m^{s^*_m}_{\mathrm{dn}}, m^{s^*_m}_{\mathrm{up}}))).$$

Assume not. Then we find $q' \geq q^*$ and $m > m^*$ such that

$$q' \Vdash s^*_m \notin W(m^{s^*_m}_{\mathrm{dn}}, m^{s^*_m}_{\mathrm{up}}, \dot{\eta}{\restriction}[m^{s^*_m}_{\mathrm{dn}}, m^{s^*_m}_{\mathrm{up}}))).$$

Of course we may assume that $\ell g(w^{q'}) \geq m^{t^{q^*}_{i(m)}}_{\mathrm{up}}$. Let k^0, k^1, k^2 be such that $m^{t_{k^\ell}}_{\mathrm{dn}} = m^{s_{\ell(4i(m)+\ell)}}_{\mathrm{dn}}$ and let $v = w^{q'}{\restriction} m^{t^q_{k^0}}_{\mathrm{dn}}$. Clearly

$$m^{t^{q^*}_{i(m)}}_{\mathrm{dn}} < m^{t^q_{k^0}}_{\mathrm{up}} < m^{t^q_{k^1}}_{\mathrm{up}} < m^{t^q_{k^2}}_{\mathrm{up}} \leq m^{t^q_{k(m)}}_{\mathrm{up}} < m^{t^q_{x(i(m))}}_{\mathrm{up}} = m^{t^{q^*}_{i(m)}}_{\mathrm{up}},$$

$v \in \mathrm{pos}(w^q, t^q_0, \ldots, t^q_{k^0-1})$ (by smoothness) and $w^{q'}{\restriction}[m^{t^q_{k^0}}_{\mathrm{dn}}, m^{t^q_{x(i(m))}}_{\mathrm{dn}}) \equiv \mathbf{0}$ (by the definition of $t^{q^*}_{i(m)}$, remember (K, Σ) is monotonic). Consequently for each $k \in [k^0, x(i(m)))$ we have $v^k \trianglelefteq w^{q'}$. Since $k^0 < k(m) < x(i(m))$ we conclude that the condition q' forces

$$\text{“ } W(m^{t_{k(m)}}_{\mathrm{dn}}, m^{t_{k(m)}}_{\mathrm{up}}, \dot{\eta}{\restriction}[m^{t_{k(m)}}_{\mathrm{dn}}, m^{t_{k(m)}}_{\mathrm{up}})) = W(m^{t_{k(m)}}_{\mathrm{dn}}, m^{t_{k(m)}}_{\mathrm{up}}, \eta(v){\restriction}[m^{t_{k(m)}}_{\mathrm{dn}}, m^{t_{k(m)}}_{\mathrm{up}})) \text{ ”}.$$

Since $m^{t_{k^0}}_{\mathrm{up}} < m^{s_{\ell(4i(m)+1)}}_{\mathrm{up}}$, for each $i \in [\ell(4i(m)+2), \ell(4i(m)+3))$ we have $s_i \in W(m^{s_i}_{\mathrm{dn}}, m^{s_i}_{\mathrm{up}}, \eta(v){\restriction}[m^{s_i}_{\mathrm{dn}}, m^{s_i}_{\mathrm{up}}))$ (remember that choice of the sequence of the $\ell(i)$'s). Now look at the condition 5.3.1(1c). Since W is omittory–compatible we have $n_W(s^*_m) = k(m) \in [k^2, x(i(m)))$ and therefore $s^*_m \in W(m^{s^*_m}_{\mathrm{dn}}, m^{s^*_m}_{\mathrm{up}}, \eta(v){\restriction}[m^{s^*_m}_{\mathrm{dn}}, m^{s^*_m}_{\mathrm{up}}))$. Hence, by 6.3.7(2$\beta$), we easily get

$$q' \Vdash \text{“ } s^*_m \in W(m^{s^*_m}_{\mathrm{dn}}, m^{s^*_m}_{\mathrm{up}}, \dot{\eta}{\restriction}[m^{s^*_m}_{\mathrm{dn}}, m^{s^*_m}_{\mathrm{up}})) \text{ ”},$$

a contradiction.

2) Repeat the proof of 1) noting that we do not have to assume that (K_0, Σ_0) is strongly finitary as we use 2.1.12 instead of 2.2.6. (Defining v^k we use fixed sequences u_n (for $n \in \omega$) such that $u_n \in \mathrm{pos}(u, t^q_n)$ for each $u \in \mathbf{basis}(t^q_n)$.)

3) Similarly (remember that (K, Σ) is finitary). □

REMARK 6.3.9. In 6.3.8(3) we need the assumption that the tree creating pair (K, Σ) is finitary. The forcing notion $\mathbb{D}_\omega$ of [**NeRo93**], in which conditions are trees $\subseteq \omega^{<\omega}$ such that each node has an extension which has all possible successors in the tree, adds a Cohen real (see [**NeRo93**, 2.1]). This forcing may be represented as $\mathbb{Q}^{\mathrm{tree}}_1(K', \Sigma')$ for some t–omittory (not finitary) tree creating pair (K', Σ').

6.4. Examples

EXAMPLE 6.4.1. Let $\mathbf{H}(m) = 2$ for $m < \omega$. We construct a creating pair $(K_{6.4.1}, \Sigma_{6.4.1})$ for $\mathbf{H}$, $\bar{t} \in \mathrm{PC}_\infty(K_{6.4.1}, \Sigma_{6.4.1})$ and $\bar{t}$-systems W^n_L (for $n, L < \omega$) such that

1. $(K_{6.4.1}, \Sigma_{6.4.1})$ is simple except omitting, finitary, forgetful, monotonic, interesting, condensed and generates an ultrafilter, the systems W^n_L are omittory–compatible,
2. if $\Gamma \subseteq \mathbb{P}^*_\infty(\bar{t}, (K_{6.4.1}, \Sigma_{6.4.1}))$ is quasi-W^n_L-generic then $\mathcal{D}(\Gamma)$ is a filter on $\bigcup\limits_{i \in \omega} \{i\} \times (i+1)$ such that for every colouring $f : \bigcup\limits_{i \in \omega} [\{i\} \times (i+1)]^n \longrightarrow L$ there is a set $A \in \mathcal{D}(\Gamma)$ almost homogeneous for f,
3. if $\Gamma \subseteq \mathbb{P}^*_\infty(\bar{t}, (K_{6.4.1}, \Sigma_{6.4.1}))$ is $\preceq$-directed and countably closed and $\mathcal{D}(\Gamma)$ is a filter such that for every colouring $f : \bigcup\limits_{i \in \omega} [\{i\} \times (i+1)]^n \longrightarrow L$ $(n, L < \omega)$ there is a set $A \in \mathcal{D}(\Gamma)$ almost homogeneous for f then Γ is quasi-W^n_L-generic for all n, L.

CONSTRUCTION. A creature $t \in \mathrm{CR}[\mathbf{H}]$ is in $K_{6.4.1}$ if for some non-empty subset a_t of $[m^t_{\mathrm{dn}}, m^t_{\mathrm{up}})$ we have $\mathbf{nor}[t] = \log_2(|a_t|)$ and

$$\mathbf{val}[t] = \{\langle u, v\rangle \in 2^{m^t_{\mathrm{dn}}} \times 2^{m^t_{\mathrm{up}}} : (\forall n \in [m^t_{\mathrm{dn}}, m^t_{\mathrm{up}}))(v(n) = 1 \ \Rightarrow\ n \in a_t)\}.$$

We define $\Sigma_{6.4.1}$ by:
if $t_0, \ldots, t_n \in K_{6.4.1}$ are such that $m^{t_\ell}_{\mathrm{up}} = m^{t_{\ell+1}}_{\mathrm{dn}}$ (for $\ell < n$) then

$$\Sigma_{6.4.1}(t_0, \ldots, t_n) = \{s \in K_{6.4.1} : m^{t_0}_{\mathrm{dn}} = m^s_{\mathrm{dn}}\ \&\ m^{t_n}_{\mathrm{up}} = m^s_{\mathrm{up}}\ \&\ (\exists \ell \le n)(a_s \subseteq a_{t_\ell})\}.$$

It should be clear that $(K_{6.4.1}, \Sigma_{6.4.1})$ is a finitary, forgetful, monotonic and simple except omitting creating pair. It is interesting as $\mathbf{nor}[t] > 0$ implies $|a_t| > 1$ (for $t \in K_{6.4.1}$). By the definition of $\Sigma_{6.4.1}$, one easily shows that $(K_{6.4.1}, \Sigma_{6.4.1})$ is condensed and generates an ultrafilter. Moreover, $(K_{6.4.1}, \Sigma_{6.4.1})$ is strongly finitary modulo $\sim_{\Sigma_{6.4.1}}$. Now, let $\bar{t} = \langle t_i : i < \omega\rangle \in \mathrm{PC}_\infty(K_{6.4.1}, \Sigma_{6.4.1})$ be such that for $i \in \omega$:

$$m^{t_i}_{\mathrm{dn}} = \frac{i(i+1)}{2}, \quad m^{t_i}_{\mathrm{up}} = \frac{(i+1)(i+2)}{2}, \quad a_{t_i} = [m^{t_i}_{\mathrm{dn}}, m^{t_i}_{\mathrm{up}})$$

(so $\mathbf{nor}[t_i] = \log_2(i+1)$). We will identify the interval $[m^{t_i}_{\mathrm{dn}}, m^{t_i}_{\mathrm{up}})$ with $\{i\} \times (i+1)$ (for $i \in \omega$). Thus, if $\Gamma \subseteq \mathbb{P}^*_\infty(\bar{t}, (K_{6.4.1}, \Sigma_{6.4.1}))$ is quasi-W-generic for some $\bar{t}$–system W, then we may think of $\mathcal{D}(\Gamma)$ as a filter on $\bigcup\limits_{i \in \omega} \{i\} \times (i+1)$. The filter $\mathcal{D}(\Gamma)$ is interesting by 6.1.6(1).

Fix $n, L < \omega$. For each $i \in \omega$ choose a mapping

$$\psi_i : \omega^{[m^{t_i}_{\rm dn}, m^{t_i}_{\rm up})} \stackrel{\rm onto}{\longrightarrow} \{f :\ f \text{ is a function from } [[m^{t_i}_{\rm dn}, m^{t_i}_{\rm up})]^n \text{ to } L\}.$$

Next, for $i \leq j < \omega$ and $\sigma : [m^{t_i}_{\rm dn}, m^{t_j}_{\rm up}) \longrightarrow \omega$ define

$$\begin{aligned} W^n_L(m^{t_i}_{\rm dn}, m^{t_j}_{\rm up}, \sigma) = \quad & \{t \in \Sigma_{6.4.1}(t_i, \ldots, t_j) : \text{if } i \leq k \leq j,\ a_t \subseteq a_{t_k} \\ & \qquad \text{then } a_t \text{ is homogeneous for } \psi_k(\sigma{\restriction}[m^{t_k}_{\rm dn}, m^{t_k}_{\rm up}))\} \end{aligned}$$

(in all other instances we let $W^n_L(m', m'', \sigma) = \emptyset$).

Claim 6.4.1.1. *W^n_L is an omittory-compatible $\bar{t}$-system.*

Proof of the claim: The requirement 5.3.1(1a) is immediate by the definition of W^n_L. For 5.3.1(1b,c) remember the way we defined the composition operation $\Sigma_{6.4.1}$: if $s \in \Sigma_{6.4.1}(s_0, \ldots, s_k)$ then $a_s \subseteq a_{s_\ell}$ for some $0 \leq \ell \leq k$. Finally note that if $s \in \Sigma_{6.4.1}(t_k, \ldots, t_\ell)$, $k \leq \ell < \omega$, $\mathbf{nor}[s] > \log_2(R_n(L, m))$ (see 6.3.5) and $\sigma : [m^s_{\rm dn}, m^s_{\rm up}) \longrightarrow \omega$ then there is $t \in W^n_L(m^s_{\rm dn}, m^s_{\rm up}, \sigma)$ such that $t \in \Sigma_{6.4.1}(s)$ and $\mathbf{nor}[t] \geq \log_2(m)$ (by the definition of $R_n(L, m)$). This gives the suitable for 5.3.1(1d) function G. Thus we have verified that W^n_L is a $\bar{t}$–system. It should be clear that W is omittory–compatible.

Note that if $\Gamma \subseteq \mathbb{P}^*_\infty(\bar{t}, (K_{6.4.1}, \Sigma_{6.4.1}))$ is $\preceq$–directed then $\mathcal{D}(\Gamma)$ is the filter generated by all sets $A^N_{\bar{s}} \stackrel{\rm def}{=} \bigcup\limits_{N \leq n} a_{s_n}$, for $\bar{s} = \langle s_n : n < \omega\rangle \in \Gamma$ and $N < \omega$. Therefore we easily check that the systems W^n_L (for $n, L < \omega$) are as required in 2, 3 of 6.4.1, noting that each colouring $f : \bigcup\limits_{i \in \omega} [\{i\} \times (i+1)]^n \longrightarrow L$ corresponds via $\langle \psi_i : i < \omega\rangle$ to some function $\eta \in \omega^\omega$. □

Proposition 6.4.2. *Assume CH. Then there exist $\Gamma \subseteq \mathbb{P}^*_\infty(\bar{t}, (K_{6.4.1}, \Sigma_{6.4.1}))$ which is quasi-W^n_L-generic for all $n, L < \omega$ and such that $\mathcal{D}(\Gamma)$ is a semi-Ramsey ultrafilter on $\bigcup\limits_{i \in \omega} \{i\} \times (i+1)$. Consequently, $\mathcal{D}(\Gamma)$ is an interesting almost Ramsey ultrafilter on $\bigcup\limits_{i \in \omega} \{i\} \times (i+1)$.*

Proof. This is somewhat similar to 6.1.6(3) (and so to 5.3.4(2)), but we have to be more careful to ensure that $\mathcal{D}(\Gamma)$ is semi–Ramsey. For this, as a basic step of the inductive construction of Γ, we use the following observation.

Claim 6.4.2.1. *Suppose that $\bar{s} \in \mathbb{P}^*_\infty(\bar{t}, (K_{6.4.1}, \Sigma_{6.4.1}))$, $n, L < \omega$, $\eta \in \omega^\omega$ and $\varphi : \omega \longrightarrow [\bigcup\limits_{i \in \omega} \{i\} \times (i+1)]^\omega$. Then there exists $\bar{s}^* = \langle s^*_m : m < \omega\rangle \in \mathbb{P}^*_\infty(\bar{t}, (K_{6.4.1}, \Sigma_{6.4.1}))$ such that $\bar{s} \leq \bar{s}^*$ and*

1. $(\forall^\infty m)(s^*_m \in W^n_L(m^{s^*_m}_{\rm dn}, m^{s^*_m}_{\rm up}, \eta{\restriction}[m^{s^*_m}_{\rm dn}, m^{s^*_m}_{\rm up})))$,
2. $(\forall m \in \omega)(|a_{s^*_m}| \leq m+1)$,
3. *if $i^*_m < \omega$ (for $m \in \omega$) are such that $a_{s^*_m} \subseteq \{i^*_m\} \times (i^*_m + 1) \cong [m^{t_{i^*_m}}_{\rm dn}, m^{t_{i^*_m}}_{\rm up})$ then*

$$\textit{either } (\exists k \in \omega)(\forall m \in \omega)(a_{s^*_m} \cap \varphi(k) = \emptyset) \quad \textit{or} \quad (\forall m \in \omega)(a_{s^*_{m+1}} \subseteq \varphi(i^*_m)).$$

Proof of the claim: Let $\bar{s} = \langle s_m : m < \omega\rangle$ and let i_m be such that $a_{s_m} \subseteq \{i_m\} \times (i_m + 1)$ (remember $\bar{t} \leq \bar{s}$, see the definition of $(K_{6.4.1}, \Sigma_{6.4.1})$). We know that $\lim\limits_{m \to \infty} |a_{s_m}| = \infty$, so we may choose $m_0 < m_1 < m_2 < \ldots < \omega$ such that

$(\forall k\in\omega)(|a_{s_{m_k}}|\geq R_n(L,k))$ (see 6.3.5). Choose $b_k\in[a_{s_{m_k}}]^k$ homogeneous for the colouring of $[\{i_{m_k}\}\times(i_{m_k}+1)]^n$ (with values in L) coded by $\eta{\restriction}[m_{\rm dn}^{t_{i_{m_k}}},m_{\rm up}^{t_{i_{m_k}}})$. Next choose $k(0)<k(1)<\ldots<\omega$ and $c_\ell\in[b_{k(\ell)}]^{\ell+1}$ (for $\ell<\omega$) such that

$$\text{either}\quad(\exists k\in\omega)(\forall\ell\in\omega)(c_\ell\cap\varphi(k)=\emptyset)\quad\text{or}\quad(\forall m\in\omega)(c_{\ell+1}\subseteq\varphi(i_{m_{k(\ell)}})).$$

Let $s^*_0\in K_{6.4.1}$ be such that $m_{\rm dn}^{s^*_0}=m_{\rm dn}^{s_0}$, $m_{\rm up}^{s^*_0}=m_{\rm up}^{s_{m_{k(0)}}}$, $a_{s^*_0}=c_0$ and let $s^*_{\ell+1}\in K_{6.4.1}$ (for $\ell\in\omega$) be such that $m_{\rm dn}^{s^*_{\ell+1}}=m_{\rm up}^{s_{m_{k(\ell)}}}$, $m_{\rm up}^{s^*_{\ell+1}}=m_{\rm up}^{s_{m_{k(\ell+1)}}}$ and $a_{s^*_{\ell+1}}=c_{\ell+1}$. Easily, the sequence $\bar{s}^*=\langle s^*_n:n<\omega\rangle$ is as required.

Assume CH. Using 6.4.2.1, we may construct a sequence $\langle\bar{s}_\alpha:\alpha<\omega_1\rangle\subseteq\mathbb{P}^*_\infty(\bar{t},(K_{6.4.1},\Sigma_{6.4.1}))$ such that

(α) $\alpha<\beta<\omega_1\quad\Rightarrow\quad\bar{s}_\alpha\preceq\bar{s}_\beta$,

(β) for each $n,L<\omega$ we have

$$(\forall\eta\in\omega^\omega)(\exists\alpha<\omega_1)(\forall^\infty m)(s_{\alpha,m}\in W^n_L(m_{\rm dn}^{s_{\alpha,m}},m_{\rm up}^{s_{\alpha,m}},\eta{\restriction}[m_{\rm dn}^{s_{\alpha,m}},m_{\rm up}^{s_{\alpha,m}}))),$$

(γ) for each function $\varphi:\omega\longrightarrow[\bigcup\limits_{i\in\omega}\{i\}\times(i+1)]^\omega$ there is $\alpha<\omega_1$ such that if $a_{s_{\alpha,m}}\subseteq\{i_{\alpha,m}\}\times(i_{\alpha,m}+1)$ (for $m\in\omega$) then

$$\text{either}\quad(\exists k\in\omega)(\forall m\in\omega)(a_{s_{\alpha,m}}\cap\varphi(k)=\emptyset)\quad\text{or}\quad(\forall m\in\omega)(a_{s_{\alpha,m+1}}\subseteq\varphi(i_{\alpha,m})).$$

Like in 6.2.7 and 5.3.4(2) we check that $\Gamma\stackrel{\rm def}{=}\{\bar{s}_\alpha:\alpha<\omega_1\}$ is quasi-W^n_L-generic for all $n,L<\omega$ and $\mathcal{D}(\Gamma)$ is a semi–Ramsey ultrafilter on $\bigcup\limits_{i\in\omega}\{i\}\times(i+1)$. $\square$

Conclusion 6.4.3. Assume CH. Let $(K_{6.4.1},\Sigma_{6.4.1})$, $\bar{t}$, W^n_L be given by 6.4.1, and let $\Gamma\subseteq\mathbb{P}^*_\infty(\bar{t},(K_{6.4.1},\Sigma_{6.4.1}))$ be quasi-W^n_L-generic for all $n,L<\omega$ such that $\mathcal{D}(\Gamma)$ is a semi–Ramsey ultrafilter on $\bigcup\limits_{i\in\omega}\{i\}\times(i+1)$ (see 6.4.2). Suppose that δ is a limit ordinal and $\langle\mathbb{P}_\alpha,\dot{\mathbb{Q}}_\alpha:\alpha<\delta\rangle$ is a countable support iteration of proper ω^ω–bounding forcing notions such that for each $\alpha<\delta$:

$$\Vdash_{\mathbb{P}_\alpha}\text{“}\Gamma\text{ generates an interesting almost Ramsey ultrafilter”}.$$

Then $\Vdash_{\mathbb{P}_\delta}$“$\Gamma$ generates an interesting almost Ramsey ultrafilter”.

Proof. By 6.4.1(3) we have

$$\Vdash_{\mathbb{P}_\alpha}\text{“}\Gamma\text{ is quasi-}W^n_L\text{-generic for all }n,L<\omega\text{”}$$

(for each $\alpha<\delta$). Hence, by 6.1.7, we get

$$\Vdash_{\mathbb{P}_\delta}\text{“}\Gamma\text{ is quasi-}W^n_L\text{-generic for each }n,L<\omega\text{ and generates an ultrafilter”}.$$

As $\mathbb{P}_\delta$ is ω^ω–bounding (by [**Sh:f**, Ch VI, 2.3, 2.8]) we may apply 6.2.9 to conclude that

$$\Vdash_{\mathbb{P}_\delta}\text{“}\mathcal{D}(\Gamma)\text{ is a semi–Ramsey ultrafilter”}.$$

Consequently, by 6.4.1(2), we have

$$\Vdash_{\mathbb{P}_\delta}\text{“}\mathcal{D}(\Gamma)\text{ is an interesting almost Ramsey ultrafilter”}.$$

$\square$

Example 6.4.4. Let $\psi\in\omega^\omega$ be such that $(\forall n\in\omega)(\psi(n)>(n+1)^2)$. We build a tree creating pair $(K^\psi_{6.4.4},\Sigma^\psi_{6.4.4})$ which is: finitary, 2-big, rich (see 6.3.3) and of the $\mathbf{sUP}(\mathcal{D})^{\rm tree}$-type (see 6.3.1) for every Ramsey ultrafilter $\mathcal{D}$ on ω.

CONSTRUCTION. Let $\mathbf{H}(n) = [\psi(n)]^{n+1}$.
For $\nu \in \prod_{m<n} \mathbf{H}(m)$ $(n \in \omega)$ and $A \subseteq \bigcup_{m<\omega} \{m\} \times \psi(m)$ we will write $A \prec \nu$ if

$$(\forall m < n)(\forall k < \psi(m))((m,k) \in A \quad \Rightarrow \quad k \in \nu(m)).$$

Now we define $(K^{\psi}_{6.4.4}, \Sigma^{\psi}_{6.4.4})$. A tree–like creature $t \in \mathrm{TCR}_\eta[\mathbf{H}]$ is in $K^{\psi}_{6.4.4}$ if:
1. $\mathbf{val}[t]$ is finite and
2. $\mathbf{nor}[t] = \log_2(\min\{|A|\colon A \subseteq \bigcup_{m \geq \ell g(\eta)} \{m\} \times \psi(m) \;\; \& \;\; (\forall \nu \in \mathrm{pos}(t))(A \not\prec \nu)\})$.

By the definition, $K^{\psi}_{6.4.4}$ is finitary. The tree composition $\Sigma^{\psi}_{6.4.4}$ is generated similarly to Σ^{tsum} of 4.2.6 but with norms as above. Thus, if $\langle t_\nu : \nu \in \hat{T}\rangle \subseteq K^{\psi}_{6.4.4}$ is a system of tree-creatures such that T is a well founded quasi tree, $\mathrm{root}(t_\nu) = \nu$, and $\mathrm{rng}(\mathbf{val}[t_\nu]) = \mathrm{succ}_T(\nu)$ (for $\nu \in \hat{T}$) then we define $S^*(t_\nu : \nu \in \hat{T})$ as the unique creature t^* in $K^{\psi}_{6.4.4}$ with $\mathrm{rng}(\mathbf{val}[t^*]) = \max(T)$, $\mathrm{dom}(\mathbf{val}[t^*]) = \{\mathrm{root}(T)\}$ and $\mathbf{dis}[t^*] = \langle \mathbf{dis}[t_\nu] : \nu \in \hat{T}\rangle$. Now we let

$$\Sigma^{\psi}_{6.4.4}(t_\nu : \nu \in \hat{T}) = \{t \in K^{\psi}_{6.4.4} : \mathbf{val}[t] \subseteq \mathbf{val}[S^*(t_\nu : \nu \in \hat{T})]\}.$$

Clearly, $\Sigma^{\psi}_{6.4.4}$ is a tree-composition on $K^{\psi}_{6.4.4}$. Note that if $t \in K^{\psi}_{6.4.4}$ then we may identify elements of $\Sigma^{\psi}_{6.4.4}(t)$ with subsets of $\mathrm{pos}(t)$: for each non-empty $u \subseteq \mathrm{pos}(t)$, t^u is the unique creature in $K^{\psi}_{6.4.4}$ with $\mathrm{pos}(t^u) = u$ and $\mathrm{root}(t^u) = \mathrm{root}(t)$. More general, if $p \in \mathbb{Q}^{\mathrm{tree}}_\emptyset(K^{\psi}_{6.4.4}, \Sigma^{\psi}_{6.4.4})$, F is a front of T^p, $u \subseteq F$ then there is a unique creature $t(p,u) \in K^{\psi}_{6.4.4}$ such that

$$\mathrm{pos}(t(p,u)) = u, \quad \mathrm{root}(t(p,u)) = \mathrm{root}(p) \quad \text{and} \quad t(p,u) \in \Sigma^{\psi}_{6.4.4}(t^p_\nu\colon (\exists \eta \in F)(\nu \lhd \eta)).$$

CLAIM 6.4.4.1. $\mathbf{nor}[S^*(t_\nu : \nu \in \hat{T})] \geq \min\{\mathbf{nor}[t_\nu] : \nu \in \hat{T}\}$.

Proof of the claim: Suppose that $m < 2^{\min\{\mathbf{nor}[t_\nu]:\nu \in \hat{T}\}}$, but there is a set $A \subseteq \bigcup_{k \geq n_0} \{k\} \times \psi(k)$ (where $n_0 = \ell g(\mathrm{root}(T))$) such that $|A| = m$ and $A \not\prec \nu$ for each $\nu \in \max(T)$. Now we build inductively a bad $\nu \in \max(T)$: since $m < 2^{\mathbf{nor}[t_{\mathrm{root}(T)}]}$ we find $\nu_0 \in \mathrm{pos}(t_{\mathrm{root}(T)})$ such that $A \prec \nu_0$. Next we look at $A^1 = A \cap \bigcup_{k \geq \ell g(\nu_0)} \{k\} \times \psi(k)$. Since $m < 2^{\mathbf{nor}[t_{\nu_0}]}$ we find $\nu_1 \in \mathrm{pos}(t_{\nu_0})$ such that $A^1 \prec \nu_1$. Continuing in this fashion, after finitely many steps, we get $\nu_k \in \max(T)$ such that $A \prec \nu_k$, a contradiction.

CLAIM 6.4.4.2. $(K^{\psi}_{6.4.4}, \Sigma^{\psi}_{6.4.4})$ *is 2-big.*

Proof of the claim: Let $t \in K^{\psi}_{6.4.4}$, $\mathbf{nor}[t] > 0$ and let $\mathrm{pos}(t) = u_0 \cup u_1$. Take sets $A_0, A_1 \subseteq \bigcup_{m \geq n_0} \{m\} \times \psi(m)$ (where $n_0 = \ell g(\mathrm{root}(t))$) such that for $i = 0, 1$:

$$(\forall \nu \in u_i)(A_i \not\prec \nu) \quad \text{and} \quad \log_2(|A_i|) = \mathbf{nor}[t^{u_i}].$$

Look at $A = A_0 \cup A_1$. Clearly $(\forall \nu \in \mathrm{pos}(t))(A \not\prec \nu)$. Hence, for some $i < 2$:

$$\mathbf{nor}[t] \leq \log_2(|A|) \leq \log_2(|A_i|) + 1 = \mathbf{nor}[t^{u_i}] + 1.$$

CLAIM 6.4.4.3. $(K^{\psi}_{6.4.4}, \Sigma^{\psi}_{6.4.4})$ *is rich.*

Proof of the claim: Suppose that $\langle s_\nu : \nu \in \hat{T}\rangle \subseteq K^{\psi}_{6.4.4}$, $n \in \omega$ and u are such that
1. $T \subseteq \bigcup_{k \in \omega} \prod_{m<k} \mathbf{H}(m)$ is a well founded quasi tree, $u \subseteq \max(T)$,

2. $\operatorname{root}(s_\nu)=\nu$, $\operatorname{pos}(s_\nu)=\operatorname{succ}_T(\nu)$, $\mathbf{nor}[s_\nu]>n+3$,
3. there is no system $\langle s^*_\nu : \nu\in\hat{T}^*\rangle\subseteq K^\psi_{6.4.4}$ such that

$$\begin{array}{llll} T^*\subseteq T, & \max(T^*)\subseteq u, & \operatorname{root}(T^*)=\operatorname{root}(T), & \operatorname{pos}(s^*_\nu)=\operatorname{succ}_{T^*}(\nu)\\ \operatorname{root}(s^*_\nu)=\nu, & \mathbf{nor}[s^*_\nu]>n+1, & \text{and} & \\ s^*_\nu\in\Sigma^\psi_{6.4.4}(s_\eta:\eta\in\hat{T}_\nu) & \text{for some } T_\nu\subseteq T. & & \end{array}$$

Let $t=S^*(s_\nu:\nu\in\hat{T})$. By 6.4.4.1 we have $\mathbf{nor}[t]>n+3$ (remember 2. above). But now, considering t^u (defined as before) we note that necessarily $\mathbf{nor}[t^u]\leq n+1<\mathbf{nor}[t]-1$. Let $v=\operatorname{pos}(t)\setminus u$. By the bigness (see 6.4.4.3) we have $\mathbf{nor}[t^v]\geq\mathbf{nor}[t]-1$, finishing the claim (remember 6.4.4.1).

CLAIM 6.4.4.4. *Let $\mathcal{D}$ be a Ramsey ultrafilter on ω. Then $(K^\psi_{6.4.4},\Sigma^\psi_{6.4.4})$ is of the $\mathbf{sUP}(\mathcal{D})^{\rm tree}$-type.*

Proof of the claim: Assume that $1\leq m<\omega$, $p\in\mathbb{Q}^*_\emptyset(K^\psi_{6.4.4},\Sigma^\psi_{6.4.4})$, $\mathbf{nor}[t^p_\nu]>m+1$ for each $\nu\in T^p$ and $F_0,F_1,\ldots$ are fronts of T^p such that

$$(\forall n\in\omega)(\forall\nu\in F_{n+1})(\exists\eta\in F_n)(\eta\vartriangleleft\nu).$$

Further suppose that $u_n\subseteq F_n$ are such that there is no system $\langle s_\nu:\nu\in\hat{T}\rangle$ with

$$\operatorname{pos}(S^*(s_\nu:\nu\in\hat{T}))\subseteq u_n,\quad \operatorname{root}(T)=\operatorname{root}(T^p),\quad\text{ and } \mathbf{nor}[s_\nu]\geq m.$$

In particular, this means that $\mathbf{nor}[t(p,u_n)]<m$ for each $n\in\omega$ ($t(p,u_n)$ is as defined earlier). Thus, for each $n\in\omega$, we find a set

$$A_n\subseteq\bigcup_{k\geq\ell g(\operatorname{root}(p))}\{k\}\times\psi(k)$$

such that

$$|A_n|=2^m\quad\text{ and }\quad(\forall\nu\in u_n)(A_n\not\prec\nu).$$

Now we use the assumption that $\mathcal{D}$ is Ramsey: we find sets $Z_0\in\mathcal{D}$ and $A^*\subseteq\bigcup\limits_{i\in\omega}\{i\}\times\psi(i)$ such that

$$(\forall n_0,n_1\in Z_0)(n_0<n_1\quad\Rightarrow\quad A_{n_0}\cap A_{n_1}=A^*).$$

[How? Just consider the colouring f of $[\omega]^2$ such that $f(n_0,n_1)$ codes the trace of A_{n_0} on A_{n_1} (for $n_0<n_1$, in the canonical enumerations of A_n's) and take an f-homogeneous set.] Now choose $Z\subseteq Z_0$, $Z\in\mathcal{D}$ such that if $n_0<n_1$, $n_0,n_1\in Z$ then

$$\min\{i:(A_{n_1}\setminus A^*)\cap\{i\}\times\psi(i)\neq\emptyset\}>\max\{\ell g(\nu):\nu\in F_{n_0}\}.$$

[How? Consider the following strategy for Player I in the game $G^R(\mathcal{D})$:
at stage $k+1$ of the game he looks at the last move i_k of the second player and he chooses N such that if $n\geq N$, $n\in Z_0$ then

$$(A_n\setminus A^*)\cap\bigcup\{\{i\}\times\psi(i):i\leq\max\{\ell g(\nu):\nu\in F_{i_k}\}\}=\emptyset.$$

Now he plays $Z_0\cap(N,\omega)$. This strategy cannot be the winning one.] Using the set Z we build the suitable condition $q\in\mathbb{Q}^{\rm tree}_\emptyset(K^\psi_{6.4.4},\Sigma^\psi_{6.4.4})$. It will be constructed in such a way that $p\leq^1_0 q$, $T^q\subseteq\{\operatorname{root}(q)\}\cup\bigcup\limits_{\ell\in Z}F_\ell$ and each $F_\ell\cap T^q$ will be a front of T^q (for $\ell\in Z$). Let $\ell_0=\min Z$ and $v_{\ell_0}=\{\nu\in F_{\ell_0}:A_{\ell_0}\prec\nu\}$. By the choice of A_{ℓ_0} we know that $v_{\ell_0}\cap u_{\ell_0}=\emptyset$. Let $t^q_{\operatorname{root}(q)}=t(p,v_{\ell_0})$. Note that

$2^{\mathbf{nor}[t(p,v_{\ell_0})]}+2^m \geq 2^{\mathbf{nor}[t(p,F_{\ell_0})]}$, and hence, as $\mathbf{nor}[t(p,F_{\ell_0})] > m+1$ (see 6.4.4.1), we have

$$\mathbf{nor}[t(p,v_{\ell_0})] \geq \mathbf{nor}[t(p,F_{\ell_0})] - m \geq \min\{\mathbf{nor}[t^p_\nu] : \mathrm{root}(p) \trianglelefteq \nu \in T^p\} - m.$$

We put $t^q_{\mathrm{root}(q)} = t(p, v_{\ell_0})$. Suppose that we have defined T^q up to the level of F_ℓ, $\ell \in Z$ (thus we know $T^q \cap F_\ell$ already). Let $\eta \in T^q \cap F_\ell$ and let $\ell' = \min(Z \setminus (\ell+1))$. By the first step of the construction we know that $A^* \prec \eta$. By the choice of Z we have that

$$(A_{\ell'} \setminus A^*) \cap \bigcup_{i<\ell g(\eta)} \{i\} \times \psi(i) = \emptyset.$$

We take $v_{\eta,\ell'} = \{\nu \in F_{\ell'} : \eta \vartriangleleft \nu \ \&\ A_{\ell'} \prec \nu\}$. As before, $v_{\eta,\ell'} \cap u_{\ell'} = \emptyset$ and

$$\mathbf{nor}[t(p^{[\eta]}, v_{\eta,\ell'})] \geq \mathbf{nor}[t(p^{[\eta]}, F_{\ell_0} \cap T^{p^{[\eta]}})] - m \geq \min\{\mathbf{nor}[t^p_\nu] : \eta \trianglelefteq \nu \in T^p\} - m$$

(remember 6.4.4.1).

Now we easily check that the condition q constructed above is as required by 6.3.1 to show that $(K^\psi_{6.4.4}, \Sigma^\psi_{6.4.4})$ is of the $\mathbf{sUP}(\mathcal{D})^{\mathrm{tree}}$–type. □

One easily checks that the forcing notion $\mathbb{Q}^{\mathrm{tree}}_1(K^\psi_{6.4.4}, \Sigma^\psi_{6.4.4})$ is non-trivial (i.e. $K^\psi_{6.4.4}$ contains enough tree like creatures with arbitrarily large norms). Let us show another property of the tree creating pair $(K^\psi_{6.4.4}, \Sigma^\psi_{6.4.4})$.

PROPOSITION 6.4.5. *Let $(K^\psi_{6.4.4}, \Sigma^\psi_{6.4.4})$ be the tree creating pair of 6.4.4. Then:*

1. $\Vdash_{\mathbb{Q}^{\mathrm{tree}}_1(K^\psi_{6.4.4}, \Sigma^\psi_{6.4.4})} (\forall m \in \omega)(\dot{W}(m) \in [\psi(m)]^{m+1})$ *and*
2. *if h is a partial function,* $\mathrm{dom}(h) \in [\omega]^\omega$ *and* $(\forall m \in \mathrm{dom}(h))(h(m) < \psi(m))$ *then*

$$\Vdash_{\mathbb{Q}^{\mathrm{tree}}_1(K^\psi_{6.4.4}, \Sigma^\psi_{6.4.4})} (\exists^\infty m \in \mathrm{dom}(h))(h(m) \in \dot{W}(m))$$

(where $\dot{W}$ is the generic real added by $\mathbb{Q}^{\mathrm{tree}}_1(K^\psi_{6.4.4}, \Sigma^\psi_{6.4.4})$, see 1.1.13).

PROOF. 1) Should be clear.
2) Let $p \in \mathbb{Q}^{\mathrm{tree}}_1(K^\psi_{6.4.4}, \Sigma^\psi_{6.4.4})$ and let h be as in the assumptions. We may assume that $(\forall \eta \in T^p)(\mathbf{nor}[t^p_\eta] > 5)$. Let $m_0 = \min\big(\mathrm{dom}(h) \setminus \ell g(\mathrm{root}(p))\big)$. Take a front F of T^p such that $(\forall \eta \in F)(\ell g(\eta) > m_0)$ and look at the set $u = \{\eta \in F : h(m_0) \in \eta(m_0)\}$. Plainly, $\mathbf{nor}[t(p,u)] \geq \mathbf{nor}[t(p,F)] - 1 \geq 4$. Let q be a condition in $\mathbb{Q}^{\mathrm{tree}}_1(K^\psi_{6.4.4}, \Sigma^\psi_{6.4.4})$ such that $\mathrm{root}(q) = \mathrm{root}(p)$, $T^q \subseteq T^p$, $t^q_{\mathrm{root}(q)} = t(p,u)$ and if $\eta \in u$, $\eta \trianglelefteq \nu \in T^p$ then $\nu \in T^q$, $t^q_\nu = t^p_\nu$. Clearly $q \geq p$ and $q \Vdash h(m_0) \in \dot{W}(m_0)$. Now we may easily finish. □

CONCLUSION 6.4.6. The following is consistent with ZFC:

1. there is an almost Ramsey interesting ultrafilter on $\bigcup_{i\in\omega} \{i\} \times (i+1)$ which is generated by $\aleph_1$ elements (so $\mathfrak{m}_2 = \lambda = \aleph_1$, see the introduction to this chapter) and
2. $\mathfrak{m}_1 = \aleph_2$, and even more: for each function $\psi \in \omega^\omega$ and a family $\mathcal{F}$ of $\aleph_1$ partial infinite functions $h : \mathrm{dom}(h) \longrightarrow \omega$ such that

$$(\forall m \in \mathrm{dom}(h))(h(m) < \psi(m))$$

there is $W \in \prod_{m\in\omega} [\psi(m)]^{m+1}$ such that

$$(\forall h \in \mathcal{F})(\exists^\infty m \in \mathrm{dom}(h))(h(m) \in W(m)).$$

PROOF. Start with $\mathbf{V} \models \mathrm{CH}$. By 6.4.2 we have an interesting almost Ramsey ultrafilter $\mathcal{D} = \mathcal{D}(\Gamma)$ on $\bigcup_{i\in\omega} \{i\} \times (i+1)$ generated by a quasi generic $\Gamma \subseteq \mathbb{P}^*_\infty(\bar{t}, (K_{6.4.1}, \Sigma_{6.4.1}))$ as there. Build a countable support iteration $\langle \mathbb{P}_\alpha, \dot{\mathbb{Q}}_\alpha : \alpha < \omega_2\rangle$ and a list $\langle \dot{\psi}_\alpha : \alpha < \omega_2\rangle$ such that for each $\alpha < \omega_2$:

1. $\dot{\psi}_\alpha$ is a $\mathbb{P}_\alpha$–name for a function in ω^ω such that $(\forall n \in \omega)(\dot{\psi}_\alpha(n) > (n+1)^2)$,
2. $\dot{\mathbb{Q}}_\alpha$ is the $\mathbb{P}_\alpha$–name for the forcing notion $\mathbb{Q}^{\mathrm{tree}}_1(K^{\dot{\psi}_\alpha}_{6.4.4}, \Sigma^{\dot{\psi}_\alpha}_{6.4.4})$,
3. $\langle \dot{\psi}_\beta : \beta < \omega_2\rangle$ lists with ω_2–repetitions all (canonical) $\mathbb{P}_{\omega_2}$–names for functions $\psi \in \omega^\omega$ such that $(\forall n \in \omega)(\psi(n) > (n+1)^2)$.

We claim that if $G \subseteq \mathbb{P}_{\omega_2}$ is a generic filter over $\mathbf{V}$, then, in $\mathbf{V}[G]$, the two sentences of the conclusion hold true. Why? One can inductively show that for each $\alpha \leq \omega_2$:

$$\Vdash_{\mathbb{P}_\alpha} \text{“}\Gamma \text{ generates an almost Ramsey interesting ultrafilter on } \bigcup_{i\in\omega}\{i\} \times (i+1)\text{”}$$

(at successor stages use 6.3.4, 6.3.6 and 6.4.4; at limit stages use 6.4.3). Hence, in $\mathbf{V}[G]$, the ultrafilter $\mathcal{D}(\Gamma)$ witnesses the first property. For the second assertion use 6.4.5. □

CHAPTER 7

Friends and relatives of PP

In this chapter we answer a question of Balcerzak and Plewik, showing that the cardinal number $\kappa_{\rm BP}$ (see 7.1.1) may be smaller than the continuum (7.5.3) and that it may be larger than the dominating number (7.5.2). As this cardinal turns out to be bounded by a cardinal number related to the strong PP–property, we take this opportunity to have a look at several properties close to the PP–property.

7.1. Balcerzak–Plewik number

For an ideal $\mathcal{J}$ of subsets of 2^ω it is natural to ask if it has the following property **(P)**:

$(\mathbf{P})_{\mathcal{J}}$ every perfect subset of 2^ω contains a perfect set from $\mathcal{J}$.

The property **(P)** has numerous consequences and applications (see e.g. Balcerzak [**Ba91**], some related results and references may be found in Balcerzak Rosłanowski [**BaRo95**]) and it is usually easy to decide if $(\mathbf{P})_{\mathcal{J}}$ holds. However, that was not clear for some of Mycielski's ideals $\mathcal{M}^*_{2,\mathcal{K}}$.

Suppose that $\mathcal{K} \subseteq [\omega]^\omega$ is a non-empty family such that

$(\oplus)$ $(\forall X \in \mathcal{K})(\exists X_0, X_1 \in \mathcal{K})(X_0, X_1 \subseteq X \ \ \& \ \ X_0 \cap X_1 = \emptyset)$.

Let $\mathcal{M}^*_{2,\mathcal{K}}$ consist of these sets $A \subseteq 2^\omega$ that

$$(\forall X \in \mathcal{K})(\exists f : X \longrightarrow 2)(\forall g \in A)(\neg f \subseteq g).$$

It is easy to check that $\mathcal{M}^*_{2,\mathcal{K}}$ is a σ–ideal of subsets of 2^ω. These ideals are relatives of the ideals from Mycielski [**My69**] and were studied e.g. in Cichoń Rosłanowski Steprāns Węglorz [**CRSW93**] and [**Ro94**]. If the family $\mathcal{K}$ is countable than easily the ideal $\mathcal{M}^*_{2,\mathcal{K}}$ determined by it has the property **(P)**. Dębski, Kleszcz and Plewik [**DKP92**] showed that the ideal $\mathcal{M}^*_{2,[\omega]^\omega}$ does not satisfy **(P)**. Then Balcerzak and Plewik defined the following cardinal number $\kappa_{\rm BP}$ (see Balcerzak Plewik [**BaPl96**]).

DEFINITION 7.1.1. The Balcerzak–Plewik number $\kappa_{\rm BP}$ is the minimal size of a family $\mathcal{K} \subseteq [\omega]^\omega$ such that

for some perfect set $Q \subseteq 2^\omega$, for every perfect subset P of Q there is $X \in \mathcal{K}$ such that

$$P{\restriction}X \stackrel{\rm def}{=} \{f \in 2^X : (\exists g \in P)(f \subseteq g)\} = 2^X.$$

[Note that $\kappa_{\rm BP}$ is the minimal size of $\mathcal{K} \subseteq [\omega]^\omega$ satisfying $(\oplus)$ for which the ideal $\mathcal{M}^*_{2,\mathcal{K}}$ does not have the property **(P)**.]

They proved that $\mathfrak{d} \leq \kappa_{\rm BP}$ and asked if

- it is consistent that $\kappa_{\rm BP} < \mathfrak{c}$,
- it is consistent that $\mathfrak{d} < \kappa_{\rm BP}$.

A full answer to these questions will be given in the final part of this chapter. Now we want to give an upper bound to $\kappa_{\rm BP}$.

DEFINITION 7.1.2. Let $\mathcal{X}$ be the space of all sequences $\bar{w} = \langle w_i : i \in B\rangle$ such that $B \in [\omega]^\omega$ and $(\forall i \in B)(w_i \in [\omega]^i)$. We define a relation $R^{\rm sPP} \subseteq \omega^\omega \times \mathcal{X}$ by

$$(\eta, \bar{w}) \in R^{\rm sPP} \quad \text{if and only if} \quad \eta \in \omega^\omega,\ \bar{w} \in \mathcal{X}, \text{ and } (\forall i \in \mathrm{dom}(\bar{w}))(\eta(i) \in w_i).$$

Note that the space $\mathcal{X}$ carries a natural Polish topology (inherited from the product space of all $\bar{w} = \langle w_i : i < \omega\rangle$ such that for each $i \in \omega$, either $w_i = \emptyset$ or $|w_i| = i$). The relation $R^{\rm sPP}$ describes the strong PP–property of [**Sh:f**, Ch VI, 2.12E]: a proper forcing notion $\mathbb{P}$ has the strong PP–property if and only if it has the $R^{\rm sPP}$–localization property (see 0.2.2).

THEOREM 7.1.3. $\kappa_{\rm BP} \leq \mathfrak{d}(R^{\rm sPP})$.

PROOF. Construct inductively a perfect tree $T \subseteq 2^{<\omega}$ and an increasing sequence $0 = k_0 < k_1 < k_2 < \ldots < \omega$ such that for every $i \in \omega$:

(α) $(\forall \nu \in T \cap 2^{k_i})(\exists \eta_0, \eta_1 \in T \cap 2^{k_{i+1}})(\nu \lhd \eta_0\ \&\ \nu \lhd \eta_1\ \&\ \eta_0 \neq \eta_1)$,

(β) for each colouring $f : T \cap 2^{k_i} \longrightarrow 2$ there is $n \in [k_i, k_{i+1})$ such that

$$(\forall \eta \in T \cap 2^{k_{i+1}})(\eta(n) = f(\eta\restriction k_i)).$$

The construction is straightforward. It is not difficult to check that if $P \subseteq [T]$ is a perfect set then there is $X \in [\omega]^\omega$ such that $P\restriction X = 2^X$ (or see [**DKP92**]).

Let $D \subseteq \mathcal{X}$ be such that $|D| = \mathfrak{d}(R^{\rm sPP})$ and

$$(\forall \eta \in \omega^\omega)(\exists \bar{w} \in D)((\eta, \bar{w}) \in R^{\rm sPP}).$$

Let N be an elementary submodel of $\mathcal{H}(\chi)$ such that $D, T, \langle k_i : i < \omega\rangle \in N$, $D \subseteq N$ and $|N| = |D|$. We are going to show that

for each perfect set $P \subseteq [T]$ there is $X \in N \cap [\omega]^\omega$ such that $P\restriction X = 2^X$

(what will finish the proof of the theorem). To this end suppose that $T^* \subseteq T$ is a perfect tree. Since $N \cap \omega^\omega$ is a dominating family (as the strong PP–property implies ω^ω–bounding) we may choose an increasing sequence $\langle n_i : i < \omega\rangle \in N \cap \omega^\omega$ such that

$$(\forall i \in \omega)(\forall \nu \in T^* \cap 2^{k_{n_i}})(|\{\eta \in T^* \cap 2^{k_{n_{i+1}}} : \nu \lhd \eta\}| > 2(i+1)).$$

As we may encode (in a canonical way) subsets of 2^n as integers, we may use the choice of D and N and find a sequence $\langle w_i : i \in B\rangle \in N$ such that for each $i \in B$:

(i) $0 < |w_i| \leq i$,

(ii) $a \in w_i \quad \Rightarrow \quad a \subseteq T \cap 2^{k_{n_i}}$,

(iii) $T^* \cap 2^{k_{n_i}} \in w_i$.

By shrinking each w_i if necessary, we may additionally demand that $|a| \geq 2\cdot\min(B)$ for each $a \in w_{\min(B)}$ and if $i < j$ both are from B, $a \in w_j$ then

(iv) for each $\nu \in a$ the set $\{\eta \in a : \nu\restriction k_{n_i} = \eta\restriction k_{n_i}\}$ has at least $2j$ elements

(remember the choice of the n_i's). Now, working in N, we inductively build a perfect tree $T^+ \subseteq T$. First for each $a \in w_{\min(B)}$ we choose $\eta^{\langle\rangle}_{a,0}, \eta^{\langle\rangle}_{a,1} \in a$ such that there are no repetitions in $\langle \eta^{\langle\rangle}_{a,\ell} : a \in w_{\min(B)},\ \ell < 2\rangle$ (possible as $|w_{\min(B)}| \leq \min(B)$ and $|a| \geq 2\cdot\min(B)$ for $a \in w_{\min(B)}$). We declare that

$$T^+ \cap 2^{k_{n_{\min(B)}}} = \{\eta^{\langle\rangle}_{a,\ell} : a \in w_{\min(B)},\ \ell < 2\}.$$

Suppose that we have defined $T^+ \cap 2^{k_{n_i}}$, $i \in B$ and $j = \min(B \setminus (i+1))$. For each $\nu \in T^+ \cap 2^{k_{n_i}}$ and $a \in w_j$ we choose $\eta^\nu_{a,0}, \eta^\nu_{a,1} \in T \cap 2^{k_{n_j}}$ such that

1. there are no repetitions in $\langle \eta^\nu_{a,0}, \eta^\nu_{a,1} : a \in w_j \rangle$,
2. if $a \in w_j$ is such that $(\exists \eta \in a)(\nu \lhd \eta)$ then $\eta^\nu_{a,0}, \eta^\nu_{a,1} \in a$.

Again, the choice is possible by **(iv)**. We declare that

$$T^+ \cap 2^{k_{n_j}} = \{\eta^\nu_{a,\ell} : \nu \in T^+ \cap 2^{k_{n_i}},\ a \in w_j,\ \ell < 2\}.$$

This fully describes the construction (in N) of the tree $T^+ \subseteq T$. Next, working still in N, we choose integers $m_j \in [k_{n_j}, k_{n_j+1})$ such that for each $j \in B$:

(+) *if* $\nu \in T^+ \cap 2^{k_{n_i}}$, $i \in B$ is such that $j = \min(B \setminus (i+1))$ (or $\nu = \langle\rangle$ and $j = \min(B)$) and $a \in w_j$ and $\eta^\nu_{a,\ell} \lhd \eta \in T \cap 2^{k_{n_j+1}}$ *then* $\eta(m_j) = \ell$

(possible by (β) of the choice of the tree T and the first demand of the choice of the $\eta^\nu_{a,\ell}$'s). Let $X = \{m_j : j \in B\} \in [\omega]^\omega \cap N$. Suppose $f : X \longrightarrow 2$. By clause **(iii)** we know that for each $j \in B$, $b_j \stackrel{\rm def}{=} T^* \cap 2^{k_{n_j}} \in w_j$. Consequently we may build inductively an infinite branch $\eta \in [T^*]$ such that $\eta \restriction k_{n_{\min(B)}} = \eta^{\langle\rangle}_{b_{\min(B)}, f(m_{\min(B)})}$ and for each $i \in B$, if $j = \min(B \setminus (i+1))$, $\nu_i = \eta \restriction k_{n_i}$ then $\eta \restriction k_{n_j} = \eta^{\nu_i}_{b_j, f(m_j)}$. It follows from (+) that $f \subseteq \eta$. □

Remark 7.1.4. Note that a sequence $\langle w_i : i \in B \rangle$, where $w_i \in [\omega]^i$, may be interpreted as a sequence $\langle v_i : i \in B \rangle$, where $|v_i| = i$ and members of v_i are functions from the interval $[\frac{i(i+1)}{2}, \frac{(i+1)(i+2)}{2})$ to ω. Consequently, considering suitable diagonals, we may use Bartoszyński–Miller's characterization of $\mathbf{non}(\mathcal{M})$ (see [**BaJu95**, 2.4.7]) and show that $\mathbf{non}(\mathcal{M}) \leq \mathfrak{d}(R^{\rm sPP})$. As clearly $\mathfrak{d} \leq \mathfrak{d}(R^{\rm sPP})$ we conclude that $\mathbf{cof}(\mathcal{M}) \leq \mathfrak{d}(R^{\rm sPP})$ (by [**BaJu95**, 2.2.11]). On the other hand, it follows from Bartoszyński's characterization of $\mathbf{cof}(\mathcal{N})$ (see [**BaJu95**, 2.3.9]) that $\mathfrak{d}(R^{\rm sPP}) \leq \mathbf{cof}(\mathcal{N})$ (just note that the Sacks property implies the strong PP–property).

7.2. An iterable friend of the strong PP–property

The PP–property is preserved in countably support iterations of proper forcing notions (see [**Sh:f**, Ch VI, 2.12]). However, it is not clear if the strong PP–property is preserved. Here, we introduce a property stronger then the strong PP which is preserved in countable support iterations.

Definition 7.2.1. Let $\mathcal{D}$ be a filter on ω, $x \in \omega^\omega$ be a non-decreasing function and let $\bar{\mathcal{F}} = (\mathcal{F}, <_{\mathcal{F}})$ be a partial order on $\mathcal{F} \subseteq \omega^\omega$.

1. The filter $\mathcal{D}$ is *weakly non-reducible* if it is non-principal and for every partition $\langle X_n : n < \omega \rangle$ of ω into finite sets there exists a set $Y \in [\omega]^\omega$ such that $\bigcup\limits_{n \in \omega \setminus Y} X_n \in \mathcal{D}$.

   If above we allow partitions into sets from the dual ideal $\mathcal{D}^c$ then we say that $\mathcal{D}$ is *non-reducible.*
2. The partial order $\bar{\mathcal{F}}$ is *PP–ok* if it is dense, has no maximal and minimal elements, each member of $\mathcal{F}$ is non-decreasing, the identity function belongs to $\mathcal{F}$ and

($\otimes$) if $h_0, h_1 \in \mathcal{F}$, $h_0 <_{\mathcal{F}} h_1$ then $(\forall n \in \omega)(1 \leq h_0(n) \leq h_1(n))$ and $\lim\limits_{n\to\infty} \frac{h_1(n)}{h_0(n)} = \infty$ and there is $h \in \mathcal{F}$ such that

$$(\forall^\infty n \in \omega)(h(n) \leq \frac{h_1(n)}{h_0(n)}).$$

3. We say that a proper forcing notion $\mathbb{P}$ has *the $(\mathcal{D}, x)$–strong PP–property* if

$\Vdash_{\mathbb{P}}$ " for every $\eta \in \omega^\omega$ there are $B \in \mathcal{D} \cap \mathbf{V}$ and $\langle w_i : i \in B\rangle \in \mathbf{V}$ such that $(\forall i \in B)(|w_i| \leq x(i) \ \& \ \eta(i) \in w_i)$ ".

4. A proper forcing notion $\mathbb{P}$ has *the $(\mathcal{D}, \bar{\mathcal{F}})$–strong PP–property* if it has the $(\mathcal{D}, x)$–strong PP–property for every $x \in \mathcal{F}$.

REMARK 7.2.2. 1. Each non-principal ultrafilter on ω is non-reducible. Clearly non-reducible filters are weakly non-reducible.

2. One can easily construct a countable partial order $\bar{\mathcal{F}}$ which is PP–ok and such that $x \in \mathcal{F}$ for any pregiven non-decreasing unbounded function $x \in \omega^\omega$.

THEOREM 7.2.3. *Suppose that $\mathcal{D}$ is a non-principal p-filter on ω (see 6.2.1(2)) and $\bar{\mathcal{F}} = (\mathcal{F}, <_{\mathcal{F}})$ is a PP–ok partial order. Then:*

1. *every proper forcing notion which has the $(\mathcal{D}, \bar{\mathcal{F}})$–strong PP–property has the strong PP–property,*
2. *the $(\mathcal{D}, \bar{\mathcal{F}})$–strong PP–property is preserved in countable support iterations of proper forcing notions.*

PROOF. 1) Should be clear.

2) We will apply [**Sh:f**, Ch VI, 1.13A], so we will follow the terminology of [**Sh:f**, Ch VI, §1]. However, we will not quote the conditions which we have to check, as that was done in the proof of 5.2.9 (and the proof here is parallel to the one there). We will present the proof in a slightly more complicated way than needed, but later we will be able to refer to it in a bounded context (in 7.3.6). Moreover, in this way the analogy to 5.2.9 will be more clear.

For each $m \geq 1$ we fix a function $\psi^m : \omega \xrightarrow{\text{onto}} [\omega]^{\leq m}$ and for $h \in \mathcal{F}$ we define $\psi_h : \omega^\omega \longrightarrow \prod\limits_{n\in\omega} [\omega]^{<\omega}$ by $\psi_h(\eta)(n) = \psi^{h(n)}(\eta(n))$. Further, for $h^*, h \in \mathcal{F}$, $B \in \mathcal{D}$ and $\bar{w} = \langle w_i : i \in B\rangle$ such that $h^* <_{\mathcal{F}} h$ and $(\forall i \in B)(w_i \in [\omega]^{\leq h(i)})$ we put

$$T_{\bar{w},h^*} = \{\nu \in \omega^{<\omega} : (\forall i \in B \cap \ell g(\nu))(\psi^{h^*(i)}(\nu(i)) \subseteq w_i)\}.$$

Each $T_{\bar{w},h^*}$ is a perfect subtree of $\omega^{<\omega}$. Now we define:

- $D_{\mathcal{D},\bar{\mathcal{F}}}$ is $\mathcal{H}(\aleph_1)^{\mathbf{V}}$,
- for $x, T \in D_{\mathcal{D},\bar{\mathcal{F}}}$ we say that $x \; R_{\mathcal{D},\bar{\mathcal{F}}} \; T$ if and only if $x = \langle h^*, h\rangle$ and $T = T_{\bar{w},h^*}$ for some $h^*, h \in \mathcal{F}$ and $\bar{w} = \langle w_i : i \in B\rangle \in D_{\mathcal{D},\bar{\mathcal{F}}}$ such that $h^* <_{\mathcal{F}} h$, $B \in \mathcal{D}$, and $(\forall i \in B)(w_i \in [\omega]^{\leq h(i)})$,
- for $\langle h^*, h\rangle, \langle h^{**}, h'\rangle \in \mathrm{dom}(R_{\mathcal{D},\bar{\mathcal{F}}})$ we say that $\langle h^*, h\rangle <_{\mathcal{D},\bar{\mathcal{F}}} \langle h^{**}, h'\rangle$ if and only if $h^* = h^{**} <_{\mathcal{F}} h <_{\mathcal{F}} h'$.

CLAIM 7.2.3.1. 1. *$(D_{\mathcal{D},\bar{\mathcal{F}}}, R_{\mathcal{D},\bar{\mathcal{F}}})$ is a weak covering model in* $\mathbf{V}$.

2. *In any generic extension $\mathbf{V}^*$ of $\mathbf{V}$ in which $(D_{\mathcal{D},\bar{\mathcal{F}}}, R_{\mathcal{D},\bar{\mathcal{F}}})$ covers, a forcing notion $\mathbb{P}$ is $(D_{\mathcal{D},\bar{\mathcal{F}}}, R_{\mathcal{D},\bar{\mathcal{F}}})$–preserving if and only if it has the $(\mathcal{D}, \bar{\mathcal{F}})$–strong PP–property.*

[Compare 5.2.9.1, 5.2.9.2.]

Proof of the claim: 1) Check.
2) Suppose that $(D_{\mathcal{D},\bar{\mathcal{F}}}, R_{\mathcal{D},\bar{\mathcal{F}}})$ covers in $\mathbf{V}^*$ and $\mathbb{P} \in \mathbf{V}^*$ is a forcing notion with the $(\mathcal{D},\bar{\mathcal{F}})$–strong PP–property. Let $\langle h^*, h\rangle \in \mathrm{dom}(R_{\mathcal{D},\bar{\mathcal{F}}})$ and $\eta \in \omega^\omega \cap (\mathbf{V}^*)^{\mathbb{P}}$. Choose $h_0, h_1 \in \mathcal{F}$ such that $h_0 <_{\mathcal{F}} h_1$ and $(\forall^\infty n \in \omega)(h_1(n) \leq \frac{h(n)}{h^*(n)})$ (possible by 7.2.1(2)). Take $\bar{w}^* = \langle w^*_i : i \in B^*\rangle \in \mathbf{V}^*$ such that $B^* \in \mathcal{D} \cap \mathbf{V}$ and

$$(\forall i \in B^*)(w^*_i \in [\omega]^{\leq h_0(i)} \ \ \& \ \ \eta(i) \in w^*_i).$$

Let $\eta^* \in \omega^\omega \cap \mathbf{V}^*$ be such that $\psi^{h_0(i)}(\eta^*(i)) = w^*_i$ (for $i \in B^*$). Since $(D_{\mathcal{D},\bar{\mathcal{F}}}, R_{\mathcal{D},\bar{\mathcal{F}}})$ covers in $\mathbf{V}^*$ and $\langle h_0, h_1\rangle \in \mathrm{dom}(D_{\mathcal{D},\bar{\mathcal{F}}})$, we find $\bar{w} = \langle w_i : i \in B\rangle \in D_{\mathcal{D},\bar{\mathcal{F}}}$ such that $\eta^* \in \lim(T_{\bar{w},h_0})$ and $(\forall i \in B)(w_i \in [\omega]^{\leq h_1(i)})$. Let $B^+ = B \cap B^* \in \mathcal{D} \cap \mathbf{V}$, $w^+_i = \bigcup\limits_{k \in w_i} \psi^{h^*(i)}(k)$ for $i \in B^+$. Note that for sufficiently large $i \in B^+$

$$|w^+_i| \leq |w_i| \cdot h^*(i) \leq h_1(i) \cdot h^*(i) \leq h(i)$$

and we may assume that this holds for all $i \in B^+$. Letting $\bar{w}^+ = \langle w^+_i : i \in B^+\rangle$ we will have $\langle h^*, h\rangle\ R_{\mathcal{D},\bar{\mathcal{F}}}\ T_{\bar{w}^+,h^*}$ and $\eta \in \lim(T_{\bar{w}^+,h^*})$, and hence $(D_{\mathcal{D},\bar{\mathcal{F}}}, R_{\mathcal{D},\bar{\mathcal{F}}})$ covers in $(\mathbf{V}^*)^{\mathbb{P}}$.

The converse implication is even simpler.

CLAIM 7.2.3.2. *$(D_{\mathcal{D},\bar{\mathcal{F}}}, R_{\mathcal{D},\bar{\mathcal{F}}}, <_{\mathcal{D},\bar{\mathcal{F}}})$ is a fine covering model.* *[Compare 5.2.9.3.]*

Proof of the claim: Immediately by the definition of $(D_{\mathcal{D},\bar{\mathcal{F}}}, R_{\mathcal{D},\bar{\mathcal{F}}}, <_{\mathcal{D},\bar{\mathcal{F}}})$ one sees that the demands (α), (β)(i)–(iii) of [**Sh:f**, Ch VI, 1.2(1)] are satisfied. To verify the condition (β)(iv) of [**Sh:f**, Ch VI, 1.2(1)] suppose that $\langle h^*, h\rangle <_{\mathcal{D},\bar{\mathcal{F}}} \langle h^*, h'\rangle$ and $\langle h^*, h\rangle\ R_{\mathcal{D},\bar{\mathcal{F}}}\ T_{\bar{w}^\ell,h^*}$, $\bar{w}^\ell = \langle w^\ell_i : i \in B_\ell\rangle$, $B_\ell \in \mathcal{D}$ (for $\ell = 1, 2$). Take $n \in \omega$ such that $(\forall m \geq n)(2 \cdot h(m) < h'(m))$ (possible by 7.2.1(2)) and let $B = B_1 \cap B_2 \cap [n, \omega) \in \mathcal{D}$. Put $w_i = w^1_i \cup w^2_i$ for $i \in B$ and look at the tree $T_{\bar{w},h^*}$. Clearly $\langle h^*, h'\rangle\ R_{\mathcal{D},\bar{\mathcal{F}}}\ T_{\bar{w},h^*}$ and $T_{\bar{w}^1,h^*} \cup T_{\bar{w}^2,h^*} \subseteq T_{\bar{w},h^*}$ (so more than needed).

Checking clauses (γ) and (δ) of [**Sh:f**, Ch VI, 1.2(1)] we restrict ourselves to the stronger condition (δ). So suppose that $\mathbf{V}^*$ is a generic extension (via a proper forcing notion) of $\mathbf{V}$ such that $\mathbf{V}^* \models$ "$(D_{\mathcal{D},\bar{\mathcal{F}}}, R_{\mathcal{D},\bar{\mathcal{F}}})$ covers".
(a) Assume that $x, x^+, x_n \in \mathrm{dom}(R_{\mathcal{D},\bar{\mathcal{F}}})$, $T_n \in D_{\mathcal{D},\bar{\mathcal{F}}}$ are such that for $n \in \omega$

$$x_n <_{\mathcal{D},\bar{\mathcal{F}}} x_{n+1} <_{\mathcal{D},\bar{\mathcal{F}}} x^+ <_{\mathcal{D},\bar{\mathcal{F}}} x \quad \text{and} \quad x_n\ R_{\mathcal{D},\bar{\mathcal{F}}}\ T_n.$$

Let $x = \langle h^*, h\rangle$, $x_n = \langle h^*, h_n\rangle$, $x^+ = \langle h^*, h^+\rangle$ (so $h^* <_{\mathcal{F}} h_n <_{\mathcal{F}} h_{n+1} <_{\mathcal{F}} h^+ <_{\mathcal{F}} h$) and let $\bar{w}^n = \langle w^n_i : i \in B_n\rangle \in D_{\mathcal{D},\bar{\mathcal{F}}}$ be such that $T_n = T_{\bar{w}^n,h^*}$ (so $B_n \in \mathcal{D} \cap \mathbf{V}$ and $|w^n_i| \leq h_n(i)$). Look at the sequence $\langle B_n : n \in \omega\rangle \subseteq \mathcal{D} \cap \mathbf{V}$. It does not have to belong to $\mathbf{V}$, but it may be covered by a countable set from $\mathbf{V}$ (as $\mathbf{V}^*$ is a proper forcing extension of $\mathbf{V}$). Hence, as $\mathcal{D}$ is a p-filter, we find a set $B \in \mathcal{D}$ such that $(\forall n \in \omega)(|B \setminus B_n| < \omega)$. Take $h^-_0, h^-_1 \in \mathcal{F}$ such that $h^-_0 <_{\mathcal{F}} h^-_1$ and $(\forall^\infty n \in \omega)(h^-_1(n) \leq \frac{h(n)}{h^+(n)})$ (remember $(\otimes)$ of 7.2.1(2)) and choose an increasing sequence $m_0 < m_1 < m_2 < \ldots < \omega$ such that $(\forall i \geq m_0)(h^-_1(i) \leq \frac{h(i)}{h^+(i)})$, $B \setminus B_0 \subseteq m_0$ and for $n \in \omega$:

$$B \setminus B_{m_n} \subseteq m_{n+1} \quad \text{and} \quad (\forall i \geq m_{n+1})((n+2) \cdot h_{m_n}(i) < h^+(i)).$$

Let $\eta \in \omega^\omega \cap \mathbf{V}^*$ be such that if $i \in B \cap [m_n, m_{n+1})$ then $\psi^{h^+(i)}(\eta(i)) = w^0_i \cup \bigcup\limits_{k<n} w^{m_k}_i$, and let $\eta^* \in \omega^\omega \cap \mathbf{V}^*$ be such that $\psi^{h^-_0(i)}(\eta^*(i)) = \{\eta(i)\}$ for each $i \in \omega$. Since

$(D_{\mathcal{D},\bar{\mathcal{F}}}, R_{\mathcal{D},\bar{\mathcal{F}}})$ covers in $\mathbf{V}^*$ we find $T_{\bar{w}^-,h_0^-} \in D_{\mathcal{D},\bar{\mathcal{F}}}$ such that $\eta^* \in \lim(T_{\bar{w}^-,h_0^-})$ and $\langle h_0^-, h_1^- \rangle \ R_{\mathcal{D},\bar{\mathcal{F}}} \ T_{\bar{w}^-,h_0^-}$. Let $B^* \stackrel{\rm def}{=} B \cap \mathrm{dom}(\bar{w}^-) \setminus m_0 \in \mathcal{D}$. By the choice of η, $T_{\bar{w}^-,h_0^-}$ and h_1^- we find $\bar{w}^* = \langle w_i^* : i \in B^* \rangle \in D_{\mathcal{D},\bar{\mathcal{F}}}$ such that $|w_i^*| \leq h_1^-(i) \cdot h^+(i) \leq h(i)$ and $w_i^0 \cup \bigcup\limits_{k<n} w_i^{m_k} \subseteq w_i^*$ whenever $i \in B^* \cap [m_n, m_{n+1})$, $n \in \omega$. Clearly $T_{\bar{w}^*,h^*} \in D_{\mathcal{D},\bar{\mathcal{F}}}$ and $\langle h^*, h \rangle \ R_{\mathcal{D},\bar{\mathcal{F}}} \ T_{\bar{w}^*,h^*}$. Put $W \stackrel{\rm def}{=} \{m_0, m_1, m_2, \ldots\}$ and suppose that $\rho \in \omega^\omega \cap \mathbf{V}^*$ is such that for every $n \in \omega$, $\rho{\restriction} m_{n+1} \in \bigcup\limits_{k<n} T_{m_k} \cup T_0$. If $i \in B^*$, $m_n \leq i < m_{n+1}$ then, by the assumptions on ρ, $\psi^{h^*(i)}(\rho(i)) \subseteq w_i^0 \cup \bigcup\limits_{k<n} w_i^{m_k} \subseteq w_i^*$. Hence $\rho \in \lim(T_{\bar{w}^*,h^*})$ (remember $B^* \subseteq [m_0, \omega)$).

(b) Assume that $x = \langle h^*, h \rangle \in \mathrm{dom}(R_{\mathcal{D},\bar{\mathcal{F}}})$, $\eta_n, \eta \in \omega^\omega$ are such that $\eta{\restriction} n = \eta_n{\restriction} n$ for $n \in \omega$. Take $h' \in \mathcal{F}$ such that $h^* <_{\mathcal{F}} h' <_{\mathcal{F}} h$ and choose an increasing sequence $0 = m_0 < m_1 < m_2 < \ldots < \omega$ such that for each $n \in \omega$

$$(\forall m \geq m_{n+1})((n+2) \cdot h^*(m) < h'(m)).$$

Let $\eta^* \in \omega^\omega$ (in $\mathbf{V}^*$) be such that

if $m \in [m_n, m_{n+1})$, $n \in \omega$, $0 < k < n+1$
then $\psi^{h^*(m)}(\eta_{m_k}(m)) \subseteq \psi^{h'(m)}(\eta^*(m))$

(remember the choice of the m_n's). Since $(D_{\mathcal{D},\bar{\mathcal{F}}}, R_{\mathcal{D},\bar{\mathcal{F}}})$ covers in $\mathbf{V}^*$ we find $\bar{w} = \langle w_i : i \in B \rangle$ such that $\langle h', h \rangle \ R_{\mathcal{D},\bar{\mathcal{F}}} \ T_{\bar{w},h'}$ (so in particular $B \in \mathcal{D}$ and $|w_i| \leq h(i)$) and $\eta^* \in \lim(T_{\bar{w},h'})$. But now look at the tree $T_{\bar{w},h^*}$. Clearly it satisfies $\langle h^*, h \rangle \ R_{\mathcal{D},\bar{\mathcal{F}}} \ T_{\bar{w},h^*}$. Moreover, one can inductively show that $\eta_{m_{n+1}} \in \lim(T_{\bar{w},h^*})$ for each $n \in \omega$. [Why? Plainly for each $m \in B$ we have $\psi^{h^*(m)}(\eta_{m_1}(m)) \subseteq \psi^{h'(m)}(\eta^*(m)) \subseteq w_m$, so $\eta_{m_1} \in \lim(T_{\bar{w},h^*})$. Looking at $\eta_{m_{n+1}}$, $n > 0$, note that $\eta_{m_{n+1}}{\restriction} m_n = \eta_{m_n}{\restriction} m_n$ and for each $m \geq m_n$ we have $\psi^{h^*(m)}(\eta_{m_{n+1}}(m)) \subseteq \psi^{h'(m)}(\eta^*(m))$.] Thus $T_{\bar{w},h^*}$ is as required, finishing the proof of the claim.

Finally, due to 7.2.3.1, 7.2.3.2 we may apply [**Sh:f**, Ch VI, 1.13A] to conclude the theorem. □

THEOREM 7.2.4. *Let $\mathcal{D}$ be a weakly non-reducible filter on ω, $x \in \omega^\omega$ be an unbounded non-decreasing function.*

1. *If (K, Σ) is a finitary t-omittory tree-creating pair then the forcing notion $\mathbb{Q}_1^{\rm tree}(K, \Sigma)$ has the $(\mathcal{D}, x)$-strong PP-property.*
2. *If (K, Σ) is a finitary creating pair which captures singletons then the forcing notion $\mathbb{Q}^*_{\rm w\infty}(K, \Sigma)$ has the $(\mathcal{D}, x)$-strong PP-property.*

PROOF. 1) Suppose that $\dot{\eta}$ is a $\mathbb{Q}_1^{\rm tree}(K, \Sigma)$-name for an element of ω^ω, $p \in \mathbb{Q}_1^{\rm tree}(K, \Sigma)$. Choose a condition $q \geq p$ and fronts F_n of T^q such that for each $n \in \omega$

1. if $\nu \in F_n$ then the condition $q^{[\nu]}$ decides the value of $\dot{\eta}(n)$,
2. $(\forall \nu \in F_n)(\mathbf{nor}[t_\nu^q] > n+1)$,
3. $(\forall \nu \in F_{n+1})(\exists \nu' \in F_n)(\nu' \vartriangleleft \nu)$

(possible by 2.3.7(2), 2.3.5). Next choose an increasing sequence $0 = n_0 < n_1 < n_2 < \ldots < \omega$ such that for each $k \in \omega$

$$|\bigcup\{\mathrm{pos}(t_\nu^q) : \nu \in F_{n_k}\}| < x(n_{k+1}).$$

Since $\mathcal{D}$ is weakly non-reducible we find $Y \in [\omega]^\omega$ such that $B \stackrel{\rm def}{=} \bigcup\limits_{k\in\omega\setminus Y} [n_k, n_{k+1}) \in \mathcal{D}$. Now construct inductively a condition $q^* \geq q$ such that $\mathrm{root}(q^*) = \mathrm{root}(q)$ and

(a) $T^{q^*} \subseteq \{\mathrm{root}(q)\} \cup \bigcup\{\mathrm{pos}(t^q_\nu) : \nu \in F_{n_k}\ \&\ k \in Y\}$,

(b) if $\nu \in T^{q^*}$ then $\mathrm{pos}(t^{q^*}_\nu) \subseteq \mathrm{pos}(t^q_{\nu^*})$ and $\mathbf{nor}[t^{q^*}_\nu] \geq \mathbf{nor}[t^q_{\nu^*}] - 1$ for some $\nu^* \in \bigcup\limits_{k\in Y} F_{n_k}$.

It should be clear that one can build such q^* (remember (K,Σ) is t-omittory). Note that if $k_0, k_1 \in Y$, $k_0 < k_1$ and $(k_0, k_1) \subseteq \omega \setminus Y$ then for each $n \in (n_{k_0}, n_{k_1}]$ we have

$$|\mathrm{dcl}(T^{q^*}) \cap F_n| = |\mathrm{dcl}(T^{q^*}) \cap F_{n_{k_0+1}}| \leq |\bigcup\{\mathrm{pos}(t^q_\nu) \colon \nu \in F_{n_{k_0}}\}| < x(n_{k_0+1}) \leq x(n).$$

For $n \in B$ let $w_n = \{m \in \omega : (\exists \nu \in F_n \cap \mathrm{dcl}(T^{q^*}))(q^{[\nu]} \Vdash \dot\eta(n) = m)\}$. By the above remark we have $|w_n| \leq x(n)$ (for $n \in B$) and clearly $q^* \Vdash (\forall n \in B)(\dot\eta(n) \in w_n)$.

2) Similar. □

7.3. Bounded relatives of PP

In the following definition we introduce relations which determine localization properties (see 0.2.2(2)) close to the PP–property when restricted to functions from $\prod\limits_{n\in\omega} f(n)$. Not surprisingly they include (the relation responsible for) the (f,g)–bounding property too.

DEFINITION 7.3.1. Let $f, g \in \omega^\omega$ be non-decreasing functions such that $(\forall n \in \omega)(0 < g(n) < f(n))$. Define:

1. $S_{f,g} = \prod\limits_{n\in\omega} [f(n)]^{g(n)}$, $\quad S^*_{f,g} = S_{f,g} \times [\omega]^\omega$,
2. $R^\exists_{f,g}, R^\forall_{f,g} \subseteq \prod\limits_{n\in\omega} f(n) \times S_{f,g}$ are given by

   $\eta\ R^\exists_{f,g}\ \bar A$ if and only if
   $\eta \in \prod\limits_{n\in\omega} f(n)$, $\bar A = \langle A_n : n \in \omega\rangle \in S_{f,g}$ and $(\exists^\infty n \in \omega)(\eta(n) \in A_n)$,

   $\eta\ R^\forall_{f,g}\ \bar A$ if and only if
   $\eta \in \prod\limits_{n\in\omega} f(n)$, $\bar A = \langle A_n : n \in \omega\rangle \in S_{f,g}$ and $(\forall^\infty n \in \omega)(\eta(n) \in A_n)$,
3. a relation $R^*_f \subseteq \prod\limits_{n\in\omega} f(n) \times \prod\limits_{n\in\omega} f(n)$ is defined by

   $\eta_0\ R^*_f\ \eta_1$ if and only if $\eta_0, \eta_1 \in \prod\limits_{n\in\omega} f(n)$ and $(\exists^\infty n \in \omega)(\eta_0(n) = \eta_1(n))$,
4. a relation $R^{**}_{f,g} \subseteq \prod\limits_{n\in\omega} f(n) \times S^*_{f,g}$ is such that

   $\eta\ R^{**}_{f,g}\ (\bar A, K)$ if and only if
   $\eta \in \prod\limits_{n\in\omega} f(n)$, $\bar A = \langle A_n : n \in \omega\rangle \in S_{f,g}$, $K = \{k_0, k_1, k_2, \ldots\} \in [\omega]^\omega$ (the increasing enumeration) and $(\forall m \in \omega)(\exists n \in [k_m, k_{m+1}))(\eta(n) \in A_n)$.

REMARK 7.3.2.
1. The spaces $S_{f,g}$, $S^*_{f,g}$ and $\prod\limits_{n\in\omega} f(n)$ carry natural (product) Polish topologies.
2. The relation $R^\forall_{f,g}$ corresponds to the (f,g)–bounding property, of course. The cardinal number $\mathfrak{d}(R^\forall_{f,g})$ is the $c(f,g)$ of **[GoSh 448]** (see there for various ZFC dependencies between the cardinals determined by different functions as well as for consistency results).

3. Note that the relation $R^{**}_{f,g}$ (or actually the corresponding localization property) is really very close to the PP–property. The cardinal numbers $\mathfrak{d}(R^*_f)$ and $\mathfrak{d}(R^{**}_{f,g})$ appear naturally in [**BRSh 616**].
4. There are other natural variants of relations introduced in 7.3.1. We will deal with them (and the corresponding cardinal invariants) in the continuation of this paper.

Below we list some obvious relations between the localization properties introduced in 7.3.1 and the corresponding cardinal numbers.

PROPOSITION 7.3.3. *Let $f, g, h \in \omega^\omega$ be non-decreasing functions such that $0 < g(n) < f(n)$ for each $n \in \omega$. Then:*

1. *The $R^\forall_{f,g}$–localization implies the $R^\exists_{f,g}$–localization and $\mathfrak{d}(R^\exists_{f,g}) \leq \mathfrak{d}(R^\forall_{f,g})$.*
2. *Suppose that for some increasing sequence $m_0 < m_1 < m_2 < \ldots < \omega$ we have*
$$(\forall n \in \omega)(g(n) \leq m_{n+1} - m_n \ \ \&\ \ f(n) \geq \prod_{k \in [m_n, m_{n+1})} h(k)).$$
Then the $R^\exists_{f,g}$–localization implies the R^*_h–localization and $\mathfrak{d}(R^*_h) \leq \mathfrak{d}(R^\exists_{f,g})$.
3. *The R^*_f–localization implies the $R^\exists_{f,g}$ localization and $\mathfrak{d}(R^\exists_{f,g}) \leq \mathfrak{d}(R^*_f)$.*
4. *The $R^\exists_{f,g}$–localization plus ω^ω–bounding imply the $R^{**}_{f,g}$–localization. The $R^{**}_{f,g}$–localization implies the $R^\exists_{f,g}$–localization. Hence $\mathfrak{d}(R^\exists_{f,g}) \leq \mathfrak{d}(R^{**}_{f,g}) \leq \max\{\mathfrak{d}, \mathfrak{d}(R^\exists_{f,g})\}$.*
5. *If g is unbounded then the strong PP–property implies the $R^\exists_{f,g}$–localization, and $\mathfrak{d}(R^\exists_{f,g}) \leq \mathfrak{d}(R^{\rm sPP})$.*
6. *Assume ground model reals are not meager. Then the extension has the R^*_f–localization property and thus $\mathfrak{d}(R^*_f) \leq$ **non**$(\mathcal{M})$.*
7. *The $R^{**}_{f,g}$–localization implies that there is no Cohen real over the ground model, and thus* **cov**$(\mathcal{M}) \leq \mathfrak{d}(R^{**}_{f,g})$.

For getting the $R^\forall_{f,g}$–localization (i.e. (f,g)–bounding property) for forcing notions built according to our schema see 5.1. Let us note that the other properties appear naturally too.

PROPOSITION 7.3.4. *Let $f, g \in \omega^\omega$. Suppose that $\mathbb{P}$ is a forcing notion one of the following type*

- *$\mathbb{Q}^*_{\rm s\infty}(K,\Sigma)$ for a finitary creating pair (K,Σ) which is either growing and big or omittory and omittory–big,*
- *$\mathbb{Q}^*_{\rm w\infty}(K,\Sigma)$ for a finitary creating pair which captures singletons,*
- *$\mathbb{Q}^{\rm tree}_0(K,\Sigma)$ for a finitary t-omittory tree–creating pair (K,Σ).*

Then $\mathbb{P}$ has the $R^{**}_{f,g}$–localization property.

PROOF. It should be clear, so we will sketch the proof for the first case only. Let $\dot\eta$ be a $\mathbb{Q}^*_{\rm s\infty}(K,\Sigma)$–name for a function in $\prod\limits_{n\in\omega} f(n)$ and let $p \in \mathbb{Q}^*_{\rm s\infty}(K,\Sigma)$. Using 2.2.3 or 2.2.6 construct a condition $q \in \mathbb{Q}^*_{\rm s\infty}(K,\Sigma)$, an enumeration $\langle u_k : k \in \omega\rangle$ of $\bigcup\limits_{n\in\omega} {\rm pos}(w^q, t^q_0, \ldots, t^q_{n-1})$ and a sequence $\langle m_k : k < \omega\rangle$ such that

1. $p \leq_0 q$, $\quad m_0 < m_1 < \ldots < \omega$,
2. if $u_k \in {\rm pos}(w^q, t^q_0, \ldots, t^q_{n-1})$ then the condition $(u_k, t^q_n, t^q_{n+1}, \ldots)$ decides the value of $\dot\eta(m_k)$.

Plainly the construction is possible and easily it finishes the proof. □

It may be not clear how one can preserve (in countable support iterations) the localization properties introduced in 7.3.1. To deal with the $R^{\exists}_{f,g}$–localization property we may adopt the approach of 7.2.3. It slightly changes the meaning of this notion but the change is not serious and makes dealing with compositions much easier.

DEFINITION 7.3.5. Let $h_0, h_1, f \in \omega^\omega$ be non-decreasing unbounded functions, $\mathcal{D}$ be a filter on ω and $\bar{\mathcal{F}} = (\mathcal{F}, <_{\mathcal{F}})$ be a partial order on $\mathcal{F} \subseteq \prod\limits_{n\in\omega} f(n)$. We say that a proper forcing notion $\mathbb{P}$:

1. *has the* $\mathcal{D}$-$R^{\exists}_{f,h_0,h_1}$*–localization property* if

$$\Vdash_{\mathbb{P}} \text{“ for every } \eta \in \prod_{n\in\omega} [f(n)]^{\leq h_0(n)} \text{ there are } B \in \mathcal{D} \cap \mathbf{V} \text{ and } \langle w_i : i \in B\rangle \in \mathbf{V}$$
$$\text{such that } (\forall i \in B)(|w_i| \leq h_1(i) \;\;\&\;\; \eta(i) \subseteq w_i) \text{ ”},$$

2. *has the* $(\mathcal{D}, \bar{\mathcal{F}})$–$R^{\exists}_{f}$*–localization property* if it has the $\mathcal{D}$-$R^{\exists}_{f,h_0,h_1}$–localization property for every $h_0, h_1 \in \mathcal{F}$ such that $h_0 <_{\mathcal{F}} h_1$.

PROPOSITION 7.3.6. *Suppose that $\mathcal{D}$ is a non-principal p-filter on ω, $f \in \omega^\omega$ is non-decreasing unbounded and $\bar{\mathcal{F}} = (\mathcal{F}, <_{\mathcal{F}})$ is a PP–ok partial order on $\mathcal{F} \subseteq \prod\limits_{n\in\omega} f(n)$ (see 7.2.1(2), except that it does not have to contain the identity function). Let $\langle \mathbb{P}_\alpha, \dot{\mathbb{Q}}_\alpha : \alpha < \delta\rangle$ be a countable support iteration such that for each $\alpha < \delta$*

$$\Vdash_{\mathbb{P}_\alpha} \text{“ } \dot{\mathbb{Q}}_\alpha \textit{ is a proper forcing notion which has the } (\mathcal{D}, \bar{\mathcal{F}})\text{–}R^{\exists}_{f}\textit{–localization property”.}$$

Then the forcing notion $\mathbb{P}_\alpha$ has the $(\mathcal{D}, \bar{\mathcal{F}})$–$R^{\exists}_{f}$–localization property.

PROOF. Repeat the proof of 7.2.3 making suitable adjustments to the fact that we are “below the function f”. No real changes are required. □

REMARK 7.3.7. Note that with no serious changes we may formulate and prove a variant of 7.3.6 which would be an exact reformulation of 5.2.9 for the current context.

PROPOSITION 7.3.8. *Let $f, h_0, h_1 \in \omega^\omega$ be non-decreasing unbounded functions such that $(\forall n \in \omega)(1 \leq h_0(n) \leq h_1(n) \leq f(n))$ and $\lim\limits_{n\to\infty} \frac{h_1(n)}{h_0(n)} = \infty$. Assume $\mathcal{D}$ is a non-reducible p-filter on ω. Suppose that (K, Σ) is a finitary, omittory and omittory–big creating pair. Then the forcing notion $\mathbb{Q}^*_{\mathrm{s}\infty}(K, \Sigma)$ has the $\mathcal{D}$-$R^{\exists}_{f,h_0,h_1}$–localization property.*

PROOF. Like 7.2.4 plus 7.3.4. □

REMARK 7.3.9. Note that if $x(n) \leq \frac{h_1(n)}{h_0(n)}$ for $n \in \omega$ then the $(\mathcal{D}, x)$–strong PP–property implies the $\mathcal{D}$-$R^{\exists}_{f,h_0,h_1}$–localization property. Consequently we may use 7.2.4 to get the conclusion of 7.3.8 for the two types of forcing notions specified in 7.2.4.

7.4. Weakly non-reducible p-filters in iterations

One could get an impression that 7.2.3, 7.3.6 together with 7.2.4 and 7.3.8 are everything we need: the properties involved are iterable and we may get them for various forcing notions. However, to be able to make a real use of 7.2.4 or 7.3.8 we have to know that if we start with a weakly non-reducible p-filter and then we iterate suitable forcing notions, the filter remains weakly non-reducible. One could start with a p-point and consider forcing notions which are p-point preserving only. However this is much too restrictive: we may iterate forcing notions mentioned in 7.2.4 and 7.3.8 and the iterations will preserve the fact that the filter is weakly non-reducible. The first step in proving this is the following observation.

PROPOSITION 7.4.1. *Suppose that $\mathcal{D}$ is a weakly non-reducible filter on ω. Let $\mathbb{P}$ be an almost ω^ω-bounding forcing notion. Then*

$$\Vdash_{\mathbb{P}} \text{“ (The filer generated by) } \mathcal{D} \text{ is weakly non-reducible ”.}$$

[Note that this covers ω^ω-bounding forcing notions.]

PROOF. Suppose that $\langle \dot{X}_n : n \in \omega \rangle$ is a $\mathbb{P}$–name for a partition of ω into finite sets. Let $\dot{f}$ be a $\mathbb{P}$–name for a function in ω^ω such that

$$\Vdash_{\mathbb{P}} \text{“ } (\forall n \in \omega)(\exists m \in \omega)(\dot{X}_m \subseteq [n, \dot{f}(n))) \text{ ”.}$$

Suppose $p \in \mathbb{P}$. Since $\mathbb{P}$ is almost ω^ω-bounding we find an increasing function $g \in \omega^\omega$ such that

$$(\forall A \in [\omega]^\omega)(\exists q \geq p)(q \Vdash_{\mathbb{P}} \text{“ } (\exists^\infty n \in A)(\dot{f}(n) < g(n)) \text{ ”}).$$

Let $0 = n_0 < n_1 < n_2 < \ldots < \omega$ be such that $g(n_k) < n_{k+1}$. As the filter $\mathcal{D}$ is weakly non-reducible, we find $Y \in [\omega]^\omega$ such that $Z \stackrel{\text{def}}{=} \bigcup\limits_{k \in \omega \setminus Y} [n_k, n_{k+1}) \in \mathcal{D}$. Let $A = \{n_k : k \in Y\}$. By the choice of the function g, there is a condition $q \geq p$ such that

$$q \Vdash_{\mathbb{P}} \text{“ } (\exists^\infty n \in A)(\dot{f}(n) < g(n)) \text{ ”.}$$

Now look at the choice of $\dot{f}$ – necessarily

$$q \Vdash_{\mathbb{P}} \text{“ } (\exists^\infty m \in \omega)(\dot{X}_m \cap Z = \emptyset) \text{ ”,}$$

which is enough to conclude the proposition. □

Note that 7.4.1 captures *almost* all forcing notions mentioned in 7.2.4, 7.3.8. So what is needed more is that “$\mathcal{D}$ is weakly non-reducible” is preserved at limit stages of countable support iterations of proper forcing notions. This is done like preserving unbounded families (i.e. by [**Sh:f**, Ch VI, §3]).

THEOREM 7.4.2. *Let $\mathcal{D}$ be a weakly non-reducible p-filter on ω. Suppose that $\langle \mathbb{P}_\alpha, \dot{\mathbb{Q}}_\alpha : \alpha < \delta \rangle$ is a countable support iteration of proper forcing notions such that δ is limit and for each $\alpha < \delta$*

$$\Vdash_{\mathbb{P}_\alpha} \text{“ (The filter generated by) } \mathcal{D} \text{ is weakly non-reducible ”.}$$

Then $\Vdash_{\mathbb{P}_\delta}$ “ (The filter generated by) $\mathcal{D}$ is weakly non-reducible ”.

PROOF. We will use [**Sh:f**, Ch VI, 3.13] and thus we will follow the notation there. Let $F \subseteq \omega^\omega$ be the family of all increasing enumerations of elements of $\mathcal{D}$ (i.e. $F = \{\mu_X : X \in \mathcal{D}\}$, see 4.4.4). Let R be (a definition of) the following two–place relation on ω^ω:

$g\ R\ f$ if and only if $(g, f \in \omega^\omega$ and$)$
$(\exists^\infty k)([n^g_k, n^g_{k+1}) \cap \mathrm{rng}(f) = \emptyset)$, where $n^g_0 = 0$, $n^g_{k+1} = n^g_k + g(k) + 1$ for $k \in \omega$.
As $\mathcal{D}$ is weakly non-reducible, the family F is R–bounding (i.e. $(\forall g \in \omega^\omega)(\exists f \in F)(g\ R\ f)$).

CLAIM 7.4.2.1. *(F, R) is S–nice (see [**Sh:f**, Ch VI, 3.2]; here $S \subseteq [F]^\omega$ is arbitrary).*

Proof of the claim: We have to show that for each $N \in S$ there is $g \in F$ such that for each $m_0 \in \omega$ (the n_0 of [**Sh:f**, Ch VI, 3.2.3(β)] is irrelevant here) the second player has an absolute winning strategy in the following game.

At the stage k of the game, Player I chooses $f_k \in \omega^\omega$ and $g_k \in F \cap N$ such that $f_k \restriction m_{\ell+1} = f_\ell \restriction m_{\ell+1}$ for all $0 \leq \ell < k$ and $f_k\ R\ g_k$. Then Player II answers playing an integer $m_{k+1} > m_k$.
Player II wins the game if $(\bigcup\limits_{k\in\omega} f_k \restriction m_k)\ R\ g$.

But this is easy: let $g \in F$ be such that $\mathrm{rng}(g) \subseteq^* \mathrm{rng}(f)$ for all $f \in F \cap N$ (remember $\mathcal{D}$ is a p-filter). Then

$$f_k\ R\ g_k \quad \text{implies} \quad (\exists^\infty \ell \in \omega)([n^{f_k}_\ell, n^{f_k}_{\ell+1}) \cap \mathrm{rng}(g) = \emptyset).$$

Thus, at stage k of the game, the second player may choose $m_{k+1} > m_k$ such that $[n^{f_k}_\ell, n^{f_k}_{\ell+1}) \cap \mathrm{rng}(g) = \emptyset$ for some $\ell \in (m_k, m_{k+1})$.

As we iterate proper forcing notions, countable subsets of F from $\mathbf{V}^{\mathbb{P}_\alpha}$ can be covered by countable subsets of F from $\mathbf{V}$. By our assumptions, F is R–bounding in each $\mathbf{V}^{\mathbb{P}_\alpha}$ (for $\alpha < \delta$) and it is nice there (like in the claim above). Consequently we may apply [**Sh:f**, Ch VI, 3.13(3)] and we conclude that

$$\Vdash_{\mathbb{P}_\delta} \text{“ } F \text{ is } R\text{–bounding ”}.$$

But this is exactly what we need. $\square$

7.5. Examples

EXAMPLE 7.5.1. Let $P \subseteq 2^\omega$ be a perfect set. We construct a finitary function $\mathbf{H}^P$, an $\mathbf{H}^P$–fast function $f^P : \omega \times \omega \longrightarrow \omega$ and a $\bar{2}$–big trivially meagering simple creating pair $(K^P_{7.5.1}, \Sigma^P_{7.5.1})$ for $\mathbf{H}^P$ with the (weak) Halving Property such that

$$\Vdash_{\mathbb{Q}^*_{f^P}(K^P_{7.5.1}, \Sigma^P_{7.5.1})} \text{“ there is a perfect set } Q \subseteq P \text{ such that } (\forall K \in [\omega]^\omega \cap \mathbf{V})(Q \restriction K \neq 2^K) \text{ ”}.$$

CONSTRUCTION. The creating pair $(K^P_{7.5.1}, \Sigma^P_{7.5.1})$ will be constructed in a way slightly similar to $(K^1_{2.4.12}, \Sigma^1_{2.4.12})$. For positive integers i, m let $R^i_2(2, m)$ be the minimal integer k such that for every function $\varphi : \prod\limits_{\ell<i} [k]^2 \longrightarrow 2$ there are sets $a_0, \ldots, a_{i-1} \in [k]^m$ such that $\varphi \restriction [a_0]^2 \times \ldots \times [a_{i-1}]^2$ is constant. [Thus this is the Ramsey number for polarized partition relations; $R^1_2(2, m)$ is essentially $R_2(2, m)$ of 6.3.5.]

Define inductively $Z_2^i(k)$ for $k\in\omega$ by $Z_2^i(0)=R_2^i(2,4)$, $Z_2^i(k+1)=R_2^i(2,2\cdot Z_2^i(k))$, and for a finite set X let

$$H_i(X) \stackrel{\rm def}{=} \min\{k \in \omega : |X| \leq Z_2^i(k)\}.$$

Let $T \subseteq 2^{<\omega}$ be a perfect tree such that $P = [T]$. Now construct inductively functions $\mathbf{H}^P = \mathbf{H}$ and $f^P = f$ and an increasing sequence $\bar{n} = \langle n_i : i \in \omega\rangle$ such that

(i) $f(0,\ell) = \ell+1$, $f(k+1,\ell) = 2^{\varphi_{\mathbf{H}}(\ell)+1}\cdot(f(k,\ell)+\varphi_{\mathbf{H}}(\ell)+2)$ (compare 1.1.12),

(ii) $n_0 = 0$, n_{i+1} is the first such that for every $\nu \in T\cap 2^{n_i}$

$$H_{2^i}(\{\eta \in T\cap 2^{n_{i+1}} : \nu \vartriangleleft \eta\}) \geq 2^{f(i,i)},$$

(iii) $\mathbf{H}(i)$ is the family of all non-empty subsets of $T\cap 2^{n_{i+1}}$.

It should be clear that the clauses (i)–(iii) uniquely determine $\mathbf{H}$, f and $\bar{n}$.

Call a sequence $u \in \prod\limits_{i<m} \mathbf{H}(i)$ *acceptable* if for each $i_0 < i_1 < m$

$$|u(0)| = 2,\quad u(i_0) = \{\eta{\restriction}n_{i_0+1} : \eta \in u(i_1)\},\quad \text{and}$$
$$(\forall \nu \in u(i_0))(|\{\eta \in u(i_0+1) : \nu \vartriangleleft \eta\}| = 2).$$

Note that if $W \in \prod\limits_{m\in\omega}\mathbf{H}(m)$ is such that each $W{\restriction}m$ is acceptable then the sequence W determines a perfect tree

$$T^W \stackrel{\rm def}{=} \{\nu \in T : (\exists m \in \omega)(\exists \eta \in W(m))(\nu \trianglelefteq \eta)\} \subseteq T$$

with the property that $|T^W\cap 2^{n_i}| = 2^i$ and each node from $T^W\cap 2^{n_i}$ has a ramification below n_{i+1}.

A creature $t \in \mathrm{CR}[\mathbf{H}]$ is in $K^P_{7.5.1}$ if $m^t_{\rm up} = m^t_{\rm dn}+1 = i+1$ and

- $\mathbf{dis}[t] = \langle m^t_{\rm dn}, \langle B^t_\nu : \nu \in T\cap 2^{n_i}\rangle, r^t\rangle$, where r^t is a non-negative real and $B^t_\nu \subseteq \{\eta \in T\cap 2^{n_{i+1}} : \nu \vartriangleleft \eta\}$ (for all $\nu \in T\cap 2^{n_i}$; remember $i = m^t_{\rm dn}$),
- $\mathbf{val}[t] = \{\langle u,v\rangle \in \prod\limits_{k<m^t_{\rm dn}}\mathbf{H}(k)\times\prod\limits_{k\leq m^t_{\rm dn}}\mathbf{H}(k) : u \vartriangleleft v$ both are acceptable and if $\eta \in v(m^t_{\rm dn})$ then $\eta \in B^t_{\eta{\restriction}n_i}\}$,
- $\mathbf{nor}[t] = \max\{0, \min\{H_{2^i}(B^t_\nu) : \nu \in T\cap 2^{n_i}\} - r^t\}$.

The operation $\Sigma^P_{7.5.1}$ is defined by

$$\Sigma^P_{7.5.1}(t) = \{s \in K^P_{7.5.1} : m^t_{\rm dn} = m^s_{\rm dn}\ \&\ (\forall \nu \in T\cap 2^{n_i})(B^s_\nu \subseteq B^t_\nu)\ \&\ r^s \geq r^t\}.$$

It should be clear that $(K^P_{7.5.1},\Sigma^P_{7.5.1})$ is a simple finitary creating pair and the forcing notion $\mathbb{Q}^*_f(K^P_{7.5.1},\Sigma^P_{7.5.1})$ is not trivial.

To check that $(K^P_{7.5.1},\Sigma^P_{7.5.1})$ is $\bar{2}$–big suppose that $t \in K^P_{7.5.1}$, $\mathbf{nor}[t] > 1$, $u \in \mathbf{basis}(t)$ and $c : \mathrm{pos}(u,t) \longrightarrow 2$ (note that $\mathbf{basis}(t) = \mathrm{dom}(\mathbf{val}[t])$ and $\mathrm{pos}(u,t) = \{v \in \mathrm{rng}(\mathbf{val}[t]) : u \vartriangleleft v\}$). Then u is acceptable and, if $\ell g(u) > 0$, $|u(\ell g(u)-1)| = 2^{\ell g(u)}$. Let $i = \ell g(u) = m^t_{\rm dn}$. Let $k = \min\{H_{2^i}(B^t_\nu) : \nu \in u(i-1)\}$. By the definition of the norm of t we know that $k \geq \mathbf{nor}[t] + r^t > 1$, so necessarily $k \geq 2$. Note that under natural interpretation $\prod\limits_{\nu\in u(i-1)}[B^t_\nu]^2 \subseteq \mathrm{pos}(u,t)$, so we may restrict our colouring c to this set and use the definition of H_{2^i} (and the choice of k). Thus we find sets $B^*_\nu \subseteq B^t_\nu$ (for $\nu \in u(i-1)$) such that

$$(\forall \nu \in u(i-1))(|B^*_\nu| = 2\cdot Z_2^{2^i}(k-2))\quad\text{and}\quad c{\restriction}(\prod_{\nu\in u(i-1)} B^*_\nu)\ \text{is constant.}$$

Note that $H_{2^i}(B^*_\nu) \geq k-1 \geq \mathbf{nor}[t]-1+r^t > r^t$. Let $s \in K^P_{7.5.1}$ be a creature determined by

$$m^s_{\mathrm{dn}} = m^t_{\mathrm{dn}}, \quad r^s = r^t, \quad B^s_\nu = B^*_\nu \text{ if } \nu \in u(i-1), \quad \text{and } B^s_\nu = B^t_\nu \text{ otherwise.}$$

Clearly $s \in \Sigma^P_{7.5.1}(t)$, $\mathbf{nor}[s] \geq \mathbf{nor}[t]-1$ and $c{\restriction}\mathrm{pos}(u,s)$ is constant.

Plainly $(K^P_{7.5.1}, \Sigma^P_{7.5.1})$ is trivially meagering, as if $x,y \in X$ then

$$H_i(X \setminus \{x,y\}) \geq H_i(X) - 1.$$

Let half : $K^P_{7.5.1} \longrightarrow K^P_{7.5.1}$ be such that if $\mathbf{nor}[t] \geq 2$ then

$$\mathbf{dis}(\mathrm{half}(t)) = \langle m^t_{\mathrm{dn}},\ \langle B^t_\nu : \nu \in T,\ \ell g(\nu) = n_{m^t_{\mathrm{dn}}}\rangle,\ r^t + \frac{1}{2}\mathbf{nor}[t]\rangle,$$

and $\mathrm{half}(t) = t$ otherwise. Exactly like in 2.4.12 one checks that the function half witnesses the fact that $(K^P_{7.5.1}, \Sigma^P_{7.5.1})$ has the (weak) Halving Property.

To show the last assertion of 7.5.1 we prove that

$$\Vdash_{\mathbb{Q}^*_f(K^P_{7.5.1},\Sigma^P_{7.5.1})} \text{“ } (\forall K \in [\omega]^\omega \cap \mathbf{V})([T^{\dot W}]{\restriction}K \neq 2^K) \text{ ”},$$

where $\dot W$ is the name for the generic real (see 1.1.13) and $T^{\dot W}$ is the tree defined before. To this end suppose that $p \in \mathbb{Q}^*_f(K^P_{7.5.1}, \Sigma^P_{7.5.1})$ and $K \in [\omega]^\omega$. We may assume that $\ell g(w^p) = j_0 > 0$ and $(\forall i \in \omega)(\mathbf{nor}[t^p_i] > f^P(0, m^{t^p_i}_{\mathrm{dn}}) \geq 2)$. Choose $j_1 \in \omega$ such that $|[n_{j_0}, n_{j_1}) \cap K| \geq 2^{j_0}$ and fix one-to-one mapping

$$k : w^p(j_0-1) \longrightarrow [n_{j_0}, n_{j_1}) \cap K : \nu \mapsto k(\nu)$$

(remember that w^p is acceptable, so $|w^p(j_0-1)| = 2^{j_0}$ and $w^p(j_0-1) \subseteq T \cap 2^{n_{j_0}}$).

Fix $\nu \in w^p(j_0-1)$ for a moment. Let $i(\nu) = i < j_1 - j_0$ be such that $k(\nu) \in [n_{j_0+i}, n_{j_0+i+1})$. For each $\eta \in T \cap 2^{n_{j_0+i}}$ such that $\nu \trianglelefteq \eta$ choose $c_{j_0+i}(\eta) \in 2$ such that the set $B^\nu_\eta \stackrel{\mathrm{def}}{=} \{\rho \in B^{t^p_i}_\eta : \rho(k(\nu)) = c_{j_0+i}(\nu)\}$ has at least $\frac{1}{2}|B^{t^p_i}_\eta|$ elements (so then $H_{2^{j_0+i}}(B^\nu_\eta) \geq H_{2^{j_0+i}}(B^{t^p_i}_\eta) - 1$). If $\eta = \nu$ then we finish the procedure. Otherwise, for each $\rho \in T \cap 2^{n_{j_0+i-1}}$ such that $\nu \trianglelefteq \rho$ we choose $c_{j_0+i-1}(\rho) \in 2$ such that the set $B^\nu_\rho \stackrel{\mathrm{def}}{=} \{\eta \in B^{t^p_{i-1}}_\rho : c_{j_0+i-1}(\rho) = c_{j_0+i}(\eta)\}$ has at least $\frac{1}{2}|B^{t^p_{i-1}}_\rho|$ elements (and so $H_{2^{j_0+i-1}}(B^\nu_\rho) \geq H_{2^{j_0+i-1}}(B^{t^p_{i-1}}_\rho) - 1$). Continuing this procedure downward till we arrive to ν we determine sets $\langle B^\nu_\eta : \nu \trianglelefteq \eta \in T \cap 2^{n_j},\ j_0 \leq j \leq j_0 + i(\nu)\rangle$ and $c_{j_0}(\nu) \in 2$ such that

$(\alpha)_\nu$ if $\nu \trianglelefteq \eta \in T \cap 2^{n_j}$, $j_0 \leq j \leq j_0 + i(\nu)$
then $B^\nu_\eta \subseteq B^{t^p_{j-j_0}}_\eta$ and $H_{2^j}(B^\nu_\eta) \geq H_{2^j}(B^{t^p_{j-j_0}}_\eta) - 1$,

$(\beta)_\nu$ if $\nu \trianglelefteq \eta \in T \cap 2^{n_{j_1}}$ is such that $(\forall j \in [j_0, j_0 + i(\nu)))(\eta{\restriction}n_{j+1} \in B^\nu_{\eta{\restriction}n_j})$
then $\eta(k(\nu)) = c_{j_0}(\nu)$.

For each $i < j_1 - j_0$ choose a creature $s_i \in \Sigma^P_{7.5.1}(t^p_i)$ such that $r^{s_i} = r^{t^p_i}$ and for every $\eta \in T \cap 2^{n_{j_0+i}}$

if $i(\eta{\restriction}n_{j_0}) \leq i$ then $B^{s_i}_\eta = B^{\eta{\restriction}n_{j_0}}_\eta$, otherwise $B^{s_i}_\eta = B^{t^p_i}_\eta$.

Clearly $\mathbf{nor}[s_i] \geq \mathbf{nor}[t^p_i] - 1$ and thus $q = (w^p, s_0, \ldots, s_{j_1-j_0-1}, t^p_{j_1-j_0}, \ldots)$ is a condition in $\mathbb{Q}^*_{f^P}(K^P_{7.5.1}, \Sigma^P_{7.5.1})$ stronger than p. Let $\sigma : K \cap [n_{j_0}, n_{j_1}) \longrightarrow 2$ be such that $\sigma(k(\nu)) = 1 - c_{j_0}(\nu)$ for each $\nu \in w^p(j_0-1)$. Note that

$$u \in \mathrm{pos}(w^p, s_0, \ldots, s_{j_1-j_0-1}) \quad \Rightarrow \quad (\forall \eta \in u(j_1-1))(\eta{\restriction}(K \cap [n_{j_0}, n_{j_1}) \neq \sigma)),$$

what finishes the proof. □

CONCLUSION 7.5.2. It is consistent that $\kappa_{\rm BP} = \mathbf{non}(\mathcal{M}) = \mathfrak{d}(R^{\rm sPP}) = \mathfrak{c} = \aleph_2$ and $\mathfrak{d} = \aleph_1$.

PROOF. Start with a model for CH and build inductively (with a suitable bookkeeping) a countable support iteration $\langle \mathbb{P}_\alpha, \dot{\mathbb{Q}}_\alpha : \alpha < \omega_2 \rangle$ and a sequence $\langle \dot{P}_\alpha : \alpha < \omega_2 \rangle$ such that

(α) $\langle \dot{P}_\alpha : \alpha < \omega_2 \rangle$ lists with ω_2–repetitions all $\mathbb{P}_{\omega_2}$–names for perfect subsets of 2^ω; each $\dot{P}_\alpha$ is a $\mathbb{P}_\alpha$–name,

(β) $\dot{\mathbb{Q}}_\alpha$ is a $\mathbb{P}_\alpha$–name for the forcing notion $\mathbb{Q}^*_{f\dot{P}_\alpha}(K^{\dot{P}_\alpha}_{7.5.1}, \Sigma^{\dot{P}_\alpha}_{7.5.1})$.

By 2.2.12 and 3.1.2 we know that each $\dot{\mathbb{Q}}_\alpha$ is a (name for) proper ω^ω–bounding forcing notion and hence $\Vdash_{\mathbb{P}_{\omega_2}}$ "$\mathfrak{d} = \aleph_1$". By 3.2.8(2) we easily conclude that $\Vdash_{\mathbb{P}_{\omega_2}}$ "$\mathbf{non}(\mathcal{M}) = \aleph_2$" and finally we note that by the last property of $\mathbb{Q}^*_{fP}(K^P_{7.5.1}, \Sigma^P_{7.5.1})$ stated in 7.5.1, and by the choice of $\langle \dot{P}_\alpha : \alpha < \omega_2 \rangle$, we have $\Vdash_{\mathbb{P}_{\omega_2}}$ "$\kappa_{\rm BP} = \aleph_2$". To finish remember 7.1.3. □

CONCLUSION 7.5.3. It is consistent that $\kappa_{\rm BP} = \mathfrak{d}(R^{\rm sPP}) = \aleph_1$ and $\mathbf{non}(\mathcal{N}) = \mathfrak{c} = \aleph_2$.

PROOF. Force over a model of CH with countable support iteration, ω_2 in length, of forcing notions $\mathbb{Q}^*_{\rm w\infty}(K_{5.4.3}, \Sigma_{5.4.3})$.

By 7.2.4(2) we know that the forcing notion $\mathbb{Q}^*_{\rm w\infty}(K_{5.4.3}, \Sigma_{5.4.3})$ has the $(\mathcal{D}, x)$-strong PP–property for any weakly non-reducible filter $\mathcal{D}$ on ω and an unbounded non-decreasing $x \in \omega^\omega$. Consequently, if (in $\mathbf{V}$) we take a p-point $\mathcal{D}$ and a PP–ok partial order $\bar{\mathcal{F}}$ then the iteration will have the $(\mathcal{D}, \bar{\mathcal{F}})$-strong PP–property, so in particular the strong PP–property (by 7.2.3; remember that by 7.4.1+7.4.2 the filter generated by $\mathcal{D}$ in the intermediate universes is weakly non-reducible). Hence, in the resulting model we have $\mathfrak{d}(R^{\rm sPP}) = \aleph_1$ and thus $\kappa_{\rm BP} = \aleph_1$ (by 7.1.3). Finally, it follows from 5.4.4 that in this model $\mathbf{non}(\mathcal{N}) = \mathfrak{c} = \aleph_2$.

Note that one can use the forcing notion $\mathbb{Q}^{\rm tree}_1(K^1_{5.4.5}, \Sigma^1_{5.4.5})$ of 5.4.5 as well. □

Let us recall the following notions from [**Sh 326**].

DEFINITION 7.5.4. Let $\mathcal{F} \subseteq \omega^\omega$ and $g \in \omega^\omega$.

1. We say that the family $\mathcal{F}$ is *g–closed* if

$$(\forall f \in \mathcal{F})(\exists f^+, f^* \in \mathcal{F})(\forall^\infty n \in \omega)(f(n)^{g(n)} \leq f^*(n) \ \ \& \ \ \prod_{m<n} (f(m)+1) \leq f^+(n)).$$

2. We say that a proper forcing notion $\mathbb{P}$ has the $(\mathcal{F}, g)$–bounding property if it has the (f, g^ε)–bounding property for each $\varepsilon > 0$ and $f \in \mathcal{F}$.

These notions are important when we want to iterate (f, g)–bounding forcing notions: if $\mathcal{F}$ is a g–closed family then each countable support iteration of proper $(\mathcal{F}, g)$–bounding forcing notions is $(\mathcal{F}, g)$–bounding (see [**Sh 326**, A2.5], compare to 5.2.9).

PROPOSITION 7.5.5. *Suppose that $\mathcal{F} \subseteq \omega^\omega$ is a g–closed family and $\varphi \in \omega^\omega$ is an increasing function. Let $g_\varphi = g \circ \varphi$ and let $\mathcal{F}_\varphi = \{f \circ \varphi : f \in \mathcal{F}\}$. Then $\mathcal{F}_\varphi$ is g_φ–closed.*

PROOF. Check. □

CONCLUSION 7.5.6. Let $g(n) = n^n$ for $n \in \omega$ and let $\mathcal{F} \subseteq \omega^\omega$ be a countable g–closed family. Suppose that $F \in \omega^\omega$ is an increasing function which dominates all elements of $\mathcal{F}$ (i.e. $(\forall f \in \mathcal{F})(\forall^\infty n \in \omega)(f(n) < F(n)))$ and let $\mathbf{H} = \mathbf{H}^F$, $f = f^F$ (and $\varphi_\mathbf{H}$) be as defined in 2.4.6 for F. Next, let $f_0 \in \mathcal{F}$ and $\langle m_k : k \in \omega \rangle \subseteq \omega$ and $h \in \omega^\omega$ be such that $m_0 = 0$, $m_{k+1} = m_k + \varphi_\mathbf{H}(k)^{\varphi_\mathbf{H}(k)}$, h is non-decreasing and $h(m_{k+1}) \leq f_0(\varphi_\mathbf{H}(k))$ (for $k \in \omega$). Assume that $\mathbb{P}_{\omega_2}$ is the countable support iteration of the forcing notions $\mathbb{Q}^*_f(K_{2.4.6}, \Sigma_{2.4.6})$. Then

$$\Vdash_{\mathbb{P}_{\omega_2}} \text{“ } \mathfrak{d}(R^*_\mathbf{H}) = \aleph_2 \ \ \& \ \ \mathfrak{d} = \mathfrak{d}(R^*_h) = \mathfrak{d}(R^\forall_{f_0\circ\varphi_\mathbf{H}, g\circ\varphi_\mathbf{H}}) = \aleph_1 \text{ ”}.$$

PROOF. We know that $\mathcal{F}_{\varphi_\mathbf{H}} = \{f' \circ \varphi_\mathbf{H} : f' \in \mathcal{F}\}$ is $g_{\varphi_\mathbf{H}}$–closed, where $g_{\varphi_\mathbf{H}} = g \circ \varphi_\mathbf{H}$. By 5.4.6 and the choice of g, F we have that the forcing notion $\mathbb{Q}^*_f(K_{2.4.6}, \Sigma_{2.4.6})$ is proper, ω^ω–bounding, $(\mathcal{F}_{\varphi_\mathbf{H}}, g_{\varphi_\mathbf{H}})$–bounding and so is the iteration. Hence, for each $f_1 \in \mathcal{F}$, $\Vdash_{\mathbb{P}_{\omega_2}}$ “$\mathfrak{d}(R^\forall_{f_1\circ\varphi_\mathbf{H}, g\circ\varphi_\mathbf{H}}) = \mathfrak{d} = \aleph_1$”. Next note that for the function h defined in the assumptions and for sufficiently large k we have

$$\prod_{n\in[m_k, m_{k+1})} h(n) \leq h(m_{k+1})^{\varphi_\mathbf{H}(k)} \leq ((f_0 \circ \varphi_\mathbf{H})(k))^{\varphi_\mathbf{H}(k)} \leq (f^*_0 \circ \varphi_\mathbf{H})(k),$$

where $f^*_0 \in \mathcal{F}$ is such that $(\forall^\infty n \in \omega)(f_0(n)^{g(n)} \leq f^*_0(n))$. Use 7.3.3(2) to conclude that $\Vdash_{\mathbb{P}_{\omega_2}}$ “$\mathfrak{d}(R^*_h) = \aleph_1$”. Finally, note that if $\dot{W}$ is the name for the generic real then

$$\Vdash_{\mathbb{Q}^*_f(K_{2.4.6}, \Sigma_{2.4.6})} (\forall x \in \prod_{n\in\omega} \mathbf{H}(n) \cap \mathbf{V})(\forall^\infty n \in \omega)(\dot{W}(n) \neq x(n))$$

and therefore $\Vdash_{\mathbb{P}_{\omega_2}}$ “$\mathfrak{d}(R^*_\mathbf{H}) = \aleph_2$”. □

CONCLUSION 7.5.7. It is consistent that $\mathbf{non}(\mathcal{M}) = \mathfrak{d} = \aleph_2$ and $\mathbf{cov}(\mathcal{M}) = \mathfrak{b} = \aleph_1 = \mathfrak{d}(R^\exists_{f,g})$ for every non-decreasing unbounded $g \in \omega^\omega$ and any $f \in \omega^\omega$ such that $\lim\limits_{n\to\omega} \frac{f(n)}{g(n)} = \infty$.

PROOF. Start with a model of CH and iterate ω_2 times with countable support the Blass-Shelah forcing notion $\mathbb{Q}^*_{s\infty}(K^*_{2.4.5}, \Sigma^*_{2.4.5})$. By 4.4.1 we immediately conclude that the iteration forces “$\mathbf{non}(\mathcal{M}) = \mathfrak{d} = \aleph_2$ & $\mathfrak{b} = \aleph_1$”. As each function in ω^ω appears in an intermediate model we may restrict our attention to $f, g \in \omega^\omega \cap \mathbf{V}$. By 7.3.8 and 7.3.6 we conclude that the iteration has the $R^\exists_{f,g}$–localization property (just build a suitable PP–ok partial order $\bar{\mathcal{F}}$ and take any p-point $\mathcal{D} \in \mathbf{V}$; by 7.4.2, 7.4.1 and 4.4.1 we know that $\mathcal{D}$ generates a non-reducible p-filter in the intermediate universes). Hence we get that in the resulting model $\mathbf{cov}(\mathcal{M}) = \mathfrak{d}(R^\exists_{f,g}) = \aleph_1$. □

EXAMPLE 7.5.8. We construct a finitary 2–big tree–creating pair $(K_{7.5.8}, \Sigma_{7.5.8})$ of the **NMP**–type (see 3.2.3(2)) such that the forcing notion $\mathbb{Q}^{\rm tree}_1(K_{7.5.8}, \Sigma_{7.5.8})$ does not have the strong PP–property.

CONSTRUCTION. This example is similar to that of 6.4.4 (what is not surprising if you notice some kind of duality between $\mathfrak{m}_1$ and $\mathfrak{d}(R^{\rm sPP})$).

Let $\mathbf{H}(n) = n^n$. Let $\mathcal{A}$ be the family of all pairs (n, x) such that $x \in [\mathbf{H}(n)]^n$. For $\nu \in \prod\limits_{k<m} \mathbf{H}(k)$, $m_0 < m$ and $A \subseteq \mathcal{A}$ we will write $\nu \prec^*_{m_0} A$ if

$$(\exists (n, x) \in A)(m_0 \leq n < m \ \ \& \ \ \nu(n) \in x).$$

Now we define $(K_{7.5.8}, \Sigma_{7.5.8})$. A tree–like creature $t \in \mathrm{TCR}_\eta[\mathbf{H}]$ is taken to be in $K_{7.5.8}$ if:

- $\mathbf{val}[t]$ is finite, and
- $\mathbf{nor}[t] = \log_2(\min\{|A| : A \subseteq \mathcal{A}\ \&\ (\forall \nu \in \mathrm{rng}(\mathbf{val}[t]))(\nu \prec^*_{\ell g(\eta)} A)\})$.

The tree composition $\Sigma_{7.5.8}$ is defined like in 6.4.4: if $\langle t_\nu : \nu \in \hat{T}\rangle \subseteq K_{7.5.8}$ is a system such that T is a well founded quasi tree, $\mathrm{root}(t_\nu) = \nu$, and $\mathrm{rng}(\mathbf{val}[t_\nu]) = \mathrm{succ}_T(\nu)$ (for $\nu \in \hat{T}$) then we define $S^*(t_\nu : \nu \in \hat{T})$ as the unique creature t^* in $K_{7.5.8}$ with $\mathrm{rng}(\mathbf{val}[t^*]) = \max(T)$, $\mathrm{dom}(\mathbf{val}[t^*]) = \{\mathrm{root}(T)\}$ and $\mathbf{dis}[t^*] = \langle \mathbf{dis}[t_\nu] : \nu \in \hat{T}\rangle$. Next we put

$$\Sigma_{7.5.8}(t_\nu : \nu \in \hat{T}) = \{t \in K_{7.5.8} : \mathbf{val}[t] \subseteq \mathbf{val}[S^*(t_\nu : \nu \in \hat{T})]\}.$$

It should be clear that $(K_{7.5.8}, \Sigma_{7.5.8})$ is a finitary tree–creating pair and the forcing notion $\mathbb{Q}^{\mathrm{tree}}_1(K_{7.5.8}, \Sigma_{7.5.8})$ is non-trivial.

CLAIM 7.5.8.1. *$(K_{7.5.8}, \Sigma_{7.5.8})$ is 2–big.*

Proof of the claim: Let $t \in K_{7.5.8}$, $\mathbf{nor}[t] > 0$ and suppose that $\mathrm{pos}(t) = u_0 \cup u_1$. Let $s_\ell \in \Sigma_{7.5.8}(t)$ be such that $\mathrm{pos}(s_\ell) = u_\ell$ (for $\ell = 0, 1$). Take $A_\ell \subseteq \mathcal{A}$ such that

$$|A_\ell| = 2^{\mathbf{nor}[s_\ell]} \quad \text{and} \quad (\forall \nu \in \mathrm{pos}(s_\ell))(\nu \prec^*_{m_0} A_\ell),$$

where $m_0 = \ell g(\mathrm{root}(s_\ell)) = \ell g(\mathrm{root}(t))$. Clearly $(\forall \nu \in \mathrm{pos}(t))(\nu \prec^*_{m_0} A_0 \cup A_1)$ and thus

$$\mathbf{nor}[t] \le \log_2(|A_0| + |A_1|) \le 1 + \max\{\mathbf{nor}[s_0], \mathbf{nor}[s_1]\}.$$

CLAIM 7.5.8.2. *$(K_{7.5.8}, \Sigma_{7.5.8})$ is of the* **NMP***–type (see 3.2.3(2)).*

Proof of the claim: Suppose that $\langle t_\eta : \eta \in T\rangle \in \mathbb{Q}^{\mathrm{tree}}_\emptyset(K_{7.5.8}, \Sigma_{7.5.8})$ is such that $(\forall \eta \in T)(\mathbf{nor}[t_\eta] > 1)$ and $F_0, F_1, F_2, \ldots$ are fronts of T such that

$$(\forall \nu \in F_{i+1})(\exists \nu' \in F_i)(\nu' \lhd \nu).$$

Clearly these fronts are finite (as $(K_{7.5.8}, \Sigma_{7.5.8})$ is finitary). Further suppose that $g : \bigcup\limits_{i\in\omega} F_i \longrightarrow \bigcup\limits_{i\in\omega} F_{i+1}$ is such that $\nu \lhd g(\nu) \in F_{i+1}$ for $\nu \in F_i$.

Let $r = \min\{2^{\mathbf{nor}[t_\eta]} : \eta \in T\}$. Choose increasing sequences $\langle n_k : k \le r^r + r\rangle$, $\langle i_k, j_k : k < r^r + r\rangle$ such that for each $k < r^r + r$:

(i) $i_k < j_k < i_{k+1}$ and if $k \in [r, r + r^r)$ then $j_k = i_k + 1$,
(ii) if $k < r$, $\nu \in F_{i_k}$ then $|\{\rho \in F_{j_k} : \nu \lhd \rho\}| \ge r$,
(iii) $n_k < \min\{\ell g(\nu) : \nu \in F_{i_k}\} < \max\{\ell g(\nu) : \nu \in F_{j_k}\} < n_{k+1}$.

Choose a mapping $\pi : F_{j_{r-1}} \longrightarrow r^r$ such that

$(*)_0$ if $\nu \lhd \eta_0 \in F_{j_{r-1}}$, $\nu \lhd \eta_1 \in F_{j_{r-1}}$, $\nu \in F_{j_k}$, $k < r$ then $\pi(\eta_0)(k) = \pi(\eta_1)(k)$,
$(*)_1$ for each $\nu \in F_{i_k}$, $k < r$ and $\ell < r$ there is $\eta \in F_{j_{r-1}}$ such that $\nu \lhd \eta$ and $\pi(\eta)(k) = \ell$.

(It is easy to define such a mapping if you remember clause (ii) above.) Let $\pi^* : r^r \longrightarrow r^r$ be the isomorphism of r^r equipped with the lexicographical order and r^r with the natural order of integers. Take a tree–creature $s \in \Sigma_{7.5.8}(t_\eta : (\exists \nu \in F_{j_{r^r+r-1}})(\eta \lhd \nu))$ such that

$$\mathrm{rng}(\mathbf{val}[s]) = \{\nu \in F_{j_{r^r+r-1}} : \ \text{if } \nu{\restriction}m_0 \in F_{j_{r-1}},\ m_0 \in \omega \text{ and } k = \pi^*(\pi(\nu{\restriction}m_0)) \text{ and } \nu{\restriction}m_1 \in F_{i_{r+k}},\ m_1 \in \omega \ \text{ then } g(\nu{\restriction}m_1) \lhd \nu\}.$$

(Note that we may find a suitable s by the definition of $\Sigma_{7.5.8}$.) By the choice, this s satisfies the demand $(\beta)^{\rm tree}$ of 3.2.3(2). But why does it have large enough norm? Suppose that $A \subseteq \mathcal{A}$ is such that $|A| < r$. Let $k_0 < r$ be such that

$$(n,x) \in A \quad \Rightarrow \quad n \notin [n_{k_0}, n_{k_0+1}).$$

Since $|A| < 2^{\mathbf{nor}[t_\eta]}$ for each $\eta \in T$ we may inductively build a sequence $\nu_0 \in F_{i_{k_0}}$ such that

$$(\forall (n,x) \in A)(\ell g(\mathrm{root}(T)) \leq n < \ell g(\nu_0) \Rightarrow \nu_0(n) \notin x).$$

Let $\sigma_0 : k_0 \longrightarrow r$ be such that $\sigma_0(k)$ is the value of $\pi(\eta)(k)$ for each $\eta \in F_{j_{r-1}}$, $\nu_0 \lhd \eta$. Take $\ell < r$ such that

if $\sigma \in r^r$, $\sigma_0{}^\frown\langle\ell\rangle \unlhd \sigma$

then there is no $(n,x) \in A$ with $n_{\pi^*(\sigma)+r} \leq n < n_{\pi^*(\sigma)+r+1}$

(remember the choice of π^* and that $|A| < r$). Now take $\nu_1 \in F_{j_{k_0}}$ such that $\nu_0 \lhd \nu_1$ and

$$(\forall \eta \in F_{j_{r-1}})(\nu_1 \unlhd \eta \Rightarrow \pi(\eta)(k_0) = \ell)$$

(possible by $(*)_0 + (*)_1$). By the choice of k_0 we know that

$$(\forall (n,x) \in A)(\ell g(\mathrm{root}(T)) \leq n < \ell g(\nu_1) \Rightarrow \nu_1(n) \notin x)$$

(look at (iii)). Next continue like at the beginning to get $\eta \in F_{j_{r-1}}$ such that $\nu_1 \unlhd \eta$ and $\neg(\eta \prec^*_{\ell g(\mathrm{root}(T))} A)$. We are sure that $\sigma_0{}^\frown\langle\ell\rangle \unlhd \pi(\eta)$ and therefore there is no $(n,x) \in A$ with $n_{\pi^*(\pi(\eta))+r} \leq n < n_{\pi^*(\pi(\eta))+r+1}$. Consequently we may continue the procedure applied to build η and we construct $\eta^* \in F_{j_{r^r+r-1}}$ such that

$\eta \lhd \eta^*$, if $\eta^*{\restriction}m \in F_{i_{r+\pi^*(\pi(\eta))}}$ then $g(\eta^*{\restriction}m) \lhd \eta^*$, and $\neg(\eta \prec^*_{\ell g(\mathrm{root}(T))} A)$.

Since, by its construction, the sequence η^* is in $\mathrm{rng}(\mathbf{val}[s])$, it exemplifies that A cannot witness the minimum in the definition of $\mathbf{nor}[s]$. Consequently, $\mathbf{nor}[s] \geq \log_2(r)$ and thus the tree–creature s satisfies the demand $(\alpha)^{\rm tree}$ of 3.2.3(2).

CLAIM 7.5.8.3. *The forcing notion $\mathbb{Q}^{\rm tree}_1(K_{7.5.8}, \Sigma_{7.5.8})$ does not have the strong PP–property.*

Proof of the claim: We will show that the generic real $\dot{W}$ shows that the strong PP–property fails for $\mathbb{Q}^{\rm tree}_1(K_{7.5.8}, \Sigma_{7.5.8})$. So suppose that $\langle w_i : i \in B\rangle$ is such that $B \in [\omega]^\omega$ and $(\forall i \in B)(w_i \in [\omega]^i)$ and let $p \in \mathbb{Q}^{\rm tree}_1(K_{7.5.8}, \Sigma_{7.5.8})$. We may assume that $\mathbf{nor}[t^p_\eta] > 1$ for each $\eta \in T^p$. Take $i_0 \in B \setminus \ell g(\mathrm{root}(p))$ and build inductively a condition $q \geq p$ such that for each $\eta \in T^q$ and $\nu \in \mathrm{pos}(t^q_\eta)$

$$t^q_\eta \in \Sigma_{7.5.8}(t^p_\eta) \quad \text{and} \quad \mathbf{nor}[t^q_\eta] \geq \mathbf{nor}[t^p_\eta] - 1 \quad \text{and} \quad (i_0 < \ell g(\nu) \Rightarrow \nu(i_0) \notin w_{i_0})$$

(remember the definition of the norm of elements of $K_{7.5.8}$). Now clearly $q \Vdash$ " $\dot{W}(i_0) \notin w_{i_0}$ ", finishing the proof of the claim and the construction. $\square$

CONCLUSION 7.5.9. It is consistent that $\mathbf{cof}(\mathcal{M}) < \mathfrak{d}(R^{\rm sPP})$.

PROOF. Start with a model of CH and force with countable support iteration of length ω_2 of forcing notions $\mathbb{Q}^{\rm tree}_1(K_{7.5.8}, \Sigma_{7.5.8})$. We know that $\mathbb{Q}^{\rm tree}_1(K_{7.5.8}, \Sigma_{7.5.8})$ is proper, ω^ω–bounding and Cohen–preserving (by 3.2.5 + 3.1.1). Consequently the iteration is of the same type (see [**BaJu95**, 6.3.21, 6.3.22]) and, by standard arguments, in the final model we have $\mathbf{non}(\mathcal{M}) = \mathfrak{d} = \aleph_1$. But this implies that $\mathbf{cof}(\mathcal{M}) = \aleph_1$ too (see [**BaJu95**, 2.2.11]). Finally, as $\mathbb{Q}^{\rm tree}_1(K_{7.5.8}, \Sigma_{7.5.8})$ does not have the strong PP–property we easily conclude that the iteration forces that $\mathfrak{d}(R^{\rm sPP}) = \aleph_2$. $\square$

List of definitions

3.2.1 Cohen–preserving proper forcing notions;
3.2.3 creating pairs of the **NMP**–type, tree creating pairs of the $\mathbf{NMP}^{\mathrm{tree}}$–type;
3.2.7 trivially meagering weak creating pairs;
3.3.1 weak creating pairs of the **NNP**–type;
3.3.2 gluing and weakly gluing tree creating pairs;
3.3.4 strongly finitary creating pairs;
3.4.1 when a weak creating pair (K, Σ) strongly refuses Sacks property;
4.1.2 creating pairs which are: meagering, anti–big;
4.2.1 Σ^{sum};
4.2.3 (d, u)–sum $\Sigma^{\mathrm{sum}}_{d,u}$;
4.2.4 when a creating pair is saturated with respect to a family of pre–norms;
4.2.6 Σ^{tsum};
4.3.1 decision functions, creating pairs of the **AB**-type, condensed creating pairs;
4.3.7 creating pairs of the $\mathbf{AB}^+$–type;
5.1.1 essentially f-big weak creating pairs;
5.1.6 reducible weak creating pairs;
5.1.7 h–limited weak creating pairs;
5.1.11 $(\mathbf{H}, F)$–fast function;
5.2.1 $U_h(\bar{t})$, $V^n_h(\bar{t})$, $(\bar{t}, h_1, h_2)$–bounding forcing notions;
5.2.3 creating pairs which are monotonic, strictly monotonic, spread;
5.2.5 m–additivity $\mathbf{add}_m(t)$ of a weak creature t, (g, h)-additive weak creating pairs;
5.2.8 $\bar{t}$–good families of functions, $(\bar{t}, \mathcal{F})$–bounding forcing notions;
5.3.1 $\bar{t}$–systems, regular $\bar{t}$–systems, $\mathbb{P}^*_{\mathcal{C}(\mathbf{nor})}(\bar{t}, (K, \Sigma))$ and the partial order $\preceq$ on it, quasi-W-generic Γ;
5.3.6 (Γ, W)–genericity preserving forcing notions;
5.3.8 Cohen sensitive $\bar{t}$–systems, directed $\bar{t}$–systems;
5.3.10 $(\bar{t}_0, h_1, h_2)$–coherent $\bar{t}$–systems, $(\bar{t}_0, \bar{\mathcal{F}})$–coherent sequences of $\bar{t}$–systems;
6.1.1 creating pair which generates an ultrafilter;
6.1.3 when Γ generates a filter (ultrafilter), $\mathcal{D}(\Gamma)$;
6.1.5 interesting creating pair;
6.2.1 Ramsey filter, p–point, q–point, weak q–point;
6.2.5 interesting ultrafilters, games $G^{sR}(\mathcal{D})$, $G^{aR}(\mathcal{D})$, semi–Ramsey ultrafilters, almost Ramsey ultrafilters;
6.3.1 tree creating pairs of the $\mathbf{UP}(\mathcal{D})^{\mathrm{tree}}$, $\mathbf{sUP}(\mathcal{D})^{\mathrm{tree}}$ –types;
6.3.3 rich tree creating pairs;
6.3.5 $R_n(k, m)$;
6.3.7 simple except omitting creating pairs, omittory–compatible $\bar{t}$–systems;
7.1.1 κ_{BP};
7.1.2 R^{sPP};
7.2.1 non-reducible filters, PP–ok partial orders, forcing notions with $(\mathcal{D}, x)$-strong PP–property, $(\mathcal{D}, \bar{\mathcal{F}})$-strong PP–property;
7.3.1 $R^{\exists}_{f,g}$, $R^{\forall}_{f,g}$, R^*_f, $R^{**}_{f,g}$;
7.3.5 $(\mathcal{D}, \bar{\mathcal{F}})$-$R^{\exists}_{\bar{f}}$–localization property.

Bibliography

[Ba91] Marek Balcerzak. *Some properties of ideals of sets in Polish spaces.* Habilitation Thesis. Łódź University Press, 1991.

[BaPl96] Marek Balcerzak and Szymon Plewik. Property **(P)** and game ideals. *Tatra Mountains Mathematical Publications*, **8**:153–158.

[BaRo95] Marek Balcerzak and Andrzej Rosłanowski. Coinitial families of perfect sets. *Journal of Applied Analysis*, **1**:181–204, 1995.

[Ba94] Tomek Bartoszyński. Some old and new problems. Circulated notes of November 1994.

[BaJu95] Tomek Bartoszyński and Haim Judah. *Set Theory: On the Structure of the Real Line.* A K Peters, Wellesley, Massachusetts, 1995.

[BJSh 368] Tomek Bartoszyński, Haim Judah, and Saharon Shelah. The Cichoń diagram. *Journal of Symbolic Logic*, **58**:401–423, 1993.

[BRSh 616] Tomek Bartoszyński, Andrzej Rosłanowski, and Saharon Shelah. After all, there are some inequalities which are provable in ZFC. *Journal of Symbolic Logic*, **accepted**.

[BsSh 242] Andreas Blass and Saharon Shelah. There may be simple $P_{\aleph_1}$- and $P_{\aleph_2}$-points and the Rudin-Keisler ordering may be downward directed. *Annals of Pure and Applied Logic*, **33**:213–243, 1987.

[CRSW93] Jacek Cichoń, Andrzej Rosłanowski, Juris Steprans, and Bogdan Węglorz. Combinatorial properties of the ideal $\mathfrak{P}_2$. *Journal of Symbolic Logic*, **58**:42–54, 1993.

[CiSh 653] Krzysztof Ciesielski and Saharon Shelah. A model with no magic sets. *Journal of Symbolic Logic*, **accepted**.

[DKP92] Wojciech Dębski, Jan Kleszcz, and Szymon Plewik. Perfect sets of independent functions. *Acta Universitatis Carolinae – Mathematica et Physica*, **33**:31–33, 1992.

[FrSh 406] David H. Fremlin and Saharon Shelah. Pointwise compact and stable sets of measurable functions. *Journal of Symbolic Logic*, **58**:435–455, 1993.

[GoSh 448] Martin Goldstern and Saharon Shelah. Many simple cardinal invariants. *Archive for Mathematical Logic*, **32**:203–221, 1993.

[J] Thomas Jech. *Set theory.* Academic Press, New York, 1978.

[JRSh 373] Haim Judah, Andrzej Rosłanowski, and Saharon Shelah. Examples for Souslin Forcing. *Fundamenta Mathematicae*, **144**:23–42, 1994.

[MaPa98] Pierre Matet and Janusz Pawlikowski. Ideals over ω and cardinal invariants of the continuum. *Journal of Symbolic Logic*, **63**:1040–1054, 1998.

[My69] Jan Mycielski. Some new ideals of sets on the real line. *Colloquium Mathematicum*, **XX**:71–76, 1969.

[NeRo93] Ludomir Newelski and Andrzej Rosłanowski. The ideal determined by the unsymmetric game. *Proceedings of the American Mathematical Society*, **117**:823–831, 1993.

[Ro94] Andrzej Rosłanowski. Mycielski ideals generated by uncountable systems. *Colloquium Mathematicum*, **LXVI**:187–200, 1994.

[RoSh 670] Andrzej Rosłanowski and Saharon Shelah. Norms on possibilities III: strange subsets of the real line. *In preparation.*

[RoSh 672] Andrzej Rosłanowski and Saharon Shelah. Norms on possibilities IV: ccc forcing notions. *In preparation.*

[RoSh 501] Andrzej Rosłanowski and Saharon Shelah. Localizations of infinite subsets of ω. *Archive for Mathematical Logic*, **35**:315–339, 1996. A special volume dedicated to Prof. Azriel Levy.

[RoSh 628] Andrzej Rosłanowski and Saharon Shelah. Norms on possibilities II: more ccc ideals on 2^ω. *Journal of Applied Analysis*, **3**:103–127, 1997.

[Sh 630] Saharon Shelah. Non-elementary proper forcing notions. *Journal of Applied Analysis*, **submitted**.

[Sh 669] Saharon Shelah. Non-elementary proper forcing notions II: iterations. *In preparation.*

[Sh:f] Saharon Shelah. *Proper and improper forcing.* Perspectives in Mathematical Logic. Springer-Verlag, 1998.

[Sh:b] Saharon Shelah. *Proper forcing*, volume 940 of *Lecture Notes in Mathematics.* Springer-Verlag, Berlin-New York, xxix+496 pp, 1982.

[Sh 207] Saharon Shelah. On cardinal invariants of the continuum. In *Axiomatic set theory (Boulder, Colo., 1983)*, volume 31 of *Contemp. Mathematics*, pages 183–207. Amer. Math. Soc., Providence, RI, 1984. Proceedings of the Conference in Set Theory, Boulder, June 1983; ed. Baumgartner J., Martin, D. and Shelah, S.

[Sh 326] Saharon Shelah. Viva la difference I: Nonisomorphism of ultrapowers of countable models. In *Set Theory of the Continuum*, volume 26 of *Mathematical Sciences Research Institute Publications*, pages 357–405. Springer Verlag, 1992.

INSTITUTE OF MATHEMATICS
THE HEBREW UNIVERSITY OF JERUSALEM
91904 JERUSALEM, ISRAEL
AND MATHEMATICAL INSTITUTE OF WROCLAW UNIVERSITY
50384 WROCLAW, POLAND

INSTITUTE OF MATHEMATICS
THE HEBREW UNIVERSITY OF JERUSALEM
91904 JERUSALEM, ISRAEL
AND DEPARTMENT OF MATHEMATICS
RUTGERS UNIVERSITY
NEW BRUNSWICK, NJ 08854, USA

Editorial Information

To be published in the *Memoirs*, a paper must be correct, new, nontrivial, and significant. Further, it must be well written and of interest to a substantial number of mathematicians. Piecemeal results, such as an inconclusive step toward an unproved major theorem or a minor variation on a known result, are in general not acceptable for publication. *Transactions* Editors shall solicit and encourage publication of worthy papers. Papers appearing in *Memoirs* are generally longer than those appearing in *Transactions* with which it shares an editorial committee.

As of May 31, 1999, the backlog for this journal was approximately 5 volumes. This estimate is the result of dividing the number of manuscripts for this journal in the Providence office that have not yet gone to the printer on the above date by the average number of monographs per volume over the previous twelve months, reduced by the number of issues published in four months (the time necessary for preparing an issue for the printer). (There are 6 volumes per year, each containing at least 4 numbers.)

A Copyright Transfer Agreement is required before a paper will be published in this journal. By submitting a paper to this journal, authors certify that the manuscript has not been submitted to nor is it under consideration for publication by another journal, conference proceedings, or similar publication.

Information for Authors and Editors

Memoirs are printed by photo-offset from camera copy fully prepared by the author. This means that the finished book will look exactly like the copy submitted.

The paper must contain a *descriptive title* and an *abstract* that summarizes the article in language suitable for workers in the general field (algebra, analysis, etc.). The *descriptive title* should be short, but informative; useless or vague phrases such as "some remarks about" or "concerning" should be avoided. The *abstract* should be at least one complete sentence, and at most 300 words. Included with the footnotes to the paper, there should be the 1991 *Mathematics Subject Classification* representing the primary and secondary subjects of the article. This may be followed by a list of *key words and phrases* describing the subject matter of the article and taken from it. A list of the numbers may be found in the annual index of *Mathematical Reviews*, published with the December issue starting in 1990, as well as from the electronic service e-MATH [**telnet e-MATH.ams.org** (or **telnet 130.44.1.100**). Login and password are **e-math**]. For journal abbreviations used in bibliographies, see the list of serials in the latest *Mathematical Reviews* annual index. When the manuscript is submitted, authors should supply the editor with electronic addresses if available. These will be printed after the postal address at the end of each article.

Electronically prepared papers. The AMS encourages submission of electronically prepared papers in $\mathcal{A}\mathcal{M}\mathcal{S}$-TeX or $\mathcal{A}\mathcal{M}\mathcal{S}$-LaTeX. The Society has prepared author packages for each AMS publication. Author packages include instructions for preparing electronic papers, the *AMS Author Handbook*, samples, and a style file that generates the particular design specifications of that publication series for both $\mathcal{A}\mathcal{M}\mathcal{S}$-TeX and $\mathcal{A}\mathcal{M}\mathcal{S}$-LaTeX.

Authors with FTP access may retrieve an author package from the Society's Internet node `e-MATH.ams.org` (130.44.1.100). For those without FTP access, the author package can be obtained free of charge by sending e-mail to `pub@ams.org` (Internet) or from the Publication Division, American Mathematical Society, P.O. Box 6248, Providence, RI 02940-6248. When requesting an author package, please specify $\mathcal{A}\mathcal{M}\mathcal{S}$-TEX or $\mathcal{A}\mathcal{M}\mathcal{S}$-LATEX, Macintosh or IBM (3.5) format, and the publication in which your paper will appear. Please be sure to include your complete mailing address.

Submission of electronic files. At the time of submission, the source file(s) should be sent to the Providence office (this includes any TEX source file, any graphics files, and the DVI or PostScript file).

Before sending the source file, be sure you have proofread your paper carefully. The files you send must be the EXACT files used to generate the proof copy that was accepted for publication. For all publications, authors are required to send a printed copy of their paper, which exactly matches the copy approved for publication, along with any graphics that will appear in the paper.

TEX files may be submitted by email, FTP, or on diskette. The DVI file(s) and PostScript files should be submitted only by FTP or on diskette unless they are encoded properly to submit through e-mail. (DVI files are binary and PostScript files tend to be very large.)

Files sent by electronic mail should be addressed to the Internet address `pub-submit@ams.org`. The subject line of the message should include the publication code to identify it as a Memoir. TEX source files, DVI files, and PostScript files can be transferred over the Internet by FTP to the Internet node `e-math.ams.org` (130.44.1.100).

Electronic graphics. Figures may be submitted to the AMS in an electronic format. The AMS recommends that graphics created electronically be saved in Encapsulated PostScript (EPS) format. This includes graphics originated via a graphics application as well as scanned photographs or other computer-generated images.

If the graphics package used does not support EPS output, the graphics file should be saved in one of the standard graphics formats—such as TIFF, PICT, GIF, etc.—rather than in an application-dependent format. Graphics files submitted in an application-dependent format are not likely to be used. No matter what method was used to produce the graphic, it is necessary to provide a paper copy to the AMS.

Authors using graphics packages for the creation of electronic art should also avoid the use of any lines thinner than 0.5 points in width. Many graphics packages allow the user to specify a "hairline" for a very thin line. Hairlines often look acceptable when proofed on a typical laser printer. However, when produced on a high-resolution laser imagesetter, hairlines become nearly invisible and will be lost entirely in the final printing process.

Screens should be set to values between 15% and 85%. Screens which fall outside of this range are too light or too dark to print correctly.

Any inquiries concerning a paper that has been accepted for publication should be sent directly to the Editorial Department, American Mathematical Society, P. O. Box 6248, Providence, RI 02940-6248.

Editors

Selected Titles in This Series

(Continued from the front of this publication)

641 **W. Bulla, F. Gesztesy, H. Holden, and G. Teschl,** Algebro-geometric quasi-periodic finite-gap solutions of the Toda and Kac-van Moerbeke hierarchies, 1998

640 **Xingde Dai and David R. Larson,** Wandering vectors for unitary systems and orthogonal wavelets, 1998

639 **Joan C. Artés, Robert E. Kooij, and Jaume Llibre,** Structurally stable quadratic vector fields, 1998

638 **Gunnar Fløystad,** Higher initial ideals of homogeneous ideals, 1998

637 **Thomáš Gedeon,** Cyclic feedback systems, 1998

636 **Ching-Chau Yu,** Nonlinear eigenvalues and analytic-hypoellipticity, 1998

635 **Magdy Assem,** On stability and endoscopic transfer of unipotent orbital integrals on p-adic symplectic groups, 1998

634 **Darrin D. Frey,** Conjugacy of Alt_5 and $\mathrm{SL}(2,5)$ subgroups of $E_8(\mathbb{C})$, 1998

633 **Dikran Dikranjan and Dmitri Shakhmatov,** Algebraic structure of pseudocompact groups, 1998

632 **Shouchuan Hu and Nikolaos S. Papageorgiou,** Time-dependent subdifferential evolution inclusions and optimal control, 1998

631 **Ronnie Lee, Steven H. Weintraub, and J. William Hoffman,** The Siegel modular variety of degree two and level four/Cohomology of the Siegel modular group of degree two and level four, 1998

630 **Florin Rădulescu,** The Γ-equivariant form of the Berezin quantization of the upper half plane, 1998

629 **Richard B. Sowers,** Short-time geometry of random heat kernels, 1998

628 **Christopher K. McCord, Kenneth R. Meyer, and Quidong Wang,** The integral manifolds of the three body problem, 1998

627 **Roland Speicher,** Combinatorial theory of the free product with amalgamation and operator-valued free probability theory, 1998

626 **Mikhail Borovoi,** Abelian Galois cohomology of reductive groups, 1998

625 **George Xian-Zhi Yuan,** The study of minimax inequalities and applications to economies and variational inequalities, 1998

624 **P. Deift and K. T-R McLaughlin,** A continuum limit of the Toda lattice, 1998

623 **S. A. Adeleke and Peter M. Neumann,** Relations related to betweenness: Their structure and automorphisms, 1998

622 **Luigi Fontana, Steven G. Krantz, and Marco M. Peloso,** Hodge theory in the Sobolev topology for the de Rham complex, 1998

621 **Gregory L. Cherlin,** The classification of countable homogeneous directed graphs and countable homogeneous n-tournaments, 1998

620 **Victor Guba and Mark Sapir,** Diagram groups, 1997

619 **Kazuyoshi Kiyohara,** Two classes of Riemannian manifolds whose geodesic flows are integrable, 1997

618 **Karl H. Hofmann and Wolfgang A. F. Ruppert,** Lie groups and subsemigroups with surjective exponential function, 1997

617 **Robin Hartshorne,** Families of curves in $\mathbb{P}^3$ and Zeuthen's problem, 1997

616 **Serguei G. Bobkov and Christian Houdré,** Some connections between isoperimetric and Sobolev-type inequalities, 1997

615 **Michael A. Dritschel and Hugo J. Woerdeman,** Model theory and linear extreme points in the numerical radius unit ball, 1997

For a complete list of titles in this series, visit the AMS Bookstore at **www.ams.org/bookstore/**.